D0164200

Climatic Fluctuations of the Ice Age

Climatic Fluctuations of the Ice Age

Burkhard Frenzel

Translated by A. E. M. Nairn

The Press of Case Western Reserve University, Cleveland & London, 1973

Printed in the United States of America.
International Standard Book Number: 0–8295–0226–2.
Library of Congress Catalogue Card Number: 70–17088.

Contents

Illustrations

Tables

Foreword to the American Edition

The growing scientific interest in research into the climatic history of the Ice Age has encouraged the Press of Case Western Reserve University to undertake the translation of this book, which appeared in 1967 under the title *Die Klimaschwankungen des Eiszeitalters.* It is appropriate to add some explanatory remarks in this foreword to the American edition.

It was originally considered that in both central Europe and North America there were only three or four ice ages with intervening warmer intervals. In recent years the picture has drastically changed; currently it is assumed that in central Europe there were ten ice ages, but with the awareness that the progress of research may reveal still more of these distinctive climatic fluctuations. The oxygen isotope analysis of the calcareous shells of certain marine organisms appears to suggest some forty cold periods within the last two to three m.y. Everywhere on the earth it is becoming increasingly more difficult to define the onset of the Ice Age, to the extent that K. K. Turekian in 1971 could speak of late Cenozoic Ice Ages. This shift of accent is certainly fully justified, though our knowledge of the older climatic fluctuations is understandably far inferior to that of the younger. In addition there have been numerous attempts, particularly from the meteorological side, to model the atmospheric circulation patterns of glacial and interglacial periods. It appeared desirable to me, in view of these various approaches, to review the state of knowledge of the Pleistocene climatic fluctuations. To exclude all hypotheses and theories, well or poorly founded, I chose to take a narrow time interval which is relatively well known because of considerable research. It is for this reason that I have deliberately restricted myself to a consideration of the Pleistocene as it is presently defined, that is the last 1.0 to 1.5 m.y. It appeared to me necessary to determine the actual climate and trends of this interesting span of time, in the hope that this might provide a basis for a reconstruction of the course and cause(s) of the climatic fluctuations recorded. It was obviously impossible to do this for the whole of the Ice Age.

I am all too aware of the weaknesses of this book, which was written for European readers and naturally has Europe always as the center of interest. Other regions of the earth could not be handled in the same detail, so that the American reader may note the absence of much that he would have liked to have seen discussed in detail. I have attempted to bring the translated edition up to date with references to the most recent literature. However, it was impossible to write a climatic history of the Pleistocene of North America—the reader who seeks this is referred to that excellent book *The Quaternary of the United States,* by H. E. Wright and D. G. Frey (Princeton University Press, 1965). For this shortcoming I offer my apologies. Yet most, if not all, climatic fluctuations of the Ice Age were the result of mechanisms which affected the earth as a planet, so that a better knowledge of the climatic history of Europe may also be of value to the American scholar, and may perhaps permit him to examine his own observations from a new standpoint.

I am grateful to the Press of Case Western Reserve University for wholehearted support during the production of the American edition. I would also like to express particular thanks to Dr. A. E. M. Nairn for his care and skill in producing a good translation. I would like to think that the stimulating correspondence which passed between us in the course of the translation may lead to an international interdisciplinary research effort, for this is an extraordinarily interesting interface between biology, geology, geomorphology, and climatology.

BURKHARD FRENZEL

Stuttgart, March 1972

Translator's Preface

Contemporary interest in environmental problems has done much to focus attention on climate, one of the more obvious environmental elements. Environmentalists concerned with inadvertent weather modification by mechanisms such as thermal pollution or atmospheric changes are often without any clear understanding of the long-term nature of climate. This is indeed one area of scientific "terra incognita." Yet this region has been penetrated by explorers whose trails only serve to illustrate a general lack of knowledge. Dr. Frenzel must surely be numbered among these trail blazers. It now seems fantastic that the number of earth scientists in the late 1960's who regarded the beginning of the Ice Age as lying at least as far back as the middle Cenozoic was a mere handful. Yet the fact that this was the case, and the unnecessary obstacles it introduced, form perhaps the clearest illustration of the complexities of the study of climate.

Perhaps the study of ancient climates, more than any other interdisciplinary science, requires an extremely broadly based approach, and with data from so many disciplines synthesis becomes extremely difficult. One of the strengths of Dr. Frenzel's book is that he eschews all hypotheses and concentrates upon the presentation of data for a restricted but definable segment of time. His synthesis for the later segment of the Ice Age is extraordinarily detailed, and is one of the few truly climatic syntheses to have appeared. Inevitably there are areas covered in less detail than even the author would like; none the less the book is a considerable *tour de force,* and it would be a pity if language difficulties prevented it becoming as widely known as possible. I am glad to have had the opportunity to prepare a translation, fortified by the knowledge that Dr. Frenzel's excellent command of English would prevent serious mistranslation and by his forbearance with my limitations regarding botanical terms.

In bringing to the English-speaking scientific public Dr. Frenzel's work, the Press of Case Western Reserve University is providing the synthesis of the researches of one explorer into the climates of the recent past. His efforts will hopefully point the way for others and, one may hope, provide environmentalists with a clearer understanding of the complexities of climate and of the techniques for deriving information on climate.

In making this translation I have been aided by the author and by the Press. If the translation has any merit, much of the credit is theirs. Only errors and mistranslations are wholly my own.

A. E. M. N.

Introduction

A year with unusual weather—a rainy summer, or a particularly severe winter—quickly confirms the impression that the climate is changing. Such seasons have been recorded so often in the last decade that the present appears to be a time of great climatic change. Von Rudloff [639]* has investigated in detail the most recent climatic fluctuations. Particularly striking is the retreat of glaciers in most of the high mountainous regions of the earth. A familiar sight in the Alps is a glacier surrounded by high lateral and terminal moraines that were deposited or piled up between 1820 and 1850 (see Fig. 1). Since then, the thickness and length of the glaciers has diminished. This has not been a continuous process, for between 1890 and 1910, and between 1920 and 1925, the glaciers advanced again—or at least the retreat of the glacier front was appreciably retarded—so that most of the glaciers today are surrounded by a series of young arcuate terminal moraines.

The retreat of the glaciers is indicative of a warming of the waters of the Arctic. From the period 1893–96 and until about 1940, the annual ice thickness in the northern polar seas decreased from 2.65 meters† to 2.18 meters. During the same interval the west coast of Spitzbergen, formerly open to shipping three months a year, became usable for seven months; and the cod (*Gadus callarias*), whose first shoals on the west coast of Greenland near Godthaab were observed below 64°N in 1919, have now reached 73°N [3].

Clearly, the climate in other regions of the earth too has altered considerably during the past hundred years, to the extent that the thermal balance of the glaciers and of other large ice-covered areas has become progressively less favorable to ice accumulation.

The relatively small glacial advances of 1820 and 1850 in the Northern Hemisphere were preceded by a period of considerable retreat despite marked advances in 1600, 1640, and 1680. The glacial advances in the seventeenth century reached far down the valleys of the western Alps [402, 403]. It seems likely that the entrances to the early ore mines in the upper reaches of many Alpine valleys were blocked by advancing ice. Only recently have they begun to emerge. These ore-workings of the Middle Ages, high in the Alpine valleys, indicate that for a

---

* The numbers in brackets indicate that full references are given in the bibliography.

† According to Ahlmann [3] this figure should be 3.65 meters, but from information supplied by Dr. M. Rodewald, it appears probable that 2.65 meters is more nearly correct.

FIGURE 1. Glacial retreat in the Alps since 1850. The Vadret da Tschierva glacier approaching from the left in the picture, and the Vadret de Roseg glacier, approaching from the right, formed thick lateral moraines when they joined in about the year 1850. These are particularly distinct on both sides of the Vadret da Tschierva, and appear here as steeply piled-up ramparts. The central part of the tongue of the glacier arches itself above the lateral moraine. Since then, both glaciers have retreated and in part are in danger of being drowned in their own debris (front left, part of the Tongue of Vadret da Tschierva). Piz-Roseg group, southeast of the Engadin, Swiss Alps.

time the climate must have been mild. Yet even this milder climate was not as favorable as that during the seventh to fourth millennia B.C. when many plants spread northward, far beyond their present confines. As early as 1902 Andersson deduced from the former distribution of the hazelnut (*Corylus avellana*; see Fig. 2) that the average monthly temperatures in August and September approximately 100 kilometers to the north of the present limit of hazelnut in Sweden were 2.5°C warmer than today. It is further known that this phase of very favorable climate with respect to the humid zone of the Northern Hemisphere was preceded by a period during which the northern ice sheet reached from the Scandinavian highlands to Brandenburg in the south, to the Valdai Mountains in the southeast, and to the Kanin Peninsula in the east. In North America the continental ice had spread southward across the Great Lakes, and all the high mountains of the earth carried a greater ice cover than they do at present.

Examples of the same kind can be found in still earlier parts of the Pleistocene. Taken together, they indicate that during the youngest interval of geological time —that is to say, during the Quaternary—the climate has repeatedly and radically changed. Upon the principal climatic changes countless smaller climatic fluctuations have been superimposed, which to a greater or lesser degree have influenced

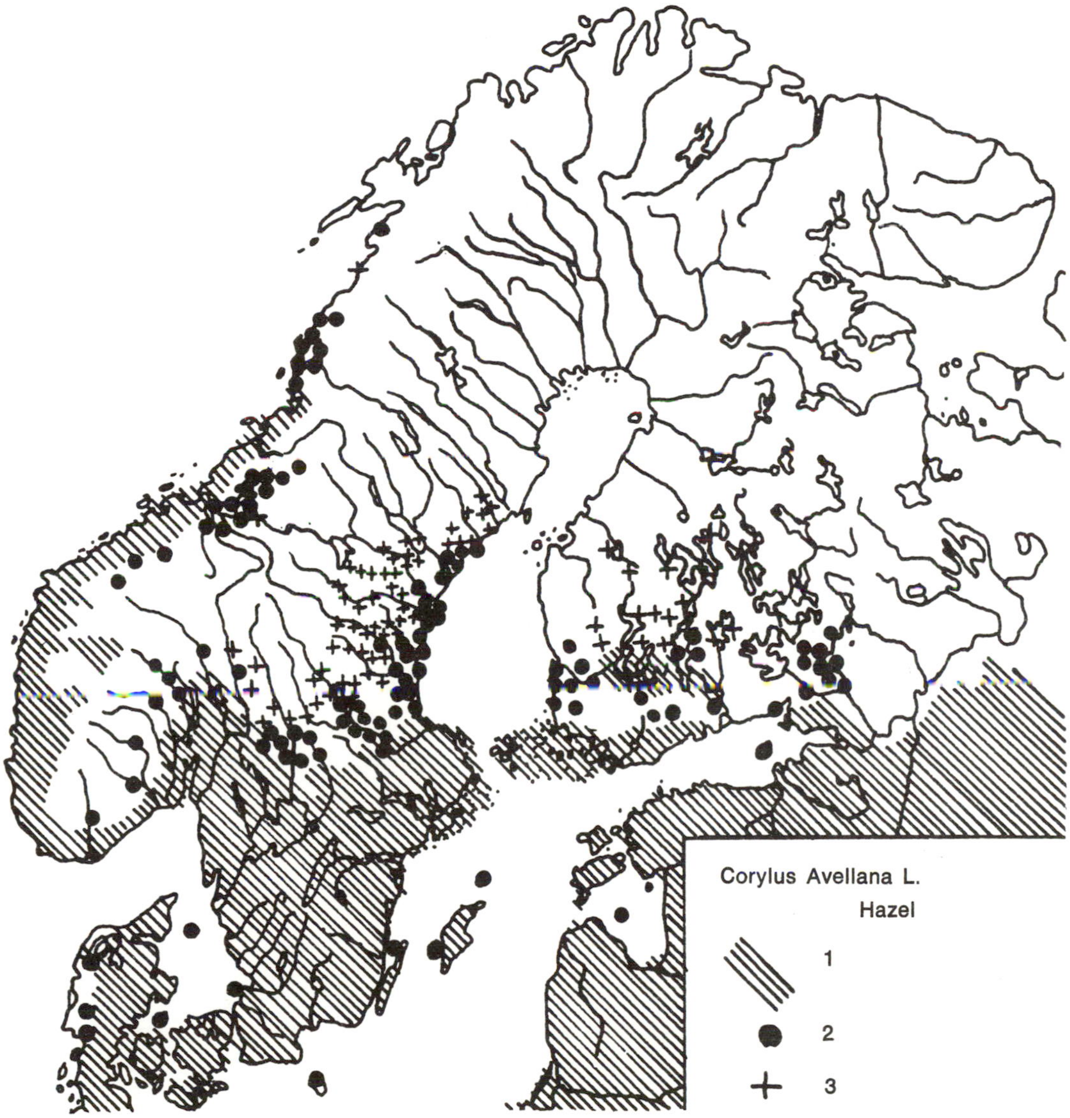

FIGURE 2. Indications of the postglacial warm period. Present and postglacial distribution of the hazelnut (*Corylus avellana*) in Scandinavia. 1. present general distribution; 2. current records of individual occurrences; 3. hazelnut fossils in sediments of the postglacial warm period. It can be seen that 5,000–6,000 years ago the hazelnut extended further north and higher in the mountainous region than at the present time. Simplified from Hulten [355].

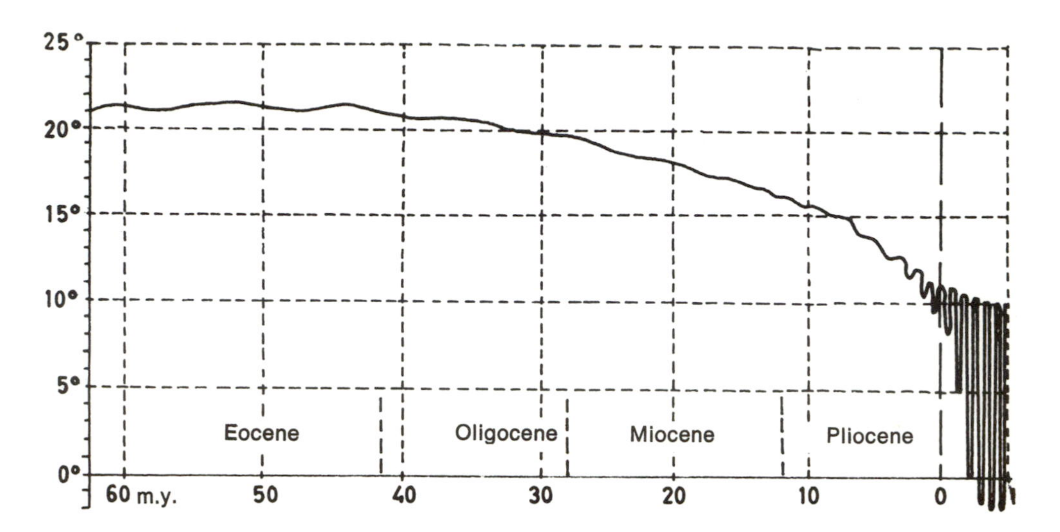

FIGURE 3. Climatic fluctuations in the Tertiary and Quaternary in central Europe. Curve of the annual average temperature of central Europe. The increase in the intensity of the fluctuations is clearly recognizable, particularly when it is observed that the time scale for the Quaternary is exaggerated five times. From Woldstedt [803].

the development of the plant and animal kingdoms. Woldstedt [803] some years ago outlined the principal characteristics of the climatic changes in the Tertiary and Quaternary (Fig. 3). His curve showing the annual average temperature of central Europe illustrates that even during the Tertiary, climatic fluctuations occurred. Their magnitude became increasingly significant toward the end of the Pliocene, until finally in the Quaternary the important climatic fluctuations set in which have so radically influenced the development of the earth's surface features, the stability of its flora and fauna, and the composition of its plant and animal communities.

Figure 3 also shows that the climatic changes within the youngest segment of geological time are qualitatively and quantitatively quite different. The slow general decrease in temperature during the latter half of the Tertiary stands in sharp contrast to the rapid fluctuations as ice age and interglacial age followed one another. During these last-named phases, climatic fluctuations on a smaller scale can also be distinguished, and the picture becomes extremely complex.

It seems advisable, in view of the preceding research and if the problems raised here are to be fully worked out, to make a distinction between climatic changes and climatic fluctuations. Under the designation of climatic *change*, slow, long-term consistent alterations in climate or in certain climatic factors are considered. Climatic *fluctuations*, on the other hand, are regarded as those alterations in climate which show a tendency to reverse or to change directions, whether once or more frequently. This definition must be made more precise, as climatic fluctuations of the past have occurred on different time scales. We shall designate as first-order climatic fluctuations those changes which occur between the high point of an interglacial and a glacial epoch, and as second-order fluctuations the lower-amplitude climatic changes such as may occur between a stadial and interstadial. Von Rudloff [639] discussed the same question from the standpoint of present-day climatic fluctuations and derived a classification which, although acceptable to meteorologists, differs from that applied here. This difference is understandable when the two hundred years of instrumental observations are contrasted with the approximately one million years of the Pleistocene.

In the history of climate, the Tertiary and Quaternary form a single unit: the climatic fluctuations of the Ice Age are only a continuation and intensification of events that had already become distinguishable during the Tertiary. It is nevertheless legitimate in the present investigation to consider only the Pleistocene and Holocene, the two divisions of the Quaternary. In justification there is, on the one hand, the lack of uniformity in treatment of the climate of earlier periods in the earth's history [see 666, 561, 562, among others], and, on the other, the nature of the climatic fluctuations of the Quaternary, which are not only particularly strong but important for the further development of the living world, the biosphere.

As noted, the Quaternary is made up of the Pleistocene and Holocene, of which the Pleistocene is much the longer. Although in many ways this division is arbitrary, it serves nicely to separate the time interval during which man has increasingly interfered with the balance of nature (the Holocene or postglacial period) from the earlier interval, during which the effects of man can be ignored. In addition, the climatic fluctuations within postglacial times—important insofar as plant and animal migrations and the development of human cultures are concerned—cannot be closely compared with the Pleistocene. As a result, in the following pages only the climatic fluctuations of the Pleistocene (that is, of the Ice Age) will be studied. They will provide some insight into the basic problems of Quaternary climatic fluctuations.

Our knowledge of the course of events during the Ice Age is naturally not equal in all regions. To the regional problems of research into the Ice Age must be added others caused by the passage of time. Although detailed information about the various stages of the Ice Age does exist, as a result of intensive study of certain regions, it cannot be affirmed with certainty that during all of the Pleistocene glaciations the glaciers actually covered a more extensive area than they do today, although clear signs of the former extraordinarily cold climate can no longer be overlooked. In view of this difficulty, it is worth adopting a proposal made by Woldstedt [804–6], who suggests that we refer not to "ice ages" or glacial epochs, but to "cold periods," and replace the term "interglacial" by the more neutral expression "warm periods." This does not imply that there are legitimate doubts about the reality of former great glacial advances; rather it avoids the use of "ice age" in cases where there is no indication of a significant glacial advance, even though the climatic character of the time may certainly be referred to as glacial in comparison with other time intervals.

There is available today a considerable volume of literature which may be drawn upon in reconstructing the climate during the Ice Age. Not only is there much interesting information, but ingenious hypotheses have been advanced. In the present work, the hypotheses will receive far less attention than the data, for the latter are needed to correct many hypothetical presentations. The factual material, however, is not equally relevant to all parts of the earth, and in consequence this volume concentrates essentially upon the already well-investigated regions of the Northern Hemisphere. On the basis of this material, we shall attempt to outline the principal climatic fluctuations during the course of the Pleistocene. The correctness of our results will be tested, so far as possible, against observations from other regions of the earth.

Climatic Fluctuations of the Ice Age

1. Methods of Investigating Ancient Climates

Every attempt to determine the climatic conditions of the past depends not only upon error-free methods of investigation but upon the measurement of time in such a way that climatic changes determined in the geological past can be dated with assurance. Since the first analyses of geological and paleontological problems, many methods have been elaborated which are said to suffice for the problems raised. They have recently been described in detail by Schwarzbach [666]. Even a short review shows that the difficulties in determining the age and ascertaining the climate of the older segments of geological time are quite different from those of the Pleistocene: the numerous and far-reaching climatic fluctuations of the Ice Age came to pass during the comparatively short interval of about one million years. It follows that exact determinations of age are needed, as well as reliable methods for determining ancient climates. Even today it is necessary to question whether such exact methods actually are available.

A further problem arises out of the very difficult starting points from which many disciplines have analyzed problems of the Ice Age. It is not always possible to define accurately the limits and applicability of every method; yet those results are relied upon. In the following section the methods of determining age and climate applicable to the Quaternary will be reviewed briefly, with some indication of the principal difficulties of each method.

THE PROBLEM OF AGE DETERMINATION

Relative Dating

In active sedimentary basins, newly deposited beds lie upon older ones, so that there is a sedimentary succession which proceeds from older to younger beds. Many times during the course of sedimentation there is a change in the type of rock being formed—a sign that the climate has changed, or that the basin in which the sediments are accumulating has been depressed or uplifted, or that there have been changes in the region of provenance, or in stream volume or velocity. The vertical succession of rock types within a given region may thus provide a useful starting point to determine the relative age of the individual beds. The rule of younger beds lying upon older does not generally apply in the case of intrusive rocks, for a rising magma may well remain at a high or a low level within a sedimentary column. Even in such cases relative dating is usually possible either by the relative stratigraphic position of the igneous rock or by the occurrence of material derived from it, in datable sediments.

The kind of sediment formed is not a function of time. Sandstone, shale, and conglomerate, for example, form in different places at the same time, or in the same place at different times. The vertical succession of varying rock types is therefore no general criterion for the establishment of relative age. The evolution of the plant and animal kingdoms is, however, time-dependent, and the development and subsequent extinction of different plant and animal groups is an irreversible and irreproducible process.

As a result, the occurrence of easily recognizable plant or animal forms which, though widely distributed have a relatively short duration in time, provides a welcome and safe means for relative dating of individual beds in a sedimentary succession, to the extent that these beds contain fossil remains. This method of relative dating of rocks by means of their "index fossils" is the basis of historical geology.

It is clear that index fossils are particularly valuable if (a) a long time interval is considered, in which marked evolution of the different plant and animal groups has occurred, or (b) when allied genera which have undergone explosive evolution during the time period are included, so that small time differences can be recognized with some certainty. It can be presumed that the new genera which appeared during the Ice Age, in a radically changing milieu, were able to occupy it and spread rapidly, so that the number of index fossils in Pleistocene sediments must have been especially great.

It is important in this respect to note an essential difference between the plant and animal kingdoms. In the animal world many new forms actually appeared and spread over wide areas. The development of the elephant and bear families are typical of numerous other developments (see Fig. 4). The rapid development of these families stands in marked contrast to the generally slow evolution of the plant groups whose fossils are known from various segments of Ice Age time. The exhaustive researches of Jentys-Szaferowa [377–80], of Firbas and Firbas [224], and of Kokawa [420–22] on the fruit and pollen of the hornbeam (*Carpinus betulus*) and the buckbean (*Menyanthes trifoliata*) in Central Europe and Japan, as well as the work of Villaret–von Rochow [773] on the seeds of the spiny water lily (*Euryale ferox*), and of Staszkiewicz [718] on the cones of the common pine (*Pinus sylvestris*) in Central Europe, have all shown that the evolution of these genera proceeded very slowly during the Ice Age. In general, no morphological changes are distinguishable.

Although this is also true of most of the remaining plant genera, it does not necessarily follow that there was no evolution at all within the plant kingdom during the Ice Age. The changes came about in ways hard to distinguish. It seems that physiologically and cytologically different subspecies were formed repeatedly, which, while strongly resembling the parent forms, possessed different physiological characteristics, permitting them to colonize different areas. Furthermore, as a result of hybridization of two different parent forms, new types continually developed which either remained close to their point or origin or left fossil remains that are exceedingly difficult to recognize.

The Robinia or false acacia (*Robinia pseudacacia*) is a good example of a change in physiological characteristics. This species was introduced into Europe and northern Asia from North America around the beginning of the seventeenth

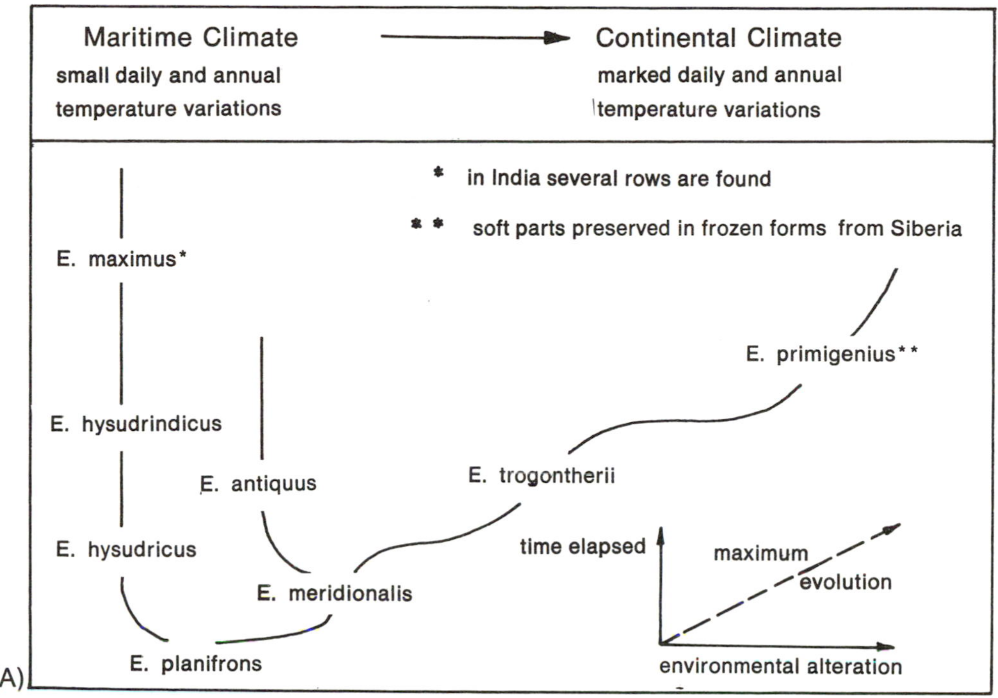

FIGURE 4. The evolution of certain animal families as a means of relative dating, illustrated by the development of the elephant (A) and the bear (B). The oldest forms are shown on the lower part of the figure. Figure A is from Adam [846], Figure B from Toepfer [847] (after Adam).

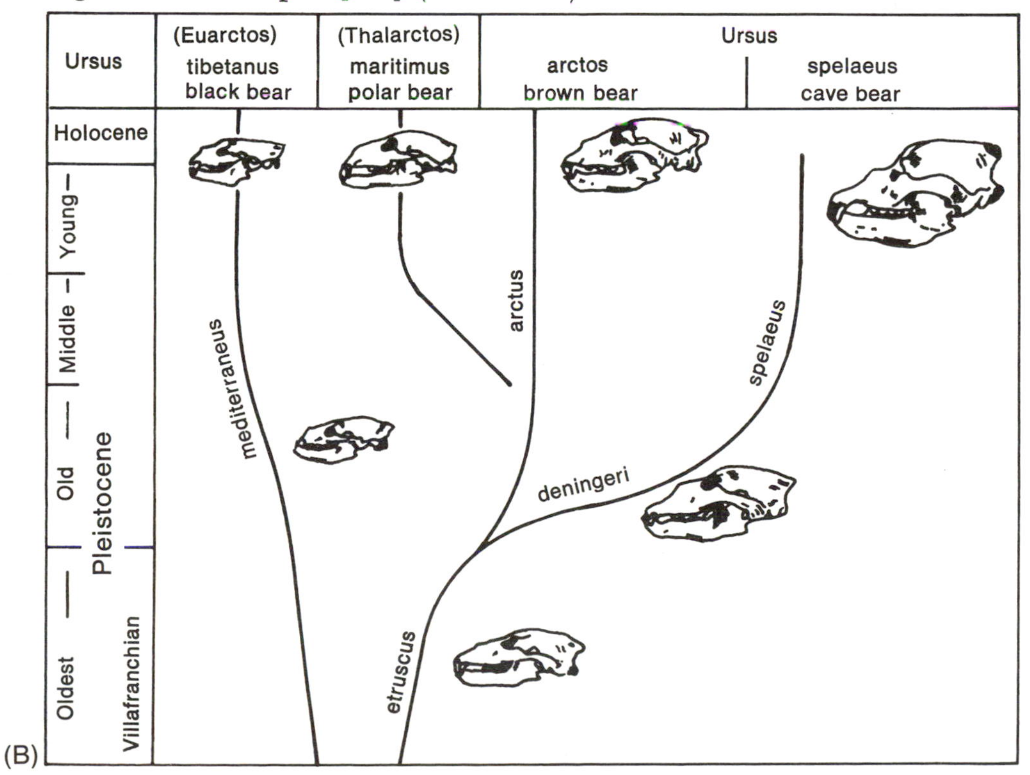

century. Since then it has formed different physiological subspecies, in the foothills of the Pamirs in central Russia and near Mosyr in White Russia, which differ in their frost resistance and daylight requirements without any apparent morphological distinctions [541].

Another example is the buckler mustard (*Biscutella laevigata*). This species represents a group of forms, some having a double chromosome number in the cell nuclei (diploid forms), and some containing a fourfold chromosome count in their nuclei (tetraploid forms). The latter occur today within the area of former glaciation in the Alps, while the former are found in the previously extraglacial region. It seems likely that the tetraploid forms, which should be regarded as hybrids of the two diploid taxa *Biscutella arvernensis* and *B. lamotti*, developed in central France. The hybrid population, because of its greater vitality, was able to colonize rapidly large areas of newly exposed surface at the close of the Ice Age, whereas the diploid parent forms, and other older diploid taxa of this group, were unable to leave their old glacial refuge when the climate improved [508, 509]. This newly formed cytological and physiological taxon can be distinguished only with difficulty from present-day material, and morphological separation on the basis of its fossil remains appears nearly hopeless.

As examples of hybridization, let us use one of the firs of the Balkan peninsula (*Abies borisii-regis*) and the hedge nettle (*Stachys germanica*). In central Greece,

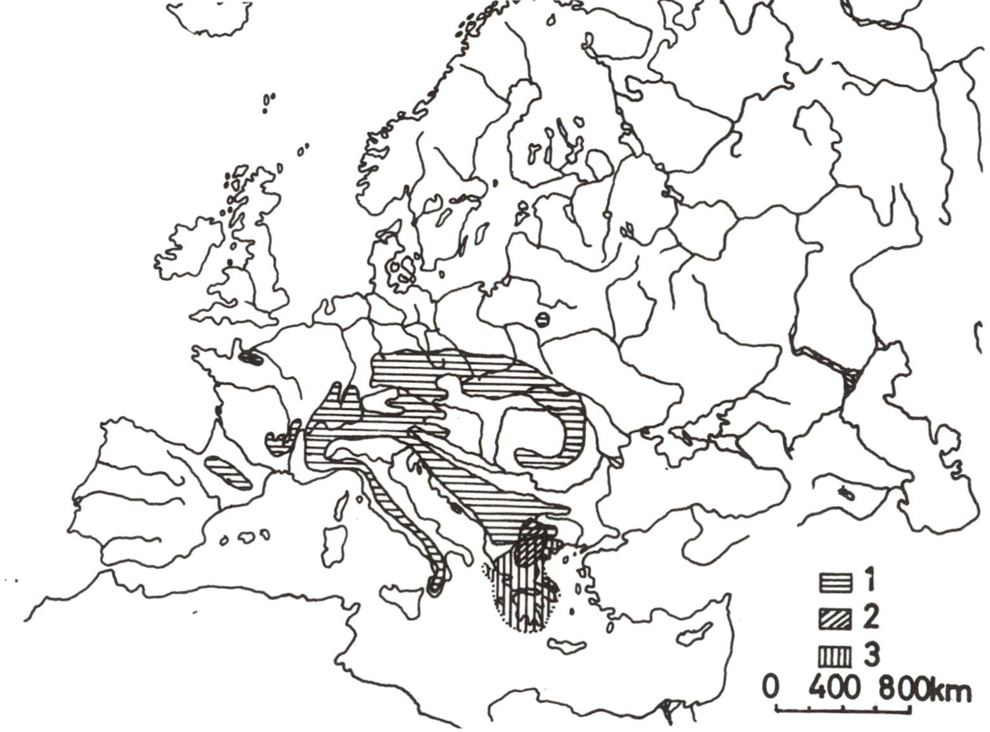

FIGURE 5. The evolution of some European firs during the Ice Age; examples of hybridization. 1. present area of *Abies alba*; 2. present area of *Abies Borisii-regis*; 3. present area of *Abies cephalonica*. The area of distribution of *Abies Borisii-regis* which developed out of the hybridization of *Abies alba* and *Abies cephalonica* has not significantly expanded beyond its original limits.

between the regions occupied by the central European silver fir (*Abies alba*) and the southern Greek fir (*Abies cephalonica*), there is a fir morphologically intermediate between them. This *Abies borisii-regis* (Fig. 5) was formed by hybridization of the other two when, as a result of Ice Age climatic fluctuations, the zone of *Abies alba* was displaced into that of *Abies cephalonica* [522]. The newly developed form, which may be used as an index fossil throughout the Quaternary sedimentary succession, still has not expanded beyond the limits of the region in which it was formed, so that it has no significance for the relative dating of Ice Age sediments.

In a similar manner the changes induced by Ice Age climatic fluctuations, in the areas occupied by two parent forms, *Stachys lanata* and *Stachys alpina*, produced the hedge nettle *Stachys germanica* [449]. The newly formed hybrid had quite different environmental requirements and was also significantly more vital than the parent species, so that it became rapidly established and widely distributed (Fig. 6). However, to date it has not proven possible to distinguish with certainty its fructifications and pollen from those of other nettles, so that despite its favorable behavior this species too can scarcely be considered an important index fossil.

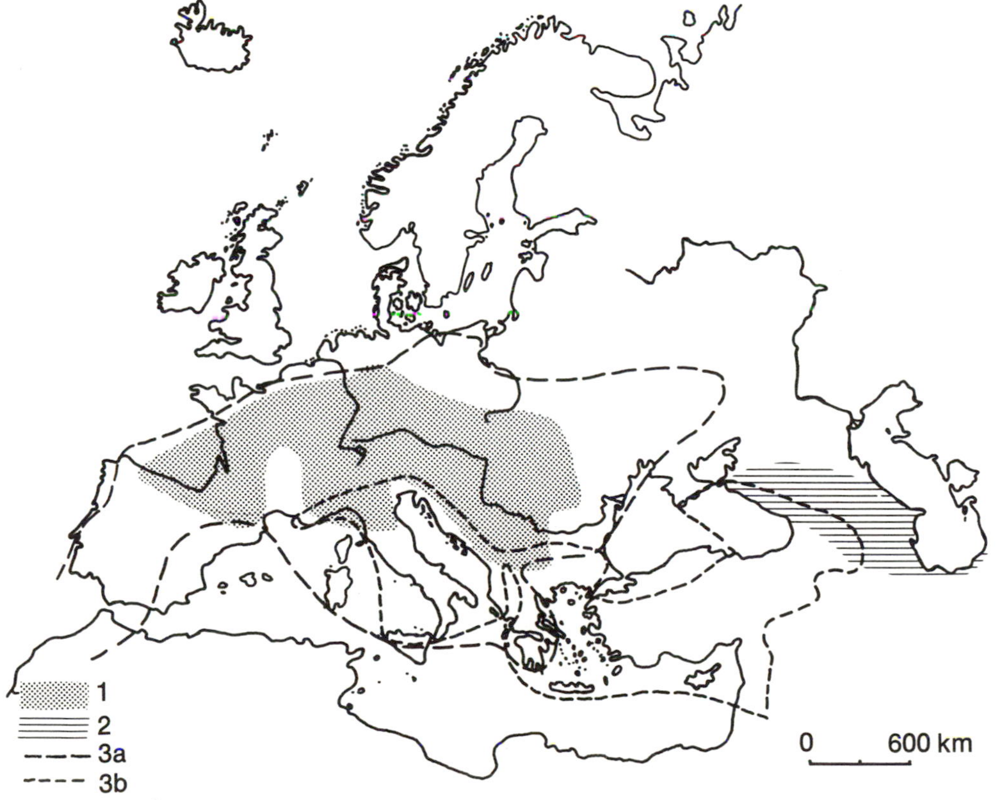

FIGURE 6. The evolution of certain European nettles (*Stachys*) during the Ice Age. An example of the development of a hybrid species with greater vitality. 1. present area of *Stachys alpina*; 2. present area of *Stachys lanata*; 3. present area of the hybrid *Stachys germanica* formed from the above parent forms (a = eugermanica group, b = cretica group). Simplified, after Lang [449].

Although further examples could be cited, the above suffice to demonstrate the important difference in the evolution of the plant and animal kingdoms. From this it might appear that the animal kingdom has greater significance for the relative dating of the Ice Age than does the plant realm, but that generalization would be incorrect. Although it is true that the evolution of certain animal groups provides a valuable basis for the relative dating of sediments which originated at widely separated times, relatively speaking, evolution requires a long period; short intervals, such as those between two successive cold or warm periods or the duration of a single cold or warm period, are usually insufficient.

It is in such cases that the great usefulness of paleobotanical research becomes evident. Even if the evolution of individual plant genera cannot be turned to good account, as much can be achieved by examination of the changing composition of a flora during long stretches of the Ice Age as by the alterations of the vegetation itself. The climatic stress of the cold periods eliminated from wide areas of the Northern Hemisphere elements of the flora which were widespread during the late Tertiary. From the decrease in the proportion of forms that prefer warm, mild climates and the concomitant increase in the proportion of those which prefer cool climates, the relative age of fossil floras from given regions can be determined with reasonable certainty (Fig. 7). Furthermore, the history of the vegetation of the plant community shows distinct differences during comparable climatic phases in different eras. There are two reasons for this. First, many of the floral elements were slowly decimated during the Ice Age in the Northern Hemisphere. Second, the unfavorable environmental conditions during the cold periods caused many forms to be displaced far to the south, only to have them slowly migrate northward again as the climate improved. The location of the Ice Age refuges in which they could survive the severe climate depended not only upon the climatic conditions in the southern refuges during the cold spells, but also upon the preceding stages in the history of the vegetation—that is, upon the available migration routes and the competitors living along the routes. These conditions, in particular the number of competitors in any one place at any one time, are unique, and the history of the vegetative cover for a given period of the Ice Age at a given place generally has its own particular characteristics, by which it can be distinguished from other intervals of Pleistocene time. This circumstance will appear very clearly in any review of the development of vegetation in the northwestern part of central Europe during even the earliest warm periods of the Pleistocene (see Table 1).

Similar distinctions in vegetal development during warm periods have been observed in many other parts of the world, so that the vegetative history of a region is particularly well suited to establishing the relative ages of the rock sequence. If, in addition, these observations can be coupled with the recognition of faunal index fossils in the succession, then a positive and unique relative dating of beds of different ages during the Quaternary is possible.

This important conclusion, however, applies only to the regions in which the climate during the Ice Age varied between two contrasting extremes, as in the region of the present oceanic and suboceanic climates of Europe, eastern Asia, North America, and possibly South America and New Zealand. In the present continental climates of the Northern Hemisphere, particularly in Siberia, the

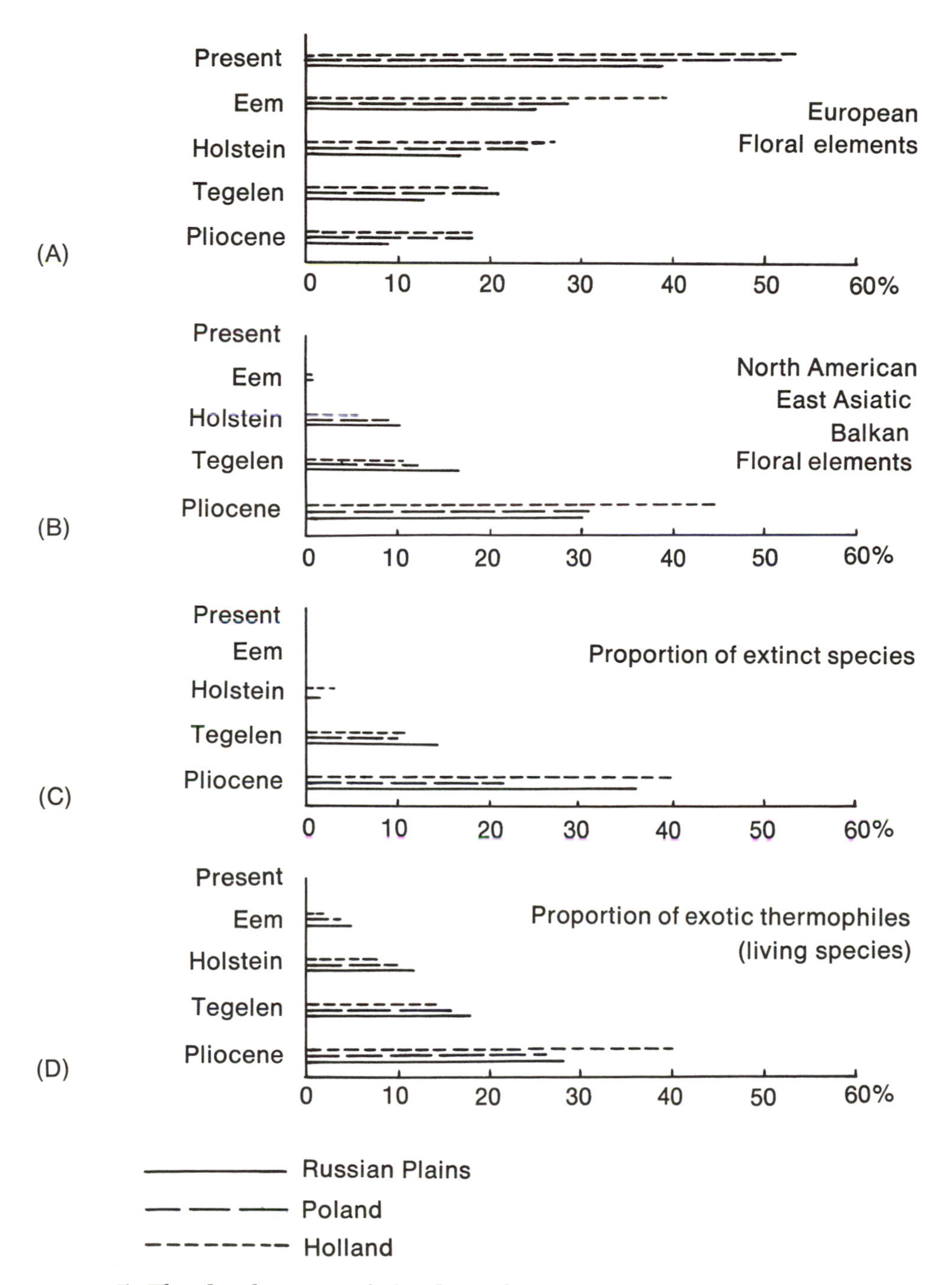

FIGURE 7. The development of the flora of a given region as an aid to relative dating. In the diagram the changes in the proportion of the individual geographic (A and B), historic (C) and ecological (D) floral elements in the floras of Holland, Poland, and the Russian plains is represented. The abscissa indicates the percentage of the total flora found during a given period; some of the warm periods are shown on the ordinate. From Grichuk [303].

TABLE 1. THE FLORAL HISTORY OF THE PLEISTOCENE WARM PERIODS IN NORTHWEST CENTRAL EUROPE [AFTER 819, 13, 14, 260] (THE PHASES ARE LISTED IN STRATIGRAPHIC ORDER)

TIGLIAN WARM PERIOD	WAALIAN WARM PERIOD	CROMERIAN WARM PERIOD	HOLSTEINIAN WARM PERIOD	EEMIAN WARM PERIOD
Pine and spruce woods with transition to a subarctic parkland. Oak-hornbeam-elm woods with many *Eucommia*, *Ostrya*, *Tsuga*, and spruce. Pine and spruce woods still with only a small proportion of thermophilic plants. Oak-hornbeam woods with numerous elm, *Ostrya*, *Eucommia*, chestnut, and evergreen oak (*Quercus* cf *Ilex*). Open woodland with wingnut, ash, hickory, alder, and vines. Alder and wingnut woods; woods of pine, spruce, and *Tsuga*; very few thermophilic species. Pine woods. Woods of beech and hornbeam, *Tsuga* and elm; spruce and fir woods; open woodland with alder, wingnut, and *Myrica*. Pine and spruce woods; oak and alder woods. Subarctic parkland.	Pine, spruce, and birch woods in transition to open vegetation. Oak woods with hornbeam and elm as well as *Ostrya* and rare *Corylus*. Reappearance of wingnut; *Tsuga* common. Few spruce. Spruce-birch phase with a high proportion of marsh grasses (*Cyperaceae*) and *Ericales* heath. Oak-hornbeam woods with many elms and later also with hazelnut and *Ostrya*. In this later phase wingnut and *Tsuga* occur. Open grassland with *Cyperaceae* vegetation and pine and spruce stands; subarctic parkland.	Tundra vegetation; pine and birch woods and spruce woods. Only a small number of thermophilic plants occur. Vegetation becomes more open. The importance of pine and birch increases. The oak woods are widespread and contain relatively numerous linden and elm. The importance of spruce decreases. Oak and elm woods with linden. Maximum spread of hazelnut. Yew occurs rarely. Strong development of yew bushes in oak and elm woods. The importance of flood plain woodland diminishes. Oak and elm woods with incoming linden. Very few hazelnuts. Elm and alder woods, with oak becoming increasingly important. Pine and birch woods.	Pine thickets and *Ericaceae* heath. Pine-spruce woods; alder flood-plain woods; widespread fir; and hornbeam woods. Pine-spruce woods; widespread alder flood-plain woods. A smaller proportion of oak and elm woods, with few linden and hazelnut. Pine-birch and spruce woods, and juniper heath. Incoming of oak and elm. Subarctic parkland.	Pine-birch woods, Sphagnum bog, and *Ericales* heath. Pine-spruce woods with birch; fir and hornbeam of decreasing importance. Strongly marked decrease in oak. Ash rare. Pine woods and spruce-fir woods, with numerous hornbeam. Hornbeam woods; also oak woods without linden and generally without elm. Hazelnut and spruce of minor importance. Fir is present. Oak-elm woods with many hazelnut and a few linden. Hornbeam and spruce increasingly important. Widespread hazel in an oak-elm wood. Hornbeam appears. Oakwood with elm. Ash-alder flood-plain woodland. The importance of pine diminishes significantly. In woodland, hazelnut important. Pine woods in which oak and

Table 1, continued

TIGLIAN WARM PERIOD	WAALIAN WARM PERIOD	CROMERIAN WARM PERIOD	HOLSTEINIAN WARM PERIOD	EEMIAN WARM PERIOD
				alder begin to spread markedly.
				Pine woods with birch, elm, and ash.
				Birch-pine woods.
				Birch-pine parkland.

NOTE: If the relative age of Pleistocene sediments can be determined by these methods, it may be possible under favorable circumstances from observations on sedimentation rates found by other means, to estabish the absolute age of the rocks under investigation. This is only very rarely possible, however, and in general it is necessary to seek other techniques to determine the absolute age of rocks.

characteristics of extreme cold and aridity were already established by the end of Tertiary times, and the unfavorable climate had already led to the development of a greatly impoverished flora and fauna comprising only those few forms able to survive under such conditions. Even the climatic amelioration of the Pleistocene warm periods was insufficient to produce any noteworthy immigration of other plant and animal communities, so that the number of reliable index fossils (animal) is small, and no marked differences are discernible in the history of the vegetation during Pleistocene warm periods.

Probably the floral and faunal changes in the present humid tropical regions and bordering zones, with alternating wet and dry seasons to the north and south, were also small. For despite the climatic fluctuations of the Ice Age, favorable environments persisted, and no single plant and animal genus was provided with selective advantages.

Reliable relative dating of various sediments from the Ice Age on the basis of the evolution of particular groups of animals and the changes in plant cover during the course of time is thus feasible only in certain regions of the earth; in others it fails or at least runs into difficulties. If the relative age of Pleistocene sediments could be determined by these methods, it might be possible, from observations on sedimentation rates made by other means, to establish the absolute age of the rocks under investigation. This is only very rarely possible; in general it is necessary to seek other techniques.

Absolute Dating

The methods of absolute dating and their applications are just as varied as are the underlying principles. Not only have the growth rates of long-lived plants been used (lichenometry and dendrochronology), and the variation in the rate and type of sedimentation in still water under a climate with a pronounced seasonal character (varve chronology), but also other methods based upon the decay rates of radioactive isotopes (radiocarbon dating and other similar methods) and the slow rate of certain chemical reactions (fluorine test) have been used as measures of time. Irrespective of the method, it is essential in every case to be aware of its applications, limitations, and inherent sources of error.

*1. **Lichenometry.*** There are many regions on the earth's surface where lichens occur on the surfaces of rocks and trees. The method of dating by lichens utilizes the growth rate of the plant body or thallus as a measure of time. The most suitable are the lichen growths on stones, such as *Rhizocarpon geographicum*, *Aspicilia cinerea*, *Diploschistes scruposus*, and *Lecidea promiscens*. Once established upon a rock, a lichen passes through a certain period before it is macroscopically visible. Then follows growth acceleration (the main growth period) until, after a few years or decades, thallus growth slows and an approximately constant growth rate is established which may persist for several centuries. Lichenometry therefore spans a maximum period of 1,000 to 1,300 years (Fig. 8). The application of the growth of lichen thalli as a measurement of time presupposes that only the same lichen species can be compared, and that these must furthermore utilize the same substrate and can there develop freely (that is, no turning or rolling of the stone, etc.) in conditions of adequate light and moisture in the same and constant local microclimates. Under differing climatic conditions the lichen thalli of the same species of the same age may be of different sizes, so that lichenometry may

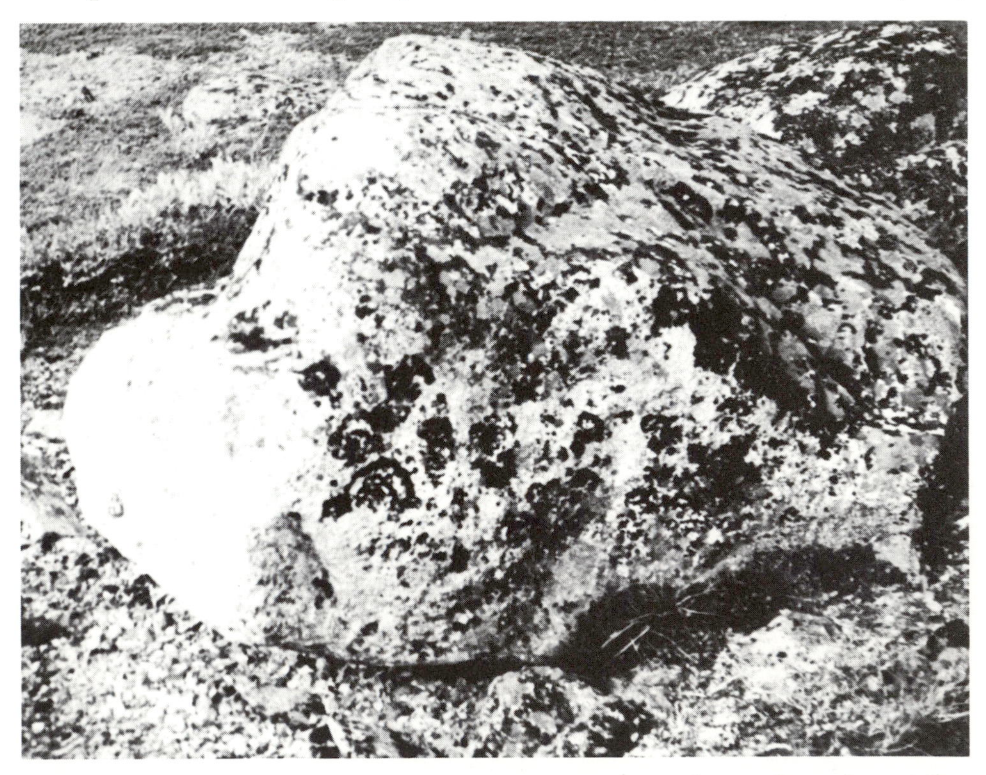

FIGURE 8. Lichens as a measure of time of exposure of a rock surface. Many lichens can be recognized upon this block approximately 150 centimeters high. They grow in a more or less concentric pattern outward from the initial starting point, as can be well seen on the lower left of the figure and on the upper surface gently dipping to the right, where the lichens appear a bright gray. Near Björkliden, Swedish Lapland.

be used as an indication of variations in microclimates, in particular of moisture conditions. Although the narrow time interval over which the method can be applied restricts its use, it affords nonetheless a means of dating the younger terminal moraines in subarctic and mountainous regions and thus performs a useful service in the dating of climatic fluctuations in historic times. It is important to stress that the method dates only exposed surfaces and not rock successions, and that thalli covered by sand or other sediments are rapidly destroyed by soil micro-organisms [48–51].

2. ***Dendrochronology.*** A great number of growing trees show an annual fluctuation in the intensity of division of cambium cells—that is, in the cells responsible for the secondary thickening of the axial elements and the differentiation of the daughter cells. This is the reason for the formation of wide-celled early wood and the small-celled late wood. Periodic difference in the activity of the living cells under the most favorable circumstances makes it possible to determine the exact age of individual trees by a count of the annual rings so formed. Though the periodic activity of the cambium and the newly formed cells depend upon endogenic rhythms of the tree, differing external factors play a critical role in the shaping of annual rings. Not only may the width of the total annual ring or the early or late wood vary from year to year, but under very unfavorable circumstances in certain years the formation of the entire ring may be suppressed, or the plant may even react with a doubling of the annual ring, depending upon how unfavorable the external factors were. Such climatically unfavorable years, together with a host of other influences—such as damage by lightning, wind, insects, or fungi—handicap the user of dendrochronology by falsifying the ring count—a problem that can be overcome only by analysis of a large number of trees or many samples from single trees. On the other hand, the fact that the kind of tree ring formed is dependent upon external conditions provides a useful means of correlating wood from trees of unknown age with those it has already been possible to date. If the effects of annual fluctuations of climatic factors about a given mean are different for individual tree species, within the same species at comparable locations similar variations between broader and narrower rings result. This is particularly true of the climatic conditions of extreme years ("Weiserjahre") [356, 357, 17, 279, 554, 79, 268, 574, 208, 349, 663, 664, 352, 353]. This characteristic sequence of broader and narrower annual rings in individuals of a particular species of tree, irrespective of the climatic interpretation, occurs over large areas, as for instance in several regions of central Europe, and in part may be traceable in differing tree species in wood of the same age [554]. As a result, using the so-called bridge method [279], it is possible to correlate trees of unknown age with already dated specimens, provided there is a minimum of 50 annual rings which exhibit the same sequence of thicker and thinner rings as well as the same increasing or decreasing trend in ring thickness and which thus possess a very low "anatropous percentage" of disagreement [354]. In such a manner, step by step, the period that may be dated by dendrochronology can be extended further into the past. This very laborious work for which special apparatus and techniques, including the use of the computer, have been developed [185, 375, 182a, 263f] has led to the establishment of a chronology based upon the oak extending back to 430 B.C. [353, 348a], and in the U.S.A. to a chronology based

upon *Pinus ponderosa* [664] extending back 2000 years and one based upon *Pinus aristata* which reaches back to 5514 B.C. [215a]. Such chronologies have already proved their worth in the dating of prehistoric buildings, and in so doing have gained recognition as a reliable means of determining absolute age.

In contrast to lichenometry, dendrochronology can also be applied to the determination of the absolute age of sediments that contain wood fragments. Moreover, it is theoretically possible to extend the dating into the past for as long as wood growth with secondary thickening has occurred. Yet in a practical sense the method has severe limitations, for the strong Ice Age climatic fluctuations have led to such far-reaching displacements of the areal distribution of tree growth that dendrochronology as a reliable measure of time, at the present moment, can under the most favorable circumstances cover only the postglacial period.

3. Varve chronology. Melt waters escape at the glacier margins. Where they enter enclosed basins the transported material is rapidly sorted according to size of grain; the coarsest material is deposited close to the entry point, with progressively finer grained material deposited increasingly further from the stream mouth. Transportation of sediment by melt water is dependent upon the volume of water to the extent that with the great increase in the melting of snow and ice during spring, much sediment is transported, with fine sand being carried further in the sedimentary basin. In autumn and winter the diminished stream flow carries in only finer material, which will be transported greater distances. The result is that in a given basin at some distant point from the ice front there is an alternation of thicker, lighter-colored sandy layers, with the thinner and darker layers of clay being formed during autumn and winter. Those sediments from the more distant part of the sedimentary basin that is receiving glacial melt waters therefore exhibit a seasonal banding. The annual clay and fine sandy layers are termed "varves," and the rock a banded or varved clay (or in some cases a banded sand).

With the melting of the glacial front, the zone in which varved clays are formed advances in the direction of the ice retreat; and in the region of formerly deposited banded clays other lacustrine sediments are deposited in which the pronounced annual banding is not normally found. If the basin is filled or drained as the result of downcutting its banks, the depositional history of that basin, and the use of varves as a measurement of time, can be carried no further. The retreating glacier in general leaves behind large numbers of hollows or basins, due either to the erosion of the former ice cover or to the melting of stranded isolated ice masses ("dead ice")—or to the erosive activity of subglacial streams. A large number of these basins are filled with melt-water sediments, so that varved sediments are deposited at very many places along the retreating ice front. As indicated, the transportation of sediment by melt water is dependent upon climate, but it is additionally dependent upon local hydrological conditions. Consequently, in adjoining regions of the ice front thicker and thinner varves may form at the same time, and as in the case of tree rings, a characteristic and unique succession of varves is formed. This makes possible the comparison of varves from different but neighboring basins. In this connection the absolute thickness of the varve is not considered, but only the nature of the variations in thickness (Fig. 9). The careful measurements of varve thicknesses from a large number of basins

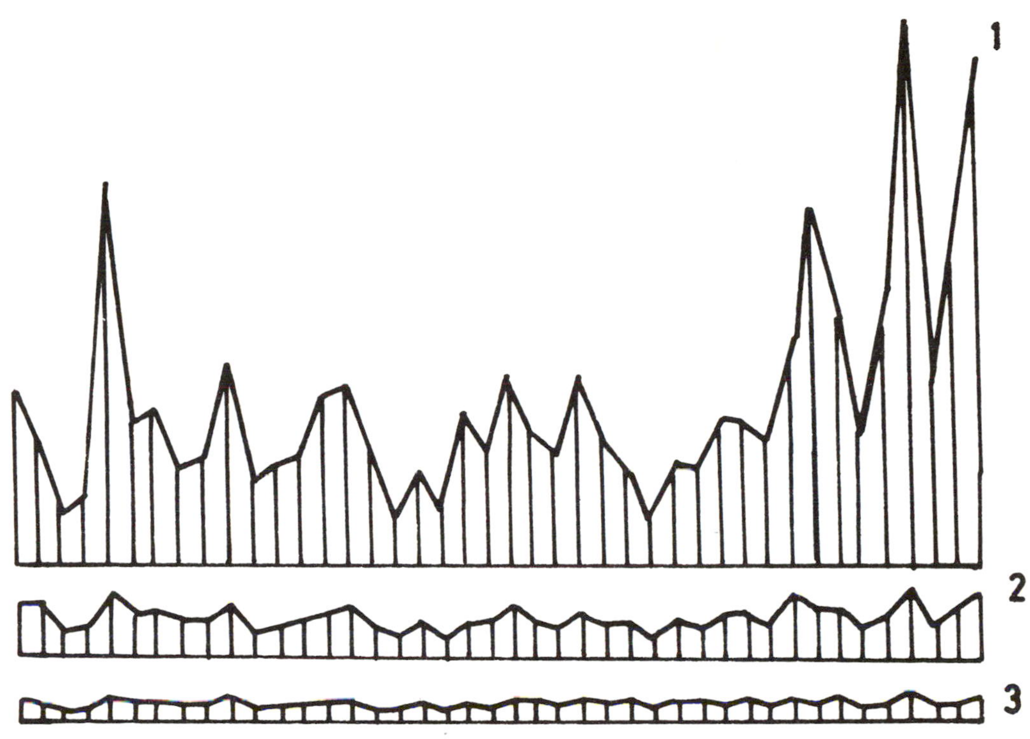

FIGURE 9. The banding of varved clays as a means of absolute dating. In the figure the thickness (ordinate) of a sequence of varved clays (abscissa) from three different glacial lakes are represented. Although clearly the varve thicknesses may be very different in the basins compared, there is good agreement in the increasing or decreasing thickness trends, so that the sections investigated must be of the same age, and the three outcrops may be correlated with confidence. After Lidén, from Woldstedt [804].

around the margins of the former Scandinavian glaciers, particularly by de Geer [269] and his co-workers, have made it possible not only to date the beginning of postglacial time but also to establish absolute ages for many periods in the late glacial development of the Scandinavian glaciers around the Baltic Sea. As a result, the chronology of postglacial time in the Nordic region was known long before the advent of radiocarbon dating, and the results of the latter method can be tested against varve chronology.

Attempts have repeatedly been made at long-distance correlation—that is, attempts to compare variations in the thickness of varves within one glacial region with those of another, in order to parallel postglacial events in Europe and North America. Such attempts must assume that the climates in the two regions under comparison agree even in the finest detail, an assumption which, in the light of the very different climatic character of these areas recently, scarcely appears to be justified [cf. the critique in 804].

A fine and apparently annual rhythmic sedimentation can be found not only in the immediate vicinity of ice fronts, but also wherever a change of seasons has produced a variation in the rate of sedimentation. This is the case wherever the climate has distinct seasons, as, for example, in a monsoon climate where the introduction of minerals varies markedly, or where the productivity rate of organisms within a body of water varies distinctly between periods of quiescence and periods of rapid growth. In this connection kieselguhr beds, composed essentially of the exoskeletons of siliceous algae (diatoms) are of interest. The observations of Follieri [251] on the kieselguhr outcrops from the penultimate warm period (Holstein) from the Valle dell'Inferno near Rome may serve as an example. The kieselguhr is horizontally bedded in such a way that an alternation of thinner and lighter with thicker and darker bands can be observed, the opposite of the situation found in basins close to the ice front. The difference arises because of the greater production of organic matter, which sediments as a darker layer, during the warm season, while during the winter months the level of organic activity is low. The kieselguhr beds near Rome are very rich in pollen, whose composition can be determined microscopically. Of the plants flourishing in the vicinity of the lake, those blossoming in early spring were the wingnut (*Pterocarya*), zelkova (*Zelkova*), elm (*Ulmus*), and alder (*Alnus*); in spring the pine (*Pinus*), spruce (*Picea*), fir (*Abies*), walnut (*Juglans*), oak (*Quercus*), maple (*Acer*), beech (*Fagus*), hornbeam (*Carpinus*), hop-hornbeam (*Ostrya*), and birch (*Betula*); in the summer the lime (*Tilia*), ash (*Fraxinus*), and Spanish chestnut (*Castanea*). Within a light and dark band respectively, tree pollens representative of blossoming forms of early spring, spring, and summer are represented in percentages as follows:

	LIGHT BAND	DARK BAND
Early spring	35	37
Spring	61	52
Summer	4	11

There is clearly an annual banding, though only very weakly developed. Although this is an example of annual banding in a lake basin remote from the former glaciated areas, it can scarcely be applied as a means of absolute dating, for the number of available basins is far too small to develop an absolute varve chronology comparable to that established along the margins of the northern ice.

Varve chronology is thus essentially restricted to the former glaciated areas, and it is doubtful whether the climatic development of two widely separated glaciated regions can be compared by this means. However, as it provides a means of dating phases in development of the vegetation in specific districts, and as the development of the plant cover is largely determined by climatic fluctuations, varve chronology seems to provide a means of assigning reliable absolute ages to the principal stages in the postglacial climatic history of large areas. It is improbable that this dating can be expanded to cover the whole of the last cold period or still earlier segments of the Ice Age, for one of the essential conditions of the method is that there should exist and be available a large number of lacustrine basins. These are first formed, or vacated by, the retreating ice. Beyond the ice front of the last cold period there were, nevertheless, enclosed basins formed during earlier cold periods, but as these were to be a large extent already filled, varved clays could be deposited there neither during the maximum phase of the last glaciation nor at its commencement.

4. Carbon 14 dating: the radiocarbon method. An extraordinarily useful dating tool for organic matter several thousand years old is the method of dating by means of the radioactive Carbon 14 isotope (^{14}C) developed by Libby [468]. This isotope is formed at altitudes between 13 and 40 kilometers by the reaction of neutrons (n) under cosmic radiation with the nuclei of the nitrogen isotope, Nitrogen 14 (^{14}N), according to the equation

$$^{14}N + n \longrightarrow {}^{14}C + {}^{1}H$$

and its subsequent decay with a half life of (5730 ± 40) years according to the expression

$$^{14}C \xrightarrow{\beta^-} {}^{14}N$$

reforming ^{14}N. The Carbon 14 is soon taken up in $^{14}C_2O$ and circulated by currents throughout the atmosphere. At the air-water interface strong absorption occurs, and $^{14}CO_2$ is also absorbed by green plants. In this connection it is important that the inclusion of the carbon dioxide and with it the establishment of an equilibrium condition in the (^{12}C + ^{13}C) Carbon 14 ratio of the plant substance on the one hand and the corresponding ratio in the atmosphere on the other should result only during the lifetime of the plant. In an extended sense this is also relevant to animals which, directly or indirectly, live upon plants. With the death of the organism free carbon exchange between the organism and the surrounding medium ceases, and the amount of Carbon 14 present decreases through radioactive decay. In consequence the proportion of Carbon 14 to the total carbon of the body substance, or the number of decaying atoms per unit time per gram of carbon, measures the age of the organic matter according to the equation:

$$A_x = \frac{T}{0.693} \cdot \ln \frac{a_0}{a_x}$$

in which A_x is the age of the sample, T the half-life of ^{14}C, a_0 the specific activity of recent plant matter and a_x that of the sample being examined [724].

The most important assumption, if it be argued that the radiocarbon method provides reliable values (excluding the precision of instrumental techniques), is that the radiocarbon content of the atmosphere during the period covered by the measurements (about 50–60,000 years) remained constant and has remained so to the present day. It follows from this that the intensity of cosmic radiation should also have remained unchanged during that time. Yet we know that even within Postglacial time the geomagnetic field intensity [191], and related to this the intensity of cosmic radiation, have repeatedly altered. Variations of this kind lead to similar fluctuations in the production of ^{14}C so that the ^{14}C age of a sample can deviate appreciably from the true age [573a]. As up to the present time the first indication of the magnitude of this source of error occurs during the last glaciation [723a], the radiocarbon dates are subject to an error whose magnitude cannot be estimated at the present time.

The radiocarbon content of the atmosphere depends not only upon the intensity of cosmic radiation, but also upon the magnitude of the organic body, in particular the plant body that absorbs $^{14}CO_2$, and upon the volume, surface area, and chemistry of water basins with whose bicarbonate and carbon dioxide content the carbon dioxide content of the atmosphere is in equilibrium. The magnitude of the plant body during the time spanned by radiocarbon measurements has certainly not altered repeatedly to any significant degree. As the proportion of the earth's total carbon dioxide locked up in plants is very small compared with that in the oceans which is freely interchangeable, it is probable that the error arising out of changes in the organic body within the last 60,000 years is of negligible importance in age determinations. Changes in the temperature and the pH of the oceans and in the oceanic surface area which occurred in the relatively short time between cold and warm periods can all have affected the radiocarbon exchange with the atmosphere, and their magnitude cannot be estimated. In addition to this, the Pleistocene sea-level fluctuations may have produced changes in oceanic currents and so in the mixing of sea water, which can have introduced alterations of unknown magnitude.

There is one further source of error to be noted. Willis, Tauber, and Münnich [799] have compared the radiocarbon dating of annual growth rings of the giant redwood (*Sequoia*) with the data from tree-ring chronology. They considered the time period from A.D. 659 to A.D. 1859. The comparison indicated that the Carbon 14 value of the atmosphere was lower than the 1859 value from A.D. 659 to A.D. 1200 but higher during the remainder of the time. The fluctuations in the original radiocarbon content follows two superimposed periodicities, one that appears to last 150 to 200 years and the other about 1,200 years. Suess [724a] recently computed a wavelength of the fluctuation in the production of ^{14}C of 200 to 400 years. It is worth recording that the fluctuations in the original radiocarbon content show a remarkable similarity to climatic fluctuations that were determined by other means for the same time period. The errors introduced by fluctuations in the original radiocarbon content of the atmosphere (deviations of around 1.5 percent) were quite large in the younger annual rings but decreased in old annual bands. Yet it is not to be concluded from this that such sources of error may be neglected; rather it should serve as a warning to treat with caution the similarity between fluctuations in the original Carbon 14 and climatic fluctuations, not least because the cause of this remarkable parallelism is not known. It must be remarked that according to Tauber [in 724a] the growth of European bogs shows a 400-year climatic period, whose wavelength is thus of the same magnitude as the fluctuation in ^{14}C production. Bray [78a] also noted that there is at least one longer period, of 2600 years duration, superposed upon this 400-year cycle.

Before applying the radiocarbon method to the determination of the ages of different objects, its usefulness and reliability were exhaustively tested upon objects dated by other means [556, 557]. They demonstrated that useful values can be obtained for Postglacial times provided extreme care is taken in sampling, and that secondary mixing of the sediments by burrowing animals can be excluded [151a, 571a, b], and provided that by use of correction available, the ^{14}C date gives a true age [573a]. The sources of error noted above, however, must still be considered if older samples are to be analyzed, for they are representative of a time in which the climate, vegetation, and possibly the magnetic field strength also were different from the present day [144a]. The number of Carbon 14 dates is extraordinarily large, and their usefulness in many fields of science can scarcely be overemphasized. Yet the principal difficulties and technical sources of error which beset the radiocarbon method are only slowly being overcome, so that the radiocarbon dates obtained at different times from the same sample or the same geological horizon are unfortunately not always comparable. The reason may be sought on the one hand in the extraordinarily sensitive measuring apparatus required in which the Null Effect is greatly reduced. On the other hand, an error may arise through contamination of the sample by inclusion of older or younger carbonaceous material during sampling, an error that can be overcome only by examining samples that have not been in contact with older or younger material. In addition, consider the following: The Stillfried B. interstadial was one of the most important interstadials during the last cold period in central Europe. Its age runs from about 30,000 to 25,000 B.P. With a half-life of about 5,700 years, this means that a sample from that age contains only about 3.1 percent of the initial amount; in the case of material dating from the first warm fluctuation of the last cold period—that is, about 60,000 years old—only about 0.05 percent of the initial Carbon 14 remains. With increasing age, therefore, the method rapidly becomes less reliable, so that even though dates of important periods are provided, they cannot replace or be used to correct established field investigations.

Regrettably, at the present time the reliability of the Carbon 14 laboratories varies considerably, and care must be used in determining dates relating to ages in excess of about 20,000 to 25,000 years.

5. Other geological methods for absolute dating. As indicated, the radiocarbon method encompasses only a few tens of thousands of years. Of the other methods that make use of radioactive decay, only the ionium method will be considered here. For further information on radiometric age-dating methods, articles by Hahn [327], Kulp [444], and Rosholt, Emiliani, Geiss, Koczy, and Wangersky [635] may be consulted. The ionium method covers the time period up to about 400,000 years and has become important for the dating of deep-sea sediments [758, 444]. On the sea floor uranium is sedimented. From this, by radioactive decay, ionium (^{230}Th) is produced. Further decay results in the formation of radium, radon, and, finally, the lead isotope ^{206}Pb. The half-life of ionium is substantially shorter than that of uranium, so that the ionium content decreases as the age of the sediments increases. Since the correspondingly small amount of ionium cannot presently be measured with sufficient accuracy, use is made of the subsequent decay products, namely radium and radon, which can be measured with greater accuracy.

The radiocarbon and the ionium methods are complementary, yet their fields of application are vastly different. Just as the radiocarbon method is entirely confined to terrestrial material, so the ionium method is most readily applicable to marine sediments. The attempts of Alekseev, Kazachevsky, Cherdyncev, and Enikeev [8] and of Cherdyncev, Strashnikov, Borisenko and Poljakova [135] to extend the latter method to terrestrial formations does not yet seem to have succeeded. In this connection it is important that the fluorine test, which was developed as a method of determining ages of terrestrial remains but provided only relative ages [cf. 149], has been so extended by Richter [626] that it can be applied to absolute age determinations bridging the gap between the radiocarbon method and the lead and helium methods. In the fluorine test the degree of alteration of the original fluorapatite in bones to P_2O_5 and fluorine is used as an indication of the time elapsed. After standardizing against the radiocarbon method, it appears to provide usable ages given suitable material. However, locally unfavorable bedding conditions of the bones considerably restrict the assertions made possible by analysis.

All the methods of absolute age determination introduced up to the present time are subject to serious errors. Yet if dating is possible by several methods and the results can be checked against field and stratigraphical relations, then at least for the youngest part of the Pleistocene and the Holocene reliable absolute age dates should be possible, even when the errors of the individual methods are not known.

QUALITATIVE AND QUANTITATIVE METHODS OF DETERMINING ANCIENT CLIMATES

Because climate is a complex phenomenon in which the various factors are functionally related one to another, an attempt to determine the character and condition of the climate of past epochs presents even greater difficulties than those involved in absolute dating. An alteration of one factor generally involves changes in one or more of the others. In addition, many of the available methods for determining past climates are based on the reactions of living organisms, on slowly unfolding physicochemical or biotic processes dependent upon many external factors (e.g., development of soils), or on the growth, movement, or disappearance of glaciers, which are themselves regulated by a series of wholly different climatic factors. These methods, because they are founded on a totality of reactions of all climatic factors, are difficult to employ. It is understandable, in this light, that a particular type of reaction may have many different causes. Only a few methods permit measurement of one dominant factor. Determining water temperatures by the ratio of the oxygen isotopes ($^{18}O/^{16}O$) and determining temperature by reconstructing the former permafrost zone are two such methods. Their applicability, however, is restricted. From the reconstructed extent of the former permafrost zone as defined by the occurrence of frozen soil structures, it is possible at most to give an indication of the average annual temperature of the different cold periods when the permafrost area spread far beyond its present limits. In contrast, only under the most favorable circumstances is it possible to delimit in the same way the area of permafrost during the Pleistocene warm periods, when the permafrost region was smaller than at present. Furthermore, the applicability of the method is restricted to the present cold, temperate latitudes, in which evidence of former permafrost occurs.

The oxygen isotope method is in general restricted to marine organisms in sedimentary rocks in regions in which there were no marked seasonal fluctuations in surface water temperatures and no significant oceanic currents that could bring and mix water from different geographic latitudes, hence different temperatures.

The former extent of glaciers, and the geomorphological forms resulting from their activity, in general will provide information about climatic conditions only within the cool temperate latitudes or in high mountains during the cold periods. The most important source of information about the climate of the Ice Age, therefore, is based on the reactions of the living world (in the widest sense, including soil formation). This base is extraordinarily susceptible to different sources of error, and in the following section an attempt will be made to outline the most important methods of determining former climates, and to present their applicability and the basic sources of error affecting them.

Physical Methods: the $^{16}O/^{18}O$ Method (Carbonate Method)

This technique was developed by Urey and his co-workers, of whom Emiliani in particular is to be noted [193, 196]. The energy of a molecule is composed of translational, vibrational, and rotational energy and electronic energy. In chemical reactions that involve several isotopes of the same element, the individual isotopes are incorporated unequally in the new compound, in such a way that the free energy of the system is at a minimum. At ordinary temperatures in all elements, with the exception of hydrogen, the specific distribution of isotopes is dependent only upon the vibrational energy. There is thus a selectivity which depends upon isotope mass. Of the oxygen isotopes ^{16}O, ^{17}O, and ^{18}O, ^{16}O is by far the most common, there being 500 atoms for every one of ^{18}O, and 2,700 for every ^{17}O atom. ^{17}O can thus be ignored in the calculations that are of concern here. From the reaction

$$CO_2 + H_20 \rightleftharpoons H_2CO_3$$

in the group $CO_2 + H_2O$ one ^{18}O atom in every 166, which may be in either the CO_2 or in H_2O, is incorporated in the H_2CO_3 formed. Upon the decomposition of H_2CO_3 into CO_2 and H_2O, the probability of ^{18}O being found in CO_2 is not twice as great as that for water but a little greater, for the total vibrational energy of the pair $C^{18}O^{16}O - H_2{}^{16}O$ is somewhat less than in the pair $C^{16}O^{16}O$ minus $H_2{}^{18}O$. Consequently, the $^{18}O/^{16}O$ "fractionation factor" between H_2O and CO_2 is 1.045 at 0°C—that is, the ratio $^{18}O/^{16}O$ in CO_2 is about 45 percent greater than in H_2O.

With increasing temperature the decrease in the free energy of the system through the incorporation of heavier isotopes by exchange reactions in relation to the total energy of the system becomes insignificant. The rise in temperature increases the free energy of the system. As a consequence, the fractionation factor decreases as the temperature increases. Between $CaCO_3$ and H_2O it is 1.025 at 0°C but 1.021 at 25°C. The temperature-dependent change of the fractionation factor of the percentage difference between the oxygen isotope ratio of the sample and some reference standard which can be measured with high accuracy on a mass spectrometer thus forms the starting point of the method for calculating the temperature at which the sample was formed.

The method has been developed to the point that sea-water temperatures of the past can be determined. To obtain these results, the oxygen isotope ratio is found by the mass-spectrographic analysis of the carbonate shells of certain marine organisms. For the results to be reliable, certain preconditions must be fulfilled. It is first necessary to establish that the oxygen isotope ratio of sea water and the carbonate and bicarbonate formed from it are solely and directly dependent upon temperature. Yet this is not the case, for $H_2{}^{16}O$ has a higher vapor pressure than $H_2{}^{18}O$. In consequence, relatively more $H_2{}^{16}O$ occurs in the atmosphere, particularly in the tropics, and it will be transported northward by air circulation, where it is precipitated as snow or rain. This means that the tropical ocean is enriched by ^{18}O, while at higher latitudes, because of rain and the influx of fresh water, the relative amount of ^{18}O is diluted below the correct thermal ratio [149a]. These deviations must have been particularly marked during the alternation of cold and warm periods, for Dansgaard and Tauber [149a] speak more of a paleoglaciation curve rather than of a paleotemperature curve. Fortunately, present geological knowledge of the Quaternary permits the necessary quantitative correction.

A subsequent source of error can arise out of the later solution or recrystallization of the original shell material, for this involves, on one or more occasions, exchange reactions that can obscure the original condition. In oxygen isotope analyses, therefore, no shell material can be used in which such processes are suspected. Naturally, each sample in which secondary deposition is found must be rejected.

A difficulty of another kind concerns the metabolic processes in organisms. For the method to be of use, one of the assumptions must be that the animals must incorporate ^{18}O in their shells in the same quantitative proportion as in sea water at the appropriate temperature. There are indications [cf. 757] that certain organisms may incorporate isotopes of a given element in nonequilibrium proportions. This is found in the Canadian water weed (*Elodea canadensis*) and in diatoms which in respiration preferentially select ^{16}O [774].

Careful observation nevertheless has shown that the foraminifera (particularly *Globigerinoïdes rubra, Gl. sacculifera, Globigerina inflata*) that have been much used in paleotemperature analysis do not show selectivity in their shell structure.

Finally, it is necessary to use only those animals whose ecological requirements are well known, in general those with a well-defined depth range throughout their life span. In the reconstruction, former temperatures of oceanic surface waters is particularly important from the paleoclimatic standpoint, so that research has concerned itself principally with animals occupying the near-surface layers. Yet in the extratropical regions it must be realized that animals may form shell only during certain periods of the year, and consequently only the average temperature of this period is determined.

Even though the method described faces appreciable difficulties, it is, because of the extensive research of the authors cited and by comparison with zoogeographic results [659 and Fig. 10] so far developed that it provides an interesting glimpse of the temperature conditions of the Oligocene and Miocene [195], the upper Cretaceous [482, 757], and even of annual temperature fluctuations during the Jurassic [757, 856]. Also, it is the only method that yields, within certain limits, reliable quantitative data on the Pleistocene and Holocene conditions of temperature of the tropical seas.

Geological Methods

1. Glacial sediments and surface morphology. The sediments and morphological forms produced by the glaciers and their melt waters in the Ice Age are among the most striking developments of that period. They were responsible for the discovery, toward the middle of the last century, of the significant climatic fluctuations of the Ice Age. They appear to rank as first-order climatic indicators. Yet the wide distribution of traces of former glaciation says very little in paleoclimatology. For the ice budget of a glacier is determined by whether the sum of melting and ablation is the same as or higher or lower than the total ice ac-

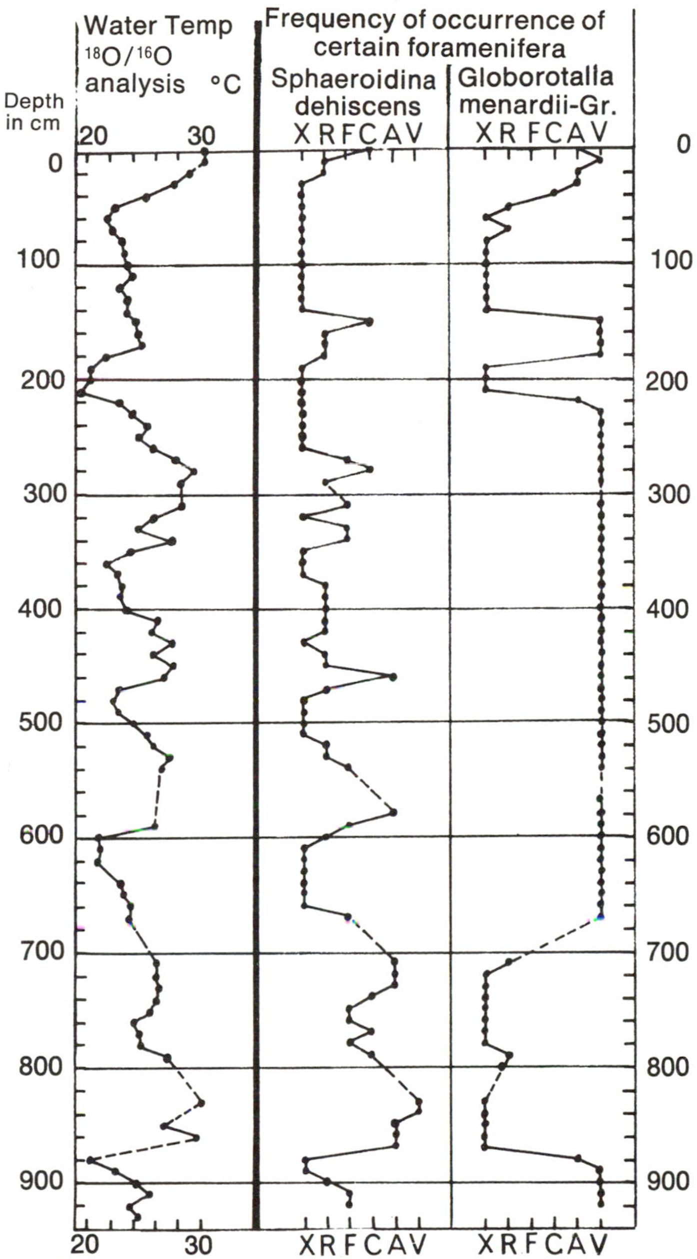

FIGURE 10. Determination of the former surface temperatures of tropical regions based upon the oxygen isotope ratio in the shells of certain foraminifera and the former occurrence of certain foraminifera species. Left-hand column: temperature determined physically; central and right-hand column: temperature determined biologically. In the left-hand column the former temperatures are recorded in centigrade; in the other two columns the scale is in terms of relative frequency data of the foraminifera investigated. In this, *V* is very abundant, *A* abundant, *C* common, *F* frequent, *R* rare and *X* absent. The alteration of the surface water temperatures in the past are reflected by only a few types (here *Sphaerodinia dehiscens*). Other species show a less sensitive reaction (here the *Globorotalia menardii* group). Simplified after Emiliani [848].

cretion for that year. The ice budget is therefore dependent on several climatic factors. Among these may be listed not only the amount of summer and winter snowfall and the magnitude of the simultaneous ablation and melting, but also the relation of the area of accumulation to the region where wasting occurs. In addition, very much depends on whether the glacier can spread horizontally or whether the ice mass is confined to long, deep valleys. Finally, the altitude of the glaciated region cannot be neglected, for in the higher mountainous regions a climatic fluctuation that would lead to a climatic amelioration at lower levels might cause a glacial advance. The most recent glacial fluctuations with whose climatic causes von Rudloff [638, 639] has concerned himself in detail show the complexity of glacial advance or retreat.

Because of the manifold difficulties, paleoclimatic conclusions can be drawn only from the distribution of traces of former glaciation, if the change of important climatic factors such as the amount of annual precipitation (or, better, in individual seasons), the average annual temperature, or the winter temperature can be determined in other ways. For all that, it is not possible to make reliable quantitative statements concerning the character of past climate.

One sure aid in the determination of past climates seems to be the height of the snow line. This is the boundary between the region in which, given a yearly average, accumulated snow does not entirely melt, and the region in which the sum of melting and ablation exceeds the snow and ice supply.

The position of the boundary is very strongly influenced by local conditions on individual mountains and mountain ranges, such as exposure to the sun's rays, the introduction of snow by avalanches, and, in particular, the relationship to the rain-bearing winds, etc. There may thus be appreciable differences in the elevation of the snow line on a single mountain, or between the flanks of a mountain range. As these local differences are strongly impressed upon the gross climate in a given mountainous region, a distinction should be made between the orographic snow line, discussed above, and a climatic snow line, whose elevation lies between the highest and lowest values of the orographic snow line in a given mountain area. The relation between the altitude of the orographic snow line to the climatic is clearly shown in Figure 11, showing the example of the Trans-Alai.

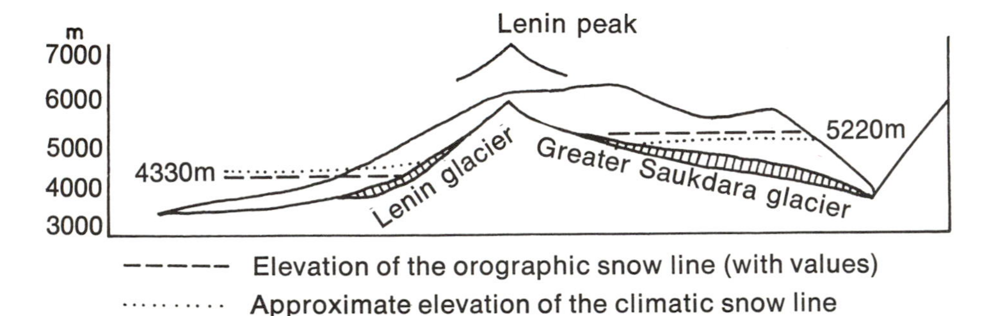

FIGURE 11. The relationship between the elevation of the orographic and climatic snow lines, illustrated by reference to the Trans-Alai Mountains. From Wissmann [801].

The climatic snow line is a theoretical concept and thus cannot be directly observed in the field. Nevertheless, it may illustrate very well the general climatic conditions with respect to snow accumulation, since local peculiarities are disregarded. Despite this, however, the elevation of the recent climatic snow line in northern Eurasia (Fig. 12) shows that its situation depends upon at least two different factors: thermal conditions, of which the summer temperature is particularly important, and the amount of the snow or firn supply in the zone of glacier accumulation above the snow line. It follows from this that the elevation of the climatic snow line rises from the foothill region toward the central part of the mountain chain, just as in similar geographic latitudes oceanic climate changes to continental away from the ocean.

It is possible to determine the elevation of not only the present climatic snow line but also those of the Pleistocene cold periods. Naturally, the most reliable data are obtained from the last cold period, for the surface forms are still fresh and more easily accessible to direct observation. The difference in elevation of the past and present climatic snow lines—that is, the depression of the snow line—appears to provide a suitable starting point for the determination of past climate. In point of fact, the oldest well-founded hypothesis of temperature conditions during the last cold period in Europe was based on this method [cf. reviews 804, 409, 801]. Yet the reliability of the method should not be overestimated, for the elevation of the present climatic snow line may go back to very different regional factors. Consequently, there is still an appreciable element of uncertainty in the calculations of the depression of the snow line. Yet it is clear that this depression, where it may be determined, is of great paleoclimatic value if the magnitude of the individual climatic factors can be found in other ways.

2. Periglacial sediments and surface forms. In front of the glaciers the vegetation cover is seldom very dense, and frost can penetrate the soil unhindered. Such a zone is referred to as the periglacial zone. Since originally the term referred to the region around the glacier, there is no statement about the outer limit of the periglacial zone, and it is in fact very difficult to define such a limit. The term is generally considered to refer to that region in which traces of frost action in the soil are frequently observed and where frost is an important element among those which shape the form of the earth's surface.

The decisive factors controlling the magnitude of the effects of frost include not only low night or winter temperatures but also the character of the vegetation, for with a dense plant cover the radiation and thermal budget of the uppermost layer of soil is affected in such a way that frost is no longer fully operative as a factor of surface morphology. The outer limit of the periglacial region thus coincides in many cases with the limits of certain types of vegetation and is in general within the tundra or alpine zone. Outside this limit frost can markedly influence soil structure or surface form where for some local reason vegetation has been destroyed or removed. Such cases are referred to as extrazonal periglacial phenomena.

In those climatic provinces in which the winter temperature is extremely low, frost is able to operate even through a thick plant cover. Examples of this are to be found in the high continental climate in the central part of northern Eurasia

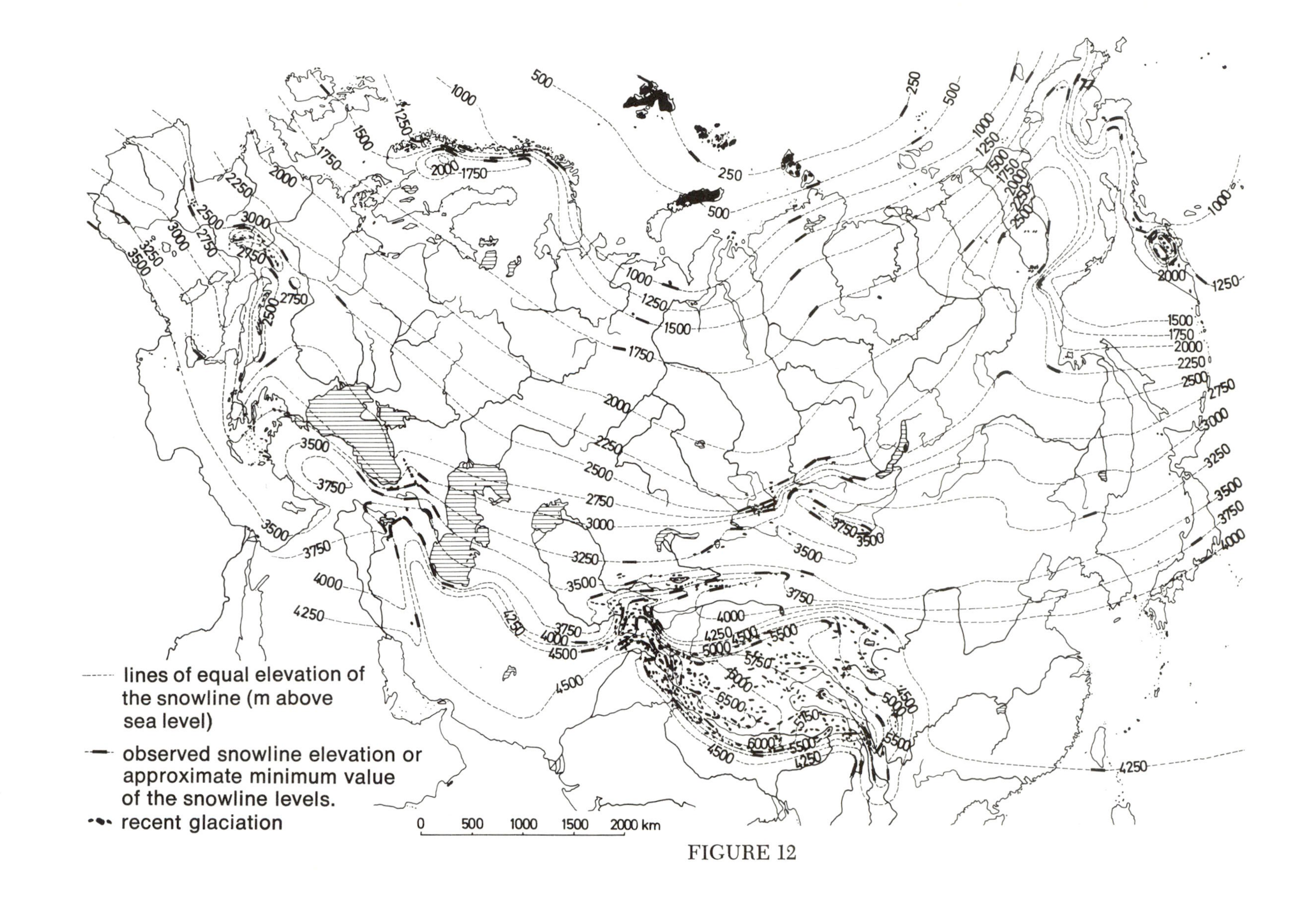
lines of equal elevation of the snowline (m above sea level)
observed snowline elevation or approximate minimum value of the snowline levels.
recent glaciation
0 500 1000 1500 2000 km

FIGURE 12

and North America. The high summer temperatures permit the vegetation there to extend far to the north notwithstanding the low winter and night temperatures which allow frost to penetrate deeply into the ground and locally inhibit vegetation, or at least are detrimental to it. In such climates the periglacial zone extends beyond the southern margin of the tundra or below the lower boundary in Alpine zones into the forest belt.

The southern margin of the periglacial zone thus appears as an important indication of the thermal budget. Yet local climatic peculiarities, soil types particularly susceptible to frost action, and the ground-water content of the substrate join to modify the influence of low temperature of the gross climate, so that from the southern limit of the periglacial zone no reliable climatic conclusions may be drawn.

In the region where prolonged and intensive frost action occurs, the soil has been perpetually frozen over several decades or even centuries. As we go south from the permafrost zone, there is a transitional region in which, in locally favored spots, nonfrozen islands of soil occur. This is the zone of thaw islands within the permafrost. As frost action lessens, this zone changes into a region where islands of permafrost appear in predominantly thawed soils, until finally the soil is only seasonally frozen or is generally unfrozen. The area in the Northern Hemisphere affected today by the various forms of permafrost is considerable (Fig. 13).

Within the area of permanently frozen soil, frost produces a series of striking surface features and disturbed bedding, which in principle derive from three processes: (1) Frost fissures open up and are enlarged by the formation of fresh ice. (2) Because of the low vapor pressure at the surface of soil ice, moisture is drawn up from lower, still unfrozen levels. As a result, the soil becomes oversaturated with water in the region of permanently frozen ground. This produces, on the one hand, compact ice lenses in the soil with an associated appreciable increase in volume and, on the other, high water content during the summer thaw of the uppermost soil horizons and, consequently, a high mobility of these oversaturated horizons and layers. (3) By the formation of permanently frozen soil, or by strong annual freezing above the water-bearing horizon, springs are plugged up, and water bodies are enclosed between two impermeable layers. Appreciable hydrostatic pressure arises, which shapes the surface into hummocks; they may be up to 50 meters high and 100 meters in diameter (pingos, bulgunnjachi, hydrolaccoliths, and analogous forms).

Often all of these processes act simultaneously, so that many different forms result.

In the Northern Hemisphere a gradual transition occurs from south to north from those surface-soil forms resulting from seasonal frost action to those of perennially frozen ground. The principal cause is the occurrence of local impermeable layers whose effects on ground-water movement are similar to those of permafrost. Thus the southern limit of the permafrost zone is not easily discerned.

FIGURE 12. The present elevation of the climatic snow line in northern Eurasia. After Frenzel [256] with small corrections incorporated from more recent data.

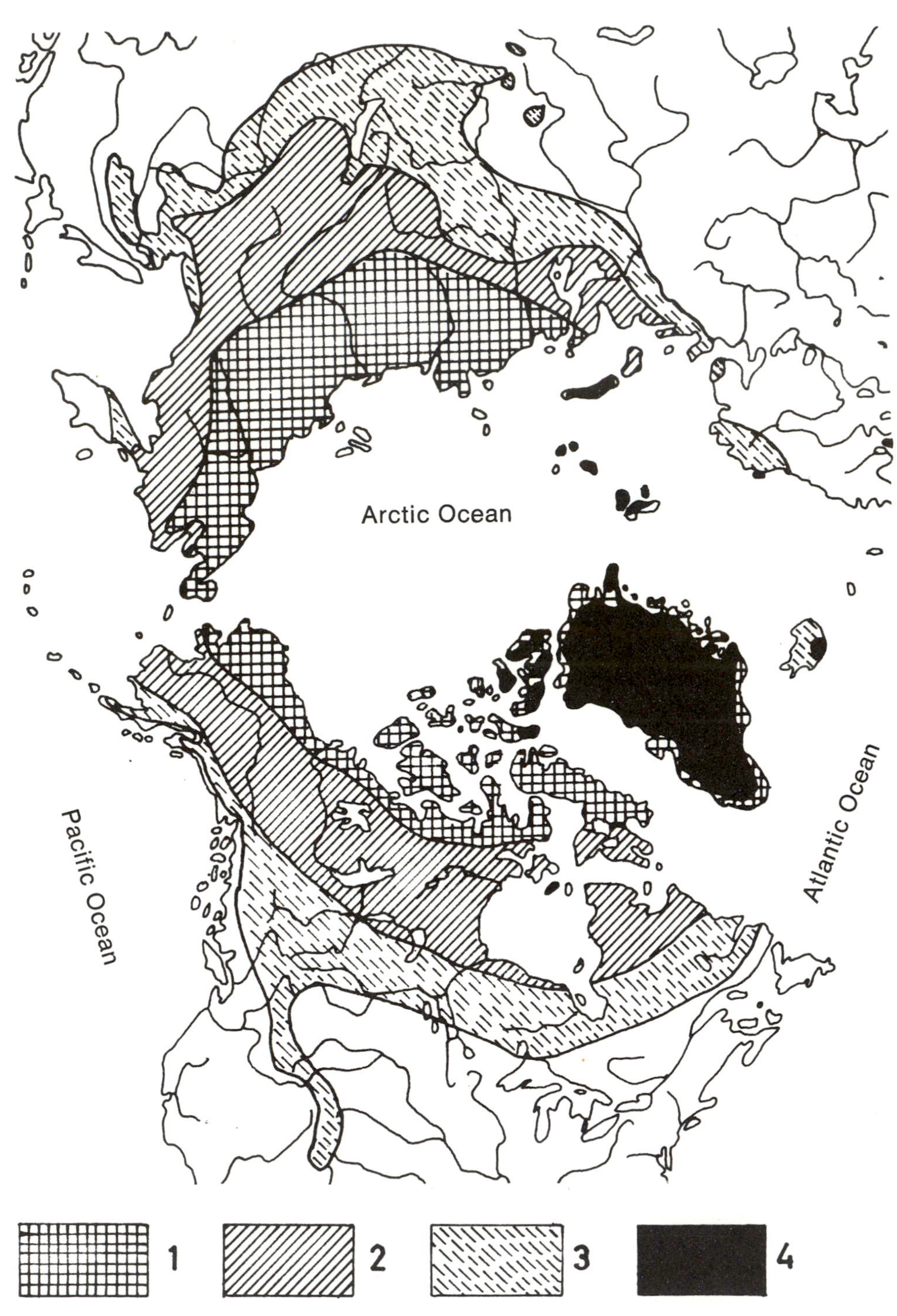

FIGURE 13. The present distribution of permafrost in the northern hemisphere: 1. perpetually frozen ground without notable regions of deep thaw; 2. large islands of thaw interspersed in a region of perpetually frozen ground; 3. islands of permafrost in a predominantly thawing soil; 4. strong recent glaciation. Modified after Black, from Woldstedt [804].

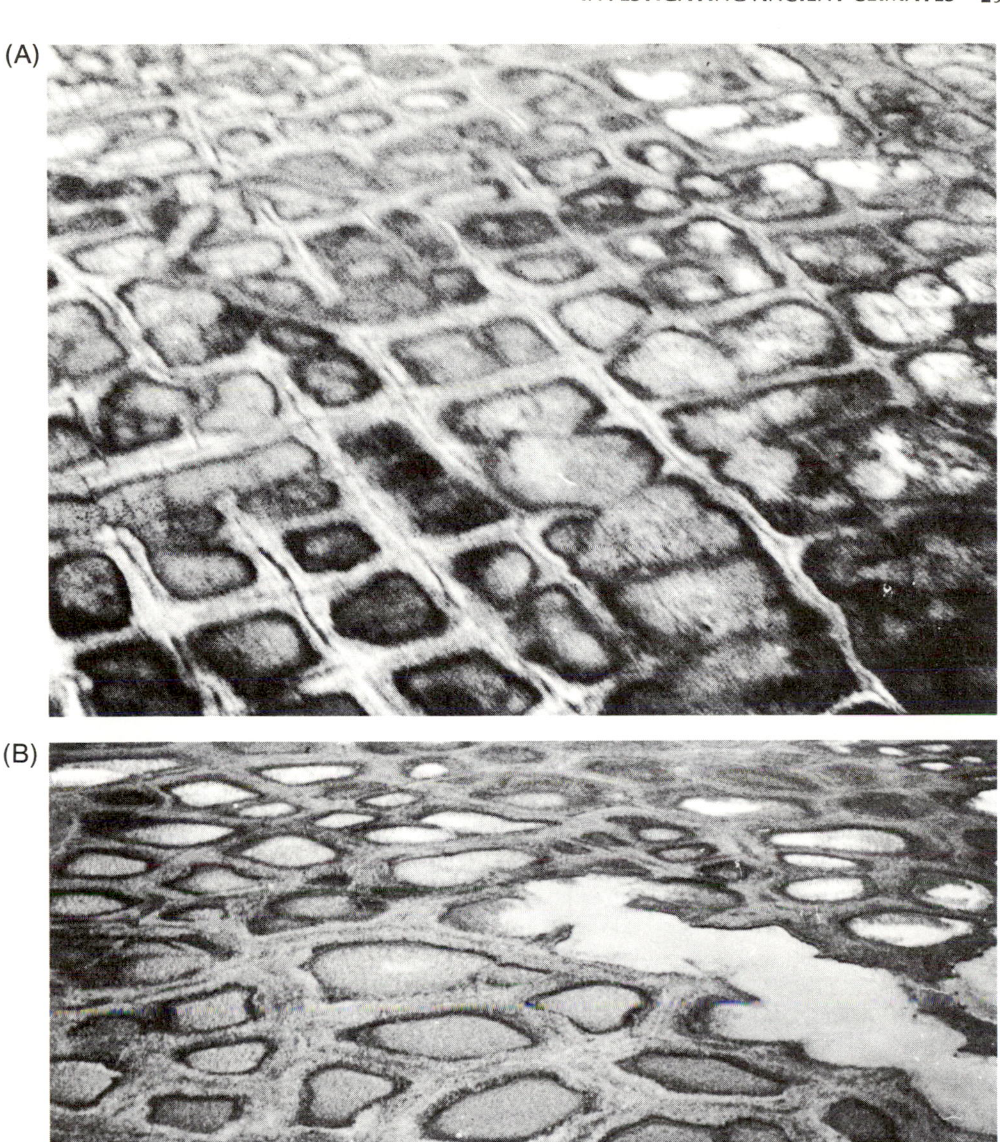

FIGURE 14. Giant polygons in the high Arctic tundra of Alaska. Upon shallowly dipping surfaces (A) a rectangular net is formed, while predominantly pentagonal and polygonal forms predominate on horizontal surfaces (B). The polygons attain a length of 10 to 20 meters on a side (majority of those in A), but are occasionally up to 150 meters in length. The polygons consist of marginal walls of upthrust soil in whose ridges central grabens (dark strip in the photos) develop that fill with water and ice wedges; the banks surround ponds or basins with swamp vegetation. Photos: Dr. F. Schenk; see also [849].

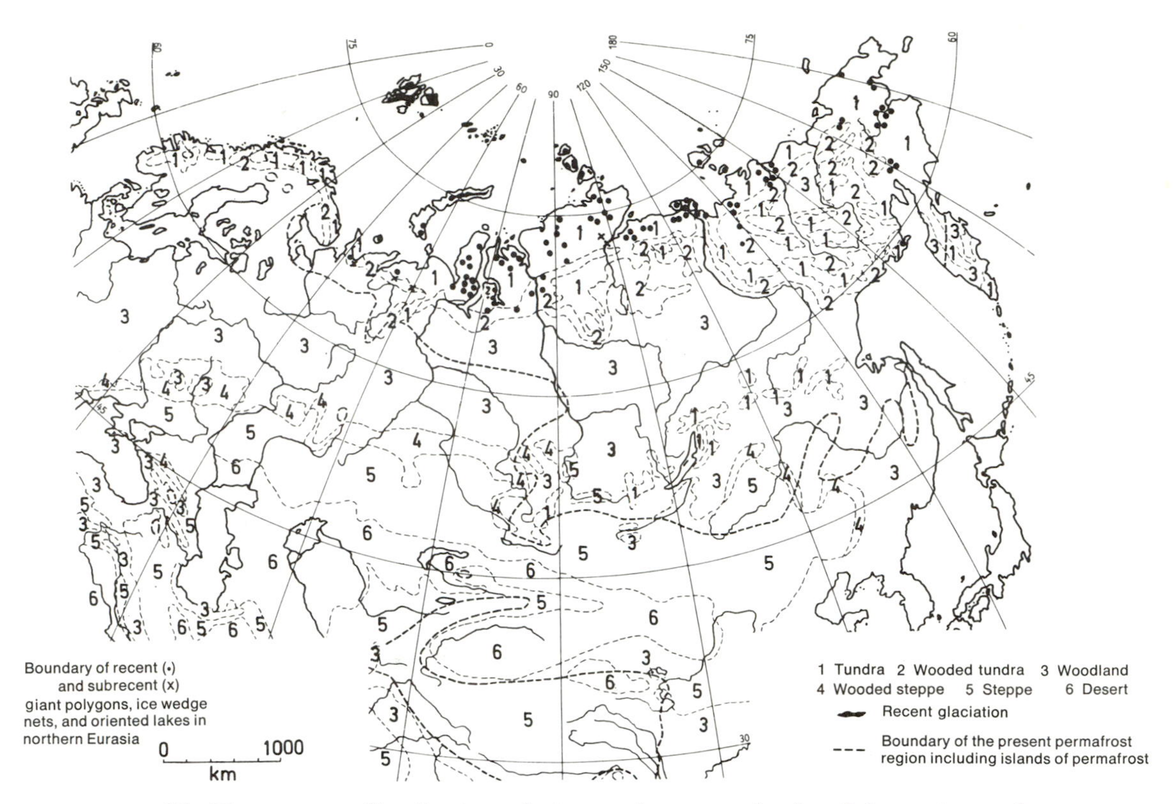

FIGURE 15. The present distribution of giant polygons and related forms in northern Eurasia in relation to the present floral zones and the area of permafrost. From Frenzel [256].

The area of the permanently frozen soil zone during former cold periods is an important paleoclimatic criterion. It is therefore necessary to investigate the closeness of the relation of superficial frost forms to the occurrence of permafrost. The investigation shows that the following forms are characteristic for the permafrost zone and are only rarely developed outside the permafrost area [256]:

(1) Giant polygons (checkered soils, tetragonal ground; Figs. 14, 15).
(2) Large pingos of many years duration (Figs. 16, 17).
(3) The so-called "palsen" in bogs (Fig. 18).
(4) Probably the hummocky tundra—that is, small earth mounds 20 to 40 centimeters high and one to one and a half meters in diameter. Their surface is generally free of vegetation, and the mounds are usually so closely spaced that the depressions between the mounds form a polygonal graben system in which herbs and dwarf shrub vegetation may be supported.
(5) Probably Golec or mountain terraces. The term covers the dense system of broader or narrower terraces on hillsides which are the result of frost weathering and soil flow under the effects of frost (solifluction) and which eat into the hillside.
(6) Finally, the true "nalyod" phenomena can be numbered here. This concerns the ice bodies in river valleys which develop in winter when the ground water flowing in the sediments in the bed of the river is trapped between

FIGURE 16. Air photo of pingos in East Greenland. Under the strong frost action in a high arctic climate the water-bearing horizon is enclosed between the upper surfaces of the permanently frozen layer and the lower surface of the winter-freezing layer under strong hydrostatic pressure. In the Siberian pingos pressures of 52 atmospheres have been measured. The high pressure causes an updoming of the uppermost layer, and so hillocks up to 40 meters in height and 100–200 meters in diameter can be formed, which have a core of ice. The core finally melts, leaving a crater-like depression whose margins are gradually reduced by frost. Commonly pingos occur in large numbers. In very favorable circumstances after the destruction of the pingo, the annular walls may still be recognizable in the field, as, for example, in the Höhen Venn in the northwest Eifel. From Hofer [850].

> the permafrost zone and the zone of seasonal frost action or some other impermeable layer. As a result, the ground water under pressure finally breaks through the annual ice and flows out, to be immediately frozen into large ice mounds (Taryn or halyod), which during the summers in the permanently frozen zone do not completely melt.

All the remaining surface forms of the periglacial region can be formed outside the permafrost zone. This is particularly true in the case of polygonal soils and honeycomb soils and includes stone stripes (Figs. 19 and 20) in high mountains and arctic regions, so that no conclusions may be drawn from their fossil occurrence regarding the former extent of the permafrost zone. With the exception of the taryns and to a certain extent the palsen, the other forms characteristic of the permafrost zone appear as fossil forms, so that the former permafrost area

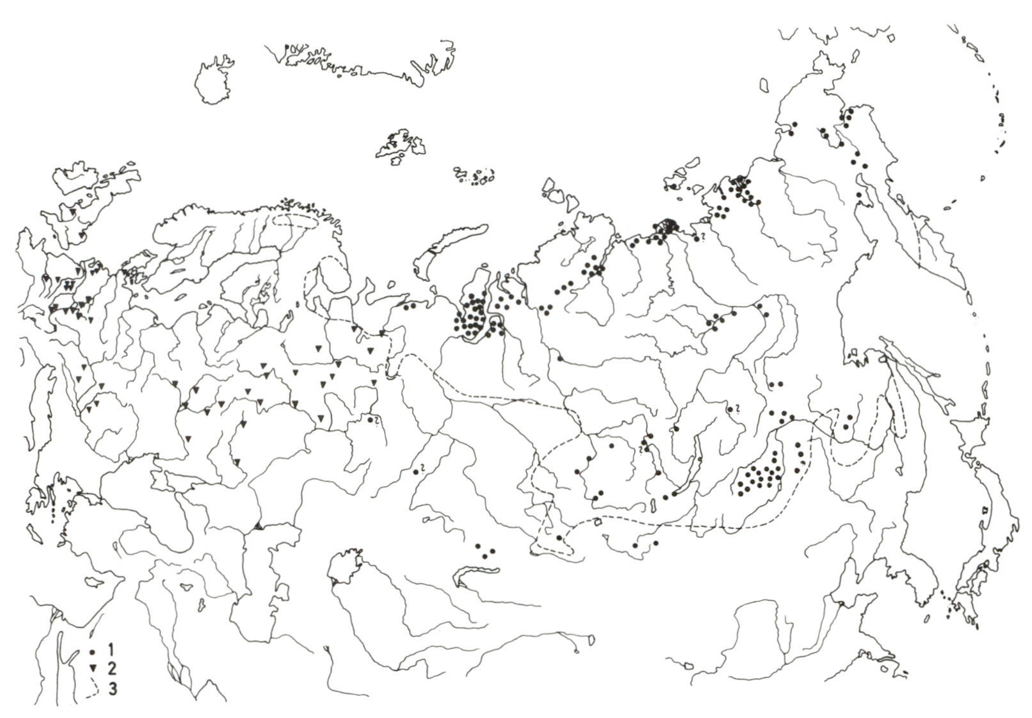

FIGURE 17. The distribution of pingos at present and during the last cold period in northern Eurasia. 1. present occurrence; 2. occurrences dating from the last cold period; 3. present furthermost limit of the occurrence of permafrost.

can be reconstructed. Unfortunately, the number of forms found in general is insufficient to enable a distinction to be made between the area of permanently frozen ground in the past and that of permanently frozen ground with islands of thaw or thawing regions with islands of permafrost. At most they may indicate the probable maximum extent of permafrost (Fig. 21).

The size of the region of permanently frozen ground is primarily dependent on the winter and summer temperatures. In addition, other factors, such as the amount of snowfall, play an important part. They have been considered by Grigor'ev [307] and combined in a formula giving the "freezing coefficient" G:

$$G = \frac{t_{neg}}{S \, . \, t_{pos} N_{pos}}$$

In this equation t_{neg} refers to the sum of subzero average monthly temperatures, S the average snow cover in centimeters for the months with the thickest snow cover, t_{pos} the sum of the average monthly temperatures above zero, and N_{pos} the sum of the average precipitation in millimeters for the months with above-zero temperatures. Temperatures are measured in degrees centigrade.

When the value of $G = -0.000095$, the ground is permanently frozen. The values between -0.000095 and -0.000085 occur in the transition zone in which the ground remains frozen for several years. Still lower values are found in regions outside the permafrost zone. Jahn [366] pointed out that two factors are not considered in the equation, the effects of microclimates and of sediment types.

FIGURE 18. Palsen in a bog near Stenbacken near Lake Torneträsk, Swedish Lapland. Palsen are peat hillocks in subarctic bogs, in which, under the shelter of *Sphagnum* peat sedges or *Sphagnum* peat may remain frozen even during summer, or a compact ice lenticle may even remain. This ice core grows by the transfer of water through the surrounding unfrozen layers, so that the palsen become increasingly larger. Consequently they are increasingly exposed to wind action, so that a clear distinction can be made between the windward and leeward sides. On the windward side (the left-hand side of the hillock and its upper surface in the photo) snow and peat are blown away, while on the leeward side (right flank of the palsen) a thick snow layer develops under whose protection the more temperature-sensitive plants may survive the hardships of winter (here willow bushes). The constant wind deflation leads to a continual thinning of the protective peat layer over the ice core, so that finally the ice melts during summer and the palsen sinks. A small pond develops (center left) which is eventually filled by sediment and the whole cycle repeats. Palsen have a duration of several decades.

Since the position of the outer boundary of the region of permanently frozen ground is a function of very different climatic and nonclimatic factors, it is scarcely significant to delimit the outer boundary of the permafrost zone by a narrowly defined annual temperature.

The formula given suffices in general to define in broad outline the southern limit of the region of permanently frozen ground at the present day in the Northern Hemisphere. It cannot be used in conjunction with the known position of this important boundary in the Pleistocene cold periods to determine former temperature conditions, as all the remaining factors in an exact equation are unknown. Jahn [366] and Büdel [100] have renewed attempts from this to give approximate values of the average annual temperature along the present southern

FIGURE 19. Stone stripes in Mount Nuolja near Abisko, Swedish Lapland. The effect of frost action on a water-bearing stratum with a sparse plant cover is to sort stones according to size and slowly force them upward. As a result, stone stripes are formed (to give some idea of scale two participants in an excursion stand at the distant end of the stripe); on horizontal surfaces they are replaced by stone polygons. Stone stripes, stone polygons, and associated forms are not restricted to the region of permanently frozen ground. Photo: Dr. H. Matthaei.

limit of the permafrost zone. Values from at least −2°C to −3°C (Jahn) to one of at least −5°C (Büdel) were obtained. These figures refer, however, to the average annual air temperature at the usual height at which meteorological measurements are taken, and not to the temperature at ground level. Jahn's conclusion [366] that lower temperatures (at least −3°C) are necessary for the development of permanently frozen ground than for its maintenance (at least −1°C to −3°C) is probably correct.

In the estimates made below (pp. 138 ff.) of annual average temperatures during the different cold periods of the Ice Age—that is, for periods in which the permafrost region was widespread in the Northern Hemisphere—an approximate annual average temperature of −2°C at the southern limit of the permafrost zone will be adopted as a minimal value of the former temperature conditions.

From the former position of the southern permafrost limit, worthwhile indications of certain characteristics of the temperature of given regions can be obtained. The magnitude of the snow cover or generally the amount of precipitation cannot, however, be reconstructed that way. In this connection, calculation of the distribution of a characteristic sediment—that is, loess—formed in the periglacial region is a useful ancillary technique. Loess is a fine-grained and

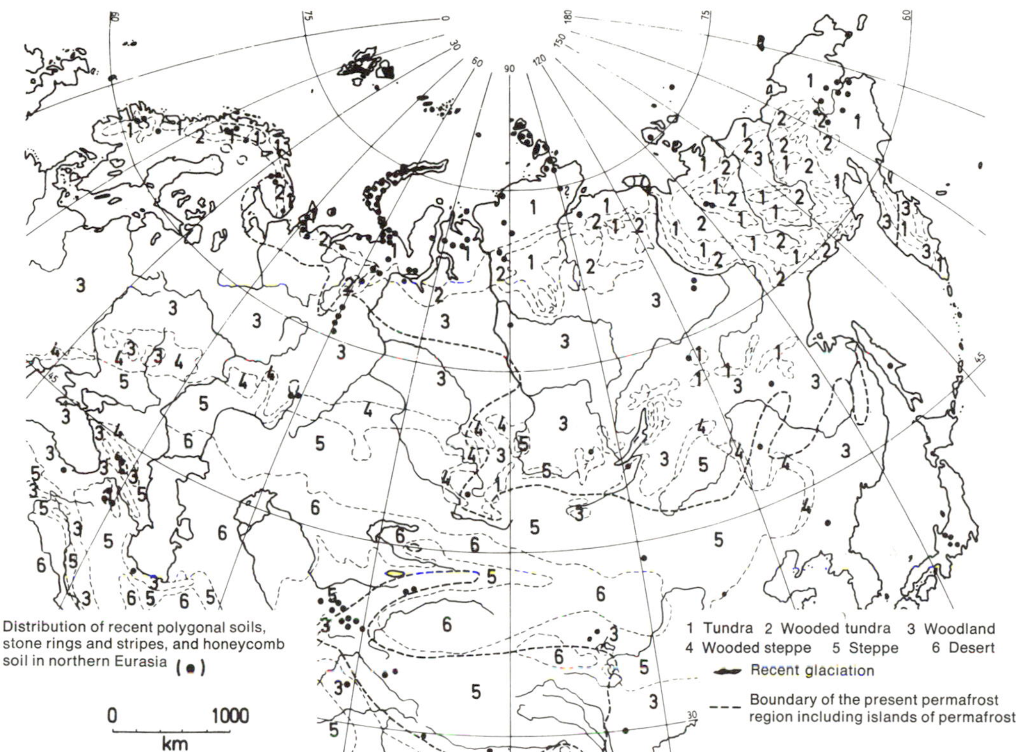

FIGURE 20. Present occurrence of stone stripes, stone polygons, and associated forms in relation to the area of permanently frozen ground and different vegetation types. From [256].

generally calcareous aeolian sediment which is usually formed in treeless steppe regions in a dry climate. It is almost always associated with severe winters. Nowhere are examples found which might indicate with certainty that loess could be formed in a moist woodland climate. The distribution of loess can thus be used as a reliable indication of the former spread of steppes. The pollen content of loess indicates the steppe plant community; its frequent richness in fossil mollusks and occasionally vertebrate fauna are further important aids in unveiling the climates of the past.

From the frozen beds or banks of rivers and streams, which carried smaller volumes of water during the cold periods than at present, and in the absence of a protective vegetation cover, sand could be blown out and form dunes. The distribution of this sand and sandy loess and its relation to the region of provenance, the current bedding, and the direction of dune movement have repeatedly been used as indications of the wind direction during the period of dune accumulation [656, 498, 602, 603, 231, et al.].

In conclusion, it can be stated that the periglacial sediments and surface forms are useful means of reconstructing the climate of the past. Generally, they may be employed in the determination of the climates of cold periods but not in the reconstruction of the climate of warm periods. For the latter segments of Ice Age time and the milder phases of the cold periods in recent years, fossil soils have been found to serve as a useful indication.

*3. **Fossil soils.*** The development of a soil requires considerable time, whose duration depends on the nature and chemistry of the sediments, the climate, and the character of the indigenous flora and fauna. Erosion and denudation as well as sedimentation can hinder soil formation when they are intense. Times of soil formation are thus *cum grano salis* periods during which there is a pause in sedimentation and deposition. This means that fossil soils can provide information

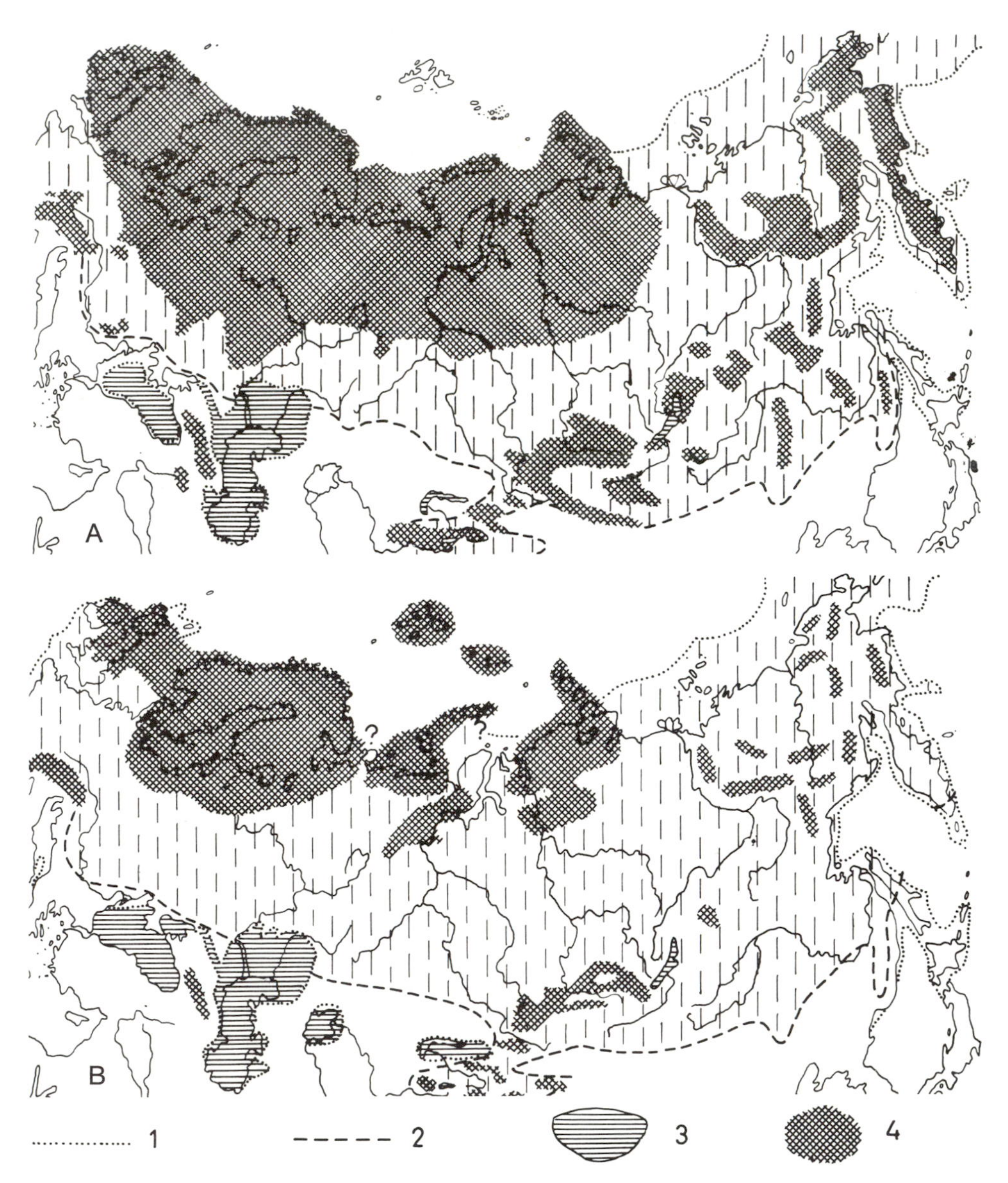

FIGURE 21. Present and past distribution of permafrost in northern Eurasia. A. maximum advance of the Saale-Dneprovsk glaciation; B. maximum advance of the Weichsel-Valdai glaciation; C. climatic optimum of the last warm (interglacial) period; D. postglacial, with the postglacial warm period indicated by widely spaced broken lines and narrowly spaced broken lines showing the present area

only about some but not all segments of Ice Age time, since during all periods of heavy sedimentation, such as the formation of loess during cold periods, sedimentation is generally more rapid than soil formation. Good examples in northern France are described in Bordes and Müller-Beck [73] and for the Black Sea coast near Constanza in Haase and Richter [324]. In both cases during the last but one cold period, warmer phases, during which soils formed, were followed by periods

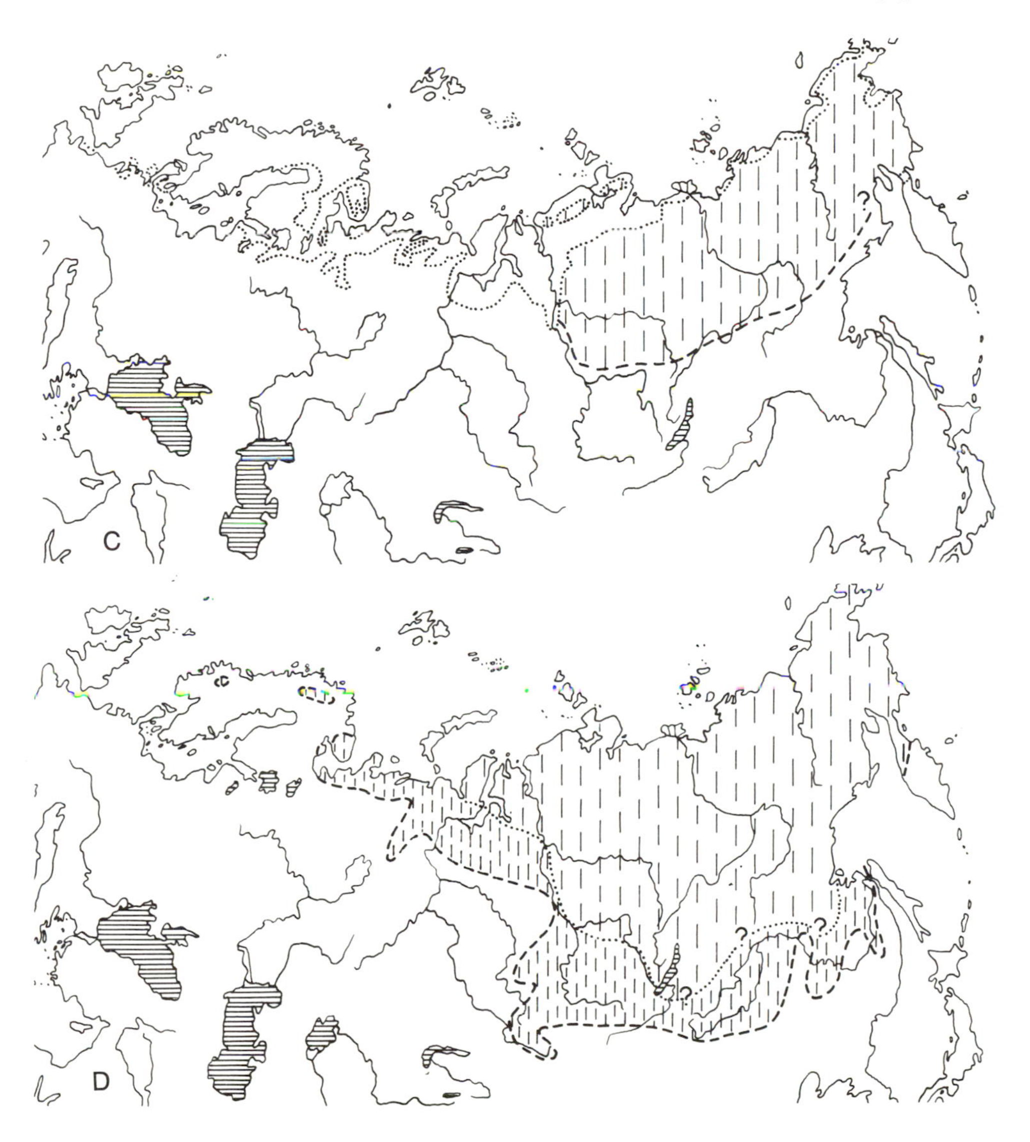

(Figure 21 caption, cont'd.)

of postglacial permafrost beyond the limits of the postglacial warm period. 1. coastline; 2. southern limit of permafrost; 3. inland lakes; 4. inland ice. The former known maximum spread of permafrost is represented. The former area of uninterrupted permanently frozen ground (Fig. 13) should be somewhat smaller. Sources of information: cf. Appendix 1, p. 245.

of heavy loess accumulation. At first, soil formation was able to keep pace with sedimentation, until finally loess drifted at a more rapid rate than could be matched by soil formation, so that traces of formation became weaker and weaker until the process ultimately halted (Fig. 22). Still, during even the most extreme phases of the cold periods, soil formation must have continued under the existing plant cover, which would provide many worthwhile glimpses of the climatic character of the time if they could but be determined.

Thus if fossil soils give information only about climatic conditions during certain time intervals, they nevertheless are a valuable criterion, and on the whole they have received little consideration up to the present time. The distinction between woodland and steppe soils, despite the ambiguity of many individual cases, provides a good indication of the climatic character of past times. More than this,

FIGURE 22. Fossil soils as climatic indicators. Profile at Litoměřice in northern Bohemia. In a profile of approximately 15 meters of loess below the present soil (top, below fence), at least four older soils can be recognized (1–4). Soils 2–4 are very probably black steppe soils. Soil 1 is difficult to type. At the time when the black earths developed, a drier steppe climate must have prevailed in place of the thermophilous deciduous woodland. The steppe climate appears to have become progressively drier and more unfavorable, so that the soil formation finally could not keep pace with the loess accumulation; and as a result the color of the fossil soil becomes weaker upward. In the case of soil 2 a short loess phase interrupted soil formation (the light band that divides the soil in two).

modern soil science is in a position not only to distinguish different fossil soil types one from another in the majority of cases, but also to recognize polygenetic soils (that is, soils formed at different times under varying conditions in the past) and resolve the individual components. The work of Brunnacker [90] in Bavaria and of Lieberoth [469] in Saxony are examples.

The soils of the last warm period in Bavaria (the Eemian warm period) are generally brown earths or parabrown earths. Soils of the same type are also being formed there today. Later there developed from the brown or parabrown earth a pseudogley. This means that as a result of the blocking of the pore spaces in the alluvial (B) horizon by the downward migration of clay minerals, the upper horizons became oversaturated. Since only in a climate with relatively high rainfall does a pseudogley develop from a parabrown or even a brown soil, it must be assumed that the rainfall was high at that time.

Alterations of other kinds are known in fossil soils in Saxony. Here during the Eemian warm period parabrown soils or pseudogleys were formed which were later overprinted by an arctic podsolgley when the climate was already arctic or subarctic. It is also possible in this case to obtain indications of the preexisting climate by a careful examination of the soil development. The results so obtained, however, are not always unequivocal. Thus, in the heart of central Europe in a climate corresponding approximately to the present, on a fine-grained substrate rich in bases and nutrients, a brown soil may develop in several stages. This soil may in turn be replaced by a parabrown earth during further stages of pedoğenesis, since the clay particles may be washed out of the upper layers and filter down to the lower horizons. They may then be retained in pores and fissures, forming a coating, and finally cause stagnation. Consequently, the soil proceeds beyond a parabrown earth and tends toward a pseudogley. This all occurs without the aid of any alteration of the conditions of precipitation of the gross climate. On the other hand, a lowering of the temperature can lead to an appreciable rise in the average humidity of the air and a reduction in the biological activity of the soil, so that in this case too a pseudogley tends to develop out of a parabrown earth. This development may take place in three different ways:

(1) As a consequence of natural soil maturity under a constant gross climate.
(2) By an increase in precipitation.
(3) A reduction in either the annual or only the summer temperatures leads to an increase in air humidity and a consequent reduction in evaporation, so that the amount of available water in pedogenetic processes appears higher and evenly spread over the year.

Thus it can be seen that the climatic causes of many developments of a polygenetic soil are unclear and must be tested by other means. Soil formation is a complex reaction of animate and inanimate nature, above and below the soil surface, to environmental conditions of which climate has a particular significance. Yet climate itself is a complex factor. Consequently, only the important trends in the development of former climates can be derived from the investigation of fossil soils. Even so, this help is extremely valuable.

Biological Methods

Among the methods applied to the interpretation of the climates of the past, ecological analysis of fossil flora and fauna has the greatest field of application and, in general, the greatest significance. In nearly all the climatic zones of the earth, remains of plants and animals from periods corresponding to both cold and warm periods of the Northern Hemisphere are known. Yet it must be made clear that every attempt to reconstruct the climate of the cold or warm periods in this manner rests upon four basic assumptions whose validity has by no means been established beyond doubt. These assumptions are:

(1) The present distribution of plant and animal species corresponds to the optimal under current climatic conditions. Changes in the regional distribution of plant and animal species investigated must be minimized, and the species must have reached the boundaries of the physiologically possible environments.

(2) The species within its area of distribution is a simple homogeneous population. Different physiological races should not occur.

(3) Extreme environmental conditions exercise a selective influence on living forms of a kind that certain unusual environments have corresponding specific morphological types—that is, favor certain constitutional factors of resistance which are independent of taxonomic relations.

(4) Once an acquired characteristic specifically suitable for a given horizon has appeared, it tends to remain over very long periods and relates not only to individual species but also to larger associations (genera, families).

It must be made clear that present conditions accord with none of these premises. All biologically oriented attempts to establish former climate consequently exhibit different inherent defects, as will be shown in a moment. Even though the character of the former climate can be accurately outlined, quantitative statements particularly necessary in reconstructing former atmospheric circulation are possible only in the rarest circumstances. The need here is much more an extension of all existing methods and a regular mutual control of the results obtained. Before the most important biological methods in paleoclimatic studies are described, some of the principal difficulties will be discussed.

Every attempt to determine the qualitative and quantitative properties of the climate of the past using biological data depends upon an analysis of the relation between present-day limits of distribution and any of the given climatic factors. This is equally valid whether the objects being discussed are, to name but a few examples, mollusks [483, 488, 459], insects [739, 340], vertebrates [141, 345], corals [211], fresh-water animals [740], plants [152, 652] (e.g. the cutsedge, *Cladium mariscus* [606], hazelnut, *Corylus avellana* [15], or *Dasycladacaean* [588, 589]), or complex features such as the polar and vertical limits of tree growth [99], or certain plant structures [607, 37] which can be explained as adaptations to environment and will not be discussed further. On the assumption that the climatic requirements of the material investigated have remained unchanged or little changed during the course of time, it is possible to extrapolate former climatic conditions from the distribution of species, genera, or families. In so doing, organisms that are relatively immobile during their lifetime—such as plants, mollusks, insects, beetles, and small rodents—are particularly important, as they best correspond to the climatic conditions of a narrowly restricted biotope. It must be stressed that they can reflect peculiarities of topoclimate to a much greater degree than the large vertebrates, which may migrate long distances, or birds with a wide habitat.

The question to what degree the climatic demands of a species have remained constant in time is difficult to answer. In many taxa the morphology and anatomy have not changed during the Quaternary at least (Danish insects [340]; European freshwater fauna [740]; hornbeam, *Carpinus betulus* [224, 377], and buckbean, *Menyanthes trifoliata* [377]). From this it is possible to conclude, admittedly with some uncertainty, that ecological conditions have remained the same. Yet it is not proven that morphology and physiological evolution are so closely linked that a significant set change in physiological efficiency goes hand in hand with a certain change in the morphology and anatomical structures of the members of a given species.

A further principal difficulty is that many times within the area of a morphological taxon, perhaps a species whose present climatic requirements have been carefully determined [201, 362, 211], several climatic races occur. These races show little or no morphological distinctions; but they have clear differences in their climatic requirements (e.g. the fruit fly, *Drosophila* [160, 161, 162, 746], the beetles *Carabus nemoralis* and *Cionus intestinalis* [525], and the buckler mustard, *Biscutella laevigata* [509]). Under climatic conditions that differ from the present, a climatic race may be preferred which is today still important but suppressed, so that an analysis of the former distribution of the whole species must lead to false conclusions about the former climate. The magnitude of this inherent difficulty can probably never be correctly estimated even when, as Erdtmann and Nordborg [204] have succeeded in showing, in the great burnet (*Sanguisorba officinalis* and *S. minor*) races with different sets of chromosomes can be distinguished morphologically by pollen. These favorable cases probably hold for only a very small number of plants, and the necessary research requirements cannot be met by routine research. Consequently, the difficulties described must be regarded as latent, and their significance must be tested in other ways.

Despite these difficulties, the biological sciences provide a series of tested methods which can be used for the determination of former climates. Because botanical methods are more highly developed than zoological, the following considerations are heavily weighted in the direction of botany.

1. Dendroclimatology. As in dendrochronology, so dendroclimatology [268] makes use of the annual growth rings in long-lived trees. The method, established by Huntington [356], uses the fact that in extratropical wood plants unfavorable environmental conditions cause thin growth rings, whereas under favorable conditions broad rings are added. Dendroclimatology is thus based upon the physiological reaction of an organism to very different external influences whose only point of contact is that they are either favorable or unfavorable to the tree in question. Herein lie the non-negligible sources of error in the method. Apart from the fact that representatives of different species react very differently to unfavorable external environments [279, 554, 349] and that local climatic and edaphic conditions—that is, peculiarities caused by local soil conditions—frequently mask regional climatic changes, every external factor that is temporarily detrimental to tree growth can lead to variations in ring thickness. A thickness may also be affected by other external factors. The variations of breadth in the growth rings are thus ambiguous and should be used in paleoclimatological interpretation only when a single environmental factor can be considered decisive—as, for example, the arid limit of the woodland (Anatolia [268]; Southwest Africa [839]; the western part of the United States [663, 279, 16, 17, 356, 357, 263a–e]), where precipitation, including ground-water resources, controls the existence of organisms; or at the polar limit and upper forest limit in mountainous regions (Lapland [208], central Norrland [186], the Alps [79, 374]), where summer temperature controls the amount of wood growth. Within the forest zone itself different species of tree react differently to unfavorable external factors or their seasonal variations [349], and only extreme, unfavorable conditions—such as extreme cold or drought or somewhat less extreme influences of cold and drought acting simultaneously or immediately following one another—lead to a similar kind of change in ring width in all types of tree, although with a time lag of one or two years. Thus, in the most favorable cases, from the comparison of the intensity of reaction of individual tree species the character of the remaining operative factors can be assessed.

It is in the nature of climatic change that there is seldom a single climatic factor which changes, but several change simultaneously, in different degrees and even in different directions. This is significant in dendroclimatology, as the growth of a tree reacts unequally to fluctuations in the same climatic factor in different climates. So the growth of the Douglas fir, *Pseudotsuga taxifolia,* in the western part of the United States is determined by the total precipitation in the preceding twelve months, of which the winter months are of greater importance [663], yet in Denmark it depends essentially upon the February–April temperatures and in part also on the May–June precipitation of the same year [349]. The thickness of the annual growth rings in the spruce (*Picea abies*) is independent of temperature fluctuations in Denmark, but it is as strongly dependent upon precipitation in the preceding year as of the May to July rainfall of the same year [349]. In contrast, in Norrland and Lapland the rainfall has no importance, but the number of days in which the maximum temperature exceeds 16°C between May and July is decisive [186, 208]. This is also true in western Norway, but here local exceptions to the rule are very important [574]. As a final example, the Anatolian pine can be considered. In the steppe climate of central Anatolia, the first part of the annual rings of the pine is caused by spring rain, whereas the deeper, more penetrating winter precipitation is necessary for the production of those cell layers of the annual ring formed during the later summer or early autumn. In the transitional climate toward the Black Sea the importance of winter precipitation for wood growth diminishes, but simultaneously that of early summer precipitation increases.

These examples suffice to show that even at the climatic limits of tree growth, the conclusions from dendroclimatology are highly uncertain; small climatic swings of a few years' duration (drought or cold) are clearly seen in variations in the tree rings. When there have been greater climatic fluctuations the method is largely inoperable.

2. Distribution of individual species as indications of climatic character. The flora of a given region is composed of species whose centers of distribution lie in different parts of the earth; it is made up of different "geographic floral elements"—a term that includes the total number of species which inhabit essentially the same region. The form of the region may differ, and often there are geographic faunal elements corresponding to geographic floral elements. Figure 23 illustrates

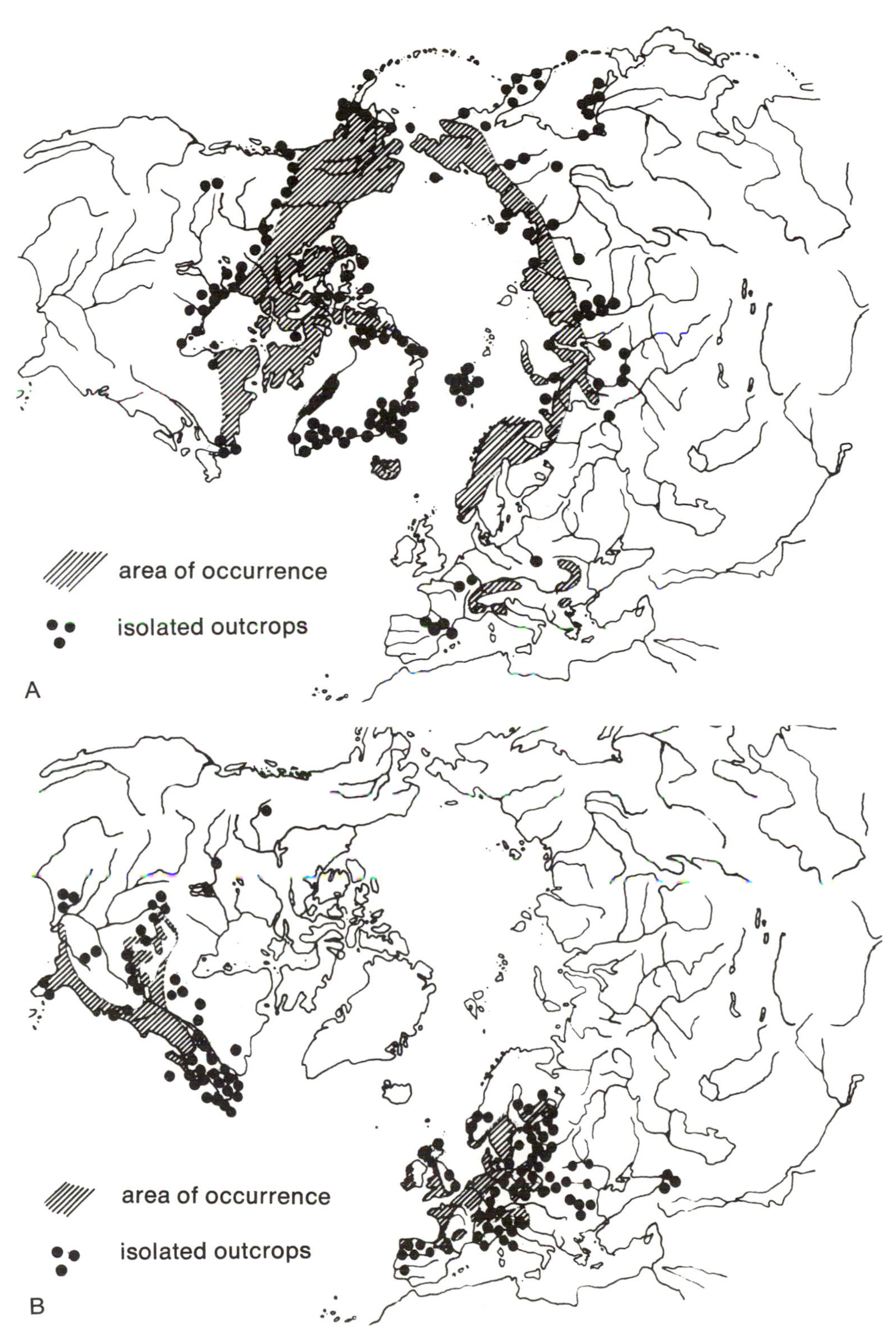

FIGURE 23. Geographic floral elements as an expression of climate. A. cotton grass, *Eriophorum Scheuchzeri*; B. sundew, *Drosera intermedia*. *Eriophorum Scheuchzeri* appears to be particularly indicative of a subarctic climate, whereas the sundew in contrast appears to reflect an oceanic climate. From [851: A] and [852: B].

examples, namely the amphi-Atlantic to suboceanic floral elements of cool temperate latitudes (sundew, *Drosera intermedia*) as well as the circumpolar subarctic-alpine floral elements of the Northern Hemisphere (cotton grass, *Eriophrum scheuchzeri*).

One may assume that the distribution of various individual plant and animal species differs according to their adaptability to the climate. This idea is basic to the paleoclimatic analysis of fossil flora using the former distribution of representatives of modern floral elements, and appears reliable if it is accepted that the critical climatic factors for the present distribution can be ascertained. Examples of this are the recent distribution of the holly (*Ilex aquifolium*) in Europe (Fig. 24), the European spruce (*Picea abies* [201], Fig. 25), as well as mistletoe (*Viscum album*), ivy (*Hedera helix*), and the holly (*Ilex aquifolium*) in Denmark [362 and Fig. 26]. The climatic conditions determined for the present position of the limit of distribution of a species investigated can by extrapolation be applied to the corresponding distribution limits for past periods where these limits can be determined. This permits quantitative statements about the climate (cf. Chapter IV).

The distribution of the spruce (Fig. 25) and the holly (Fig. 24) shows that in

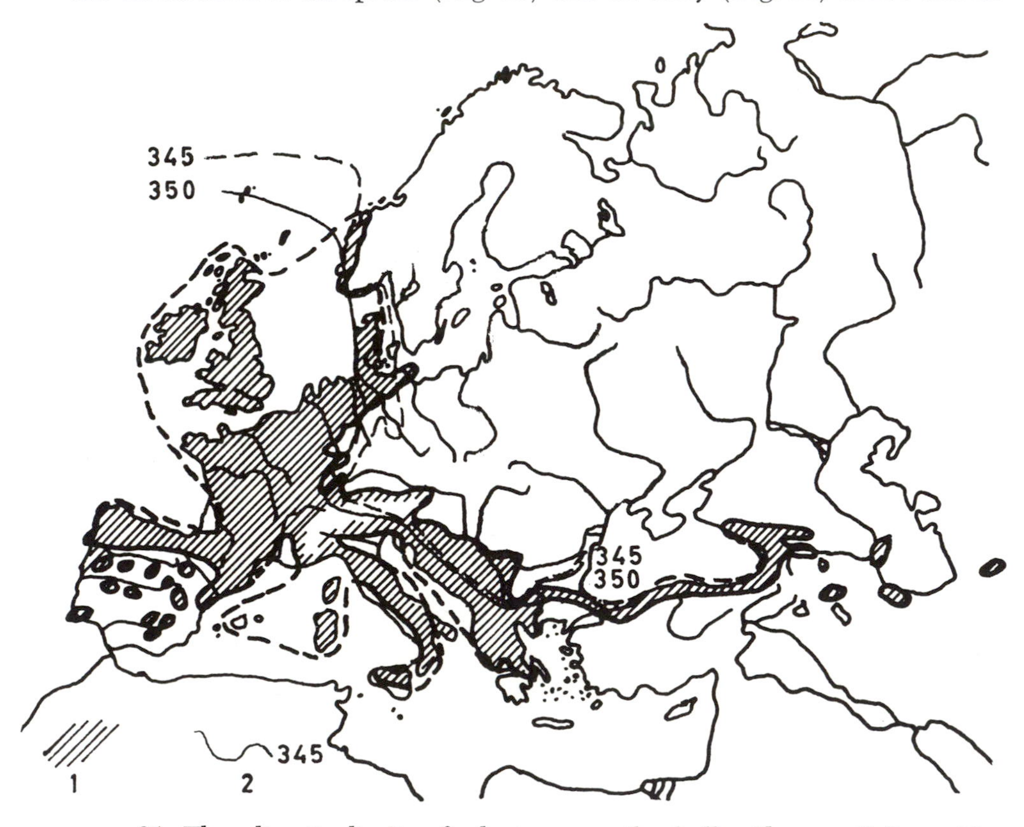

FIGURE 24. The climatic limits of plant zones; the holly, *Ilex aquifolium*. The eastern limit of the area coincides approximately with the 0°C January isotherm, or more precisely with that line which links all areas in which the number of days with a maximum temperature above zero is not less than 345. Simplified and modified after Enquist [201]. 1. present area of distribution of holly; 2. number of days with a maximum temperature above zero.

various parts of the area of distribution of a species the limit is controlled by different climatic factors (made clear in the above discussion of the difficulties facing dendroclimatology). This means that in climatic changes other factors than those present could be critical. Such a source of danger must always be considered if in the course of major climatic fluctuations extremely cold continental climate replaces a former mild oceanic climate, or vice versa. The possible uses of the method are thus restricted, and it is always necessary to test whether the climatic conditions of a given time so far resembled the present that the climatic values determined for the present limits of distribution can also be applied to the reconstruction of former climate. Even if the supposition holds, the method cannot be applied uncritically, for the distribution of plant and animal species depends not only upon the physiological capabilities of the species considered and the climate pertaining but also upon the number, kind, and vitality of competing species living in the same region. These sources of error have been clearly demonstrated in Ellenberg's work [189, 190]. It can be shown that the apparently firm association of individual plant species in given localities is essentially determined by competition. As examples of this, the "biotope demands" of erect brome grass (*Bromus erectus,* a grass characteristic of dry locations), the meadow foxtail

FIGURE 25. The climatic limits of plant zones; the European fir (*Picea abies*). At the different boundaries there are different critical climatic factors. 1. northern limit during the 65 summer days a maximum temperature of 12.5°C is necessary; 2. the western limit lies where, on fewer than 120 winter days, the minimum temperature reaches 0°C, i.e. where winters are too warm; 3. the southeastern margin lies where, on more than 65 summer days, the temperature reaches or exceeds 24°C. On the figure the distribution of the European *Picea abies* and the Siberian *Picea obovata* have been combined. After Enquist, from Meusel [853].

(*Alopecurus pratensis,* which prefers moist localities), and the smooth oats (*Arrhenatherum elatius,* generally growing on moderately wet soils) can be used (Fig. 27). When grown in the absence of other competition, all three species thrived best on soils with average water supply. When the three species compete with one another, the brome grass is found preferentially in the drier locations and the meadow foxtail in the wetter. The "biotope demands" of these species are thus a function of competition. It can be concluded that in general the environmental requirements of organisms change when different species compete with one another. When this stage of affairs is overlooked, the consequence can be significant errors in the determination of ancient climate.

An interesting example is the present and former distribution of the spruce (*Picea abies*). Figure 25 shows that the present western limit lies where the average temperature during the 120 winter days does not exceed 0°C; further to the west, the winters are warmer. During the next to last warm period (Holstein) the spruce flourished in Holland and England. If the previously mentioned temperature values were used, it would follow that the climate of England and Holland would have been much more continental (i.e., colder winters) than at the present time. Countless other observations (see below, pp. 104 ff.), however, indicate that, on the contrary, during the Holstein warm period the climate must have been more oceanic and warmer than at present. At the western limits of the spruce, then, temperature conditions must have been different from the present, with competing species different from those under present conditions.

The reason for uncertainty in determining climatic conditions that control the limits of distribution is that at the limits of a species the climate does not inhibit or block critical physiological conditions, but rather reduces the vitality of a species directly or indirectly with respect to competing species. When competition is absent, the species may spread over a much wider area. The region which is physiologically possible is significantly greater than that which is actually occupied on ecological grounds.

There are, nevertheless, essential physiological processes which can operate only if certain climatic factors reach the necessary minimum value for survival.

FIGURE 26. The climatic limits of plant zones: a. mistletoe, *Viscum album*; b. holly, *Ilex aquifolium*; c. ivy, *Hedera helix*. The ordinates of the diagrams show the average temperature of the warmest month, while the abscissa gives the average temperature of the coldest month. The diagrams thus illustrate the dependence of the plant named upon the temperature change between warmest and coldest months. In the diagrams the following symbols are used:
a) 1. mistletoe in the vicinity of the recording station; 2. a station at the limits of mistletoe distribution; 3. mistletoe absent in the vicinity of the station.
b) 1. holly in the vicinity of the climatic recording station; 2. station at the limits of holly distribution; 3. holly absent in the vicinity of the station; 4. holly which has escaped from gardens into the surrounding woods; 5. sterile holly only in the vicinity of the station; 6. the distribution limit of holly lies immediately in the vicinity of the station.
c) 1. ivy developed normally in the vicinity of the station; 2. mostly sterile with only rare ripe fruit in the tree crown; 3. ivy absent in the vicinity of the station; 4. fruits only in particularly favorable localities; 5. never bears fruit. The curves define limiting climatic values in individual regions. From Iversen [362].

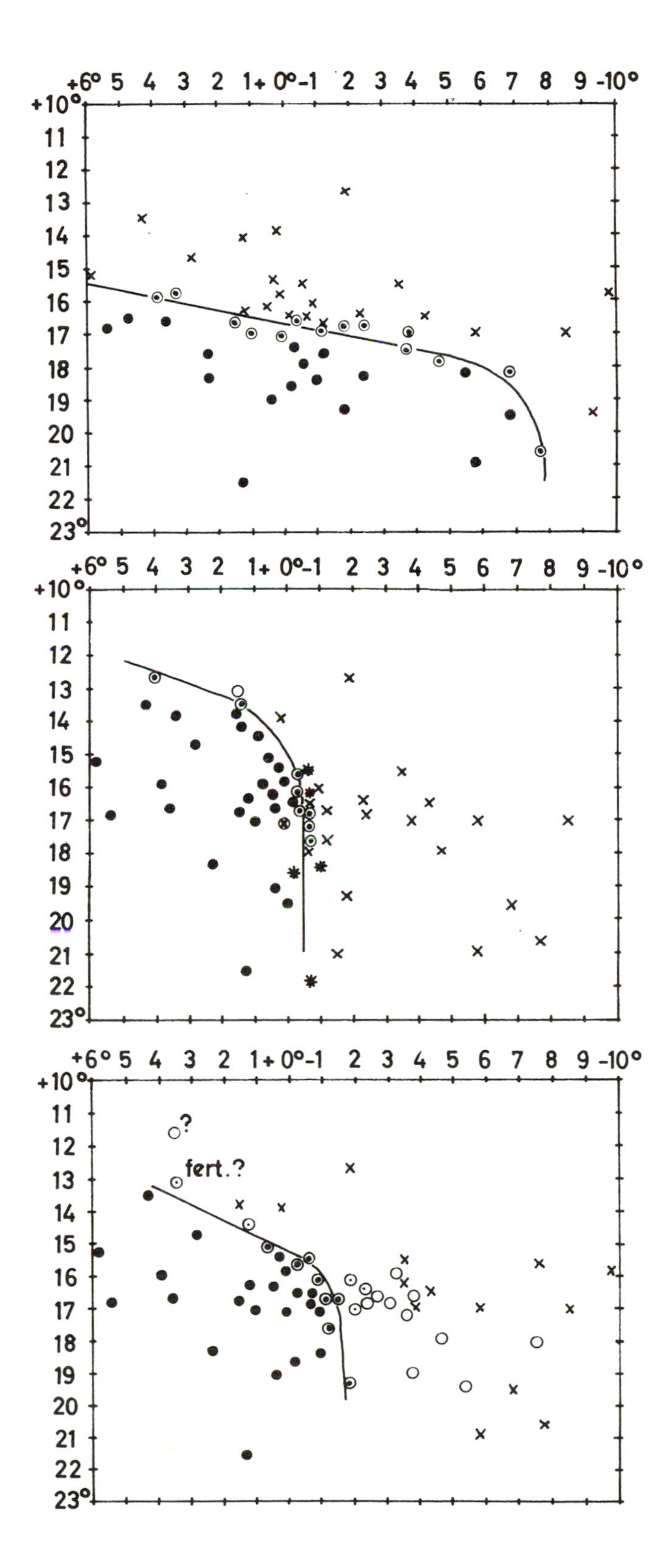
+6° 5 4 3 2 1+ 0°-1 2 3 4 5 6 7 8 9 -10°
+10°
11
12
13
14
15
16
17
18
19
20
21
22
23°
+6° 5 4 3 2 1+ 0°-1 2 3 4 5 6 7 8 9 -10°
+10°
11
12
13
14
15
16
17
18
19
20
21
22
23°
+6° 5 4 3 2 1+ 0°-1 2 3 4 5 6 7 8 9 -10°
+10°
11
12
13
14
15
16
17
18
19
20
21
22
23°
?
fert.?

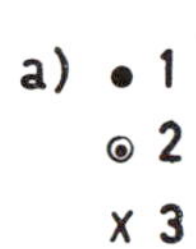
a) 1
2
3

b) 1
2
3
4
5
6

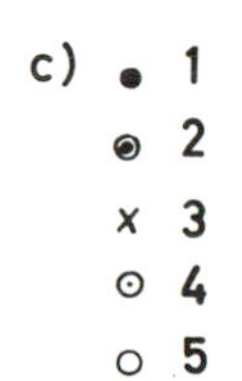
c) 1
2
3
4
5

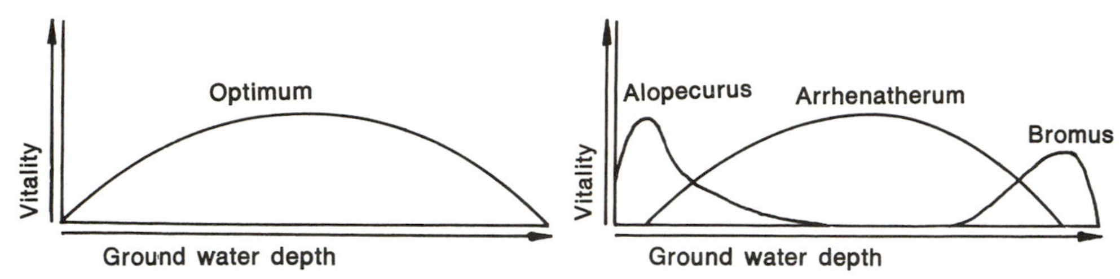

FIGURE 27. Changes in the environmental requirements of plant species with changes in competition. In the figures vitality (increasing upward) is shown on the ordinate with the depth of ground water along the abscissa (decreasing to the right). When grown alone, each of the three grasses (erect brome grass, meadow foxtail, and smooth oats) shows its optimum development in regions of average ground-water depth. Only with the mixing of all three does the meadow foxtail (*Alopecvrus pratensis*) come to inhabit the wet locations, the brome grass (*Bromus erectus*) the dry, and smooth oats (*Arnhenatherum elatius*) the central regions in which they are most commonly found in nature. From Walter [854].

For example, the development of many species depends on the length of the day and the control—acceleration or retardation—of certain reactions at certain temperatures. It is questionable, however, whether such physiological processes can acquire significance in determining past climate. Insofar as is known, the length of the day at various points on the earth's surface did not change during the Ice Age. It must also be added that physiological races whose day-length requirements deviate from the parent forms develop very rapidly, so that a plant species forced to migrate by a notable fluctuation in climate can avoid possibly deleterious climatic effects by the rapid development of new physiological races. As for the controlling effects of temperature, it is important to note that these temperature levels have been reached regularly. At least, the results so tediously won by researchers do not differ from other data less laboriously obtained.

It can be concluded that the coincidence of the limits of distribution of certain plants and animals with certain values of individual climatic factors in many cases provides a useful starting point for the reconstruction of climates of the past. Yet there are dangers in using this method which serve as a warning against too hasty an application. The difficulties are diminished if instead of considering merely the distribution of individual species, plant or animal communities are studied. Furthermore, additional useful indications are to be awaited from the resolution of the developmental and migrational history of the various plant communities. Here pollen analysis is helpful.

3. Reconstruction of the vegetation. Pollen analysis is one of the most important methods for revealing the history of vegetation. Beyond that, it can contribute to the determination of ancient climates. To be sure it provides no information about the fluctuation of individual climatic factors, only about the change in the total climate as it affects changes in the plant cover over a large area.

The pollen of wind-pollinated plants as well as the spores of ferns, horsetails, lycopods, and mosses are caught up by the wind and transported at varying distances before finally being sedimented in bogs, lake basins, rivers, or seas, or on the land surface. The membrane of pollens and spores is resistant and can persist for long periods, especially when protected from oxidation. These conditions are most commonly found in lakes or bogs. The form of pollen grains or spores (that is, sporomorphs) is characteristic for each species (Fig. 28). As a result, pollen analysis of the sporomorphs of a given bed or sedimentary horizon can yield information about the former existence or absence of wind-pollinated plant species. Detailed research into the magnitude of pollen production of different plant species [223, 203, 45, 209, 210, 338, 558, 151, 813] has shown that some species will produce more pollen than corresponds to their relative abundance in the total flora. For others the contrary is true. With some knowledge of the scale of over- or underrepresentation of pollen it is possible to obtain quantitative data on the significance of plant species in the former plant community by pollen analysis. In principle the calculation is simple: generally the proportion in the sample of tree pollen (sometimes including that of shrubs), of the so-called "nonarboreal" pollens —that is, of the nontree-forming flowering plants—and finally of spores is determined, with the spores frequently given as percentages of the tree pollens. Connected to this, the relative proportions of individual species may be determined with respect to the total sporomorph count of that group. From this a pollen spectrum is obtained of the actual sample. By determining many pollen spectra from different depths in a peat or lake basin being investigated and the reduction of percentage values to one coordinate system in which the percentage of pollen is shown against decreasing depth, this pollen diagram provides a picture of the history of the vegetation. The more sporomorphs determined, the more exact the reconstruction of the ancient vegetation. It is clear that pollen analysis in general indicates only wind-pollinated plants (anemogamous) in the former vegetation. The pollen of insect-pollinated plants (entomogamous) is, in contrast, normally very hard to find.

The fact that pollen analysis usually encompasses only the pollen of wind-pollinated plants has the distinct advantage that the pollen spectrum which results gives—by mixing pollen in the air—an average for a large region. Local peculiarities in vegetation (as, for example, in microclimatically favored localities, which are most difficult to deal with paleoclimatically), will be either poorly represented or absent.

Yet this advantage is not without two serious handicaps: (1) With the help of air currents, sporomorphs attain great altitudes [594, 575, 221, 1, 2, 745, 504, 710, and others] and may be transported to regions with a quite different vegetation, so that, for example, in deserts or arctic islands the pollens of numerous forest trees are found. The error is easily recognizable by consideration of the nontree pollens, as these are seldom transported far from the parent plant. The evaluation of the nonarboreal pollen is thus an essential of all pollen analysis work. (2) Further problems arise in pollen analysis in those localities in which the characteristic and ecologically dominant plant type is pollinated not by wind but by insects. In central and northern Europe this situation does not arise. In other places, as in the tropics, it is a serious obstacle to the successful use of the method.

A definitive climatic evaluation of pollen analysis must be based upon a sound knowledge of the ecological and plant-sociological relations of individual species with one another. Unfortunately, specific determination from the morphology of pollen is generally not possible, so that the occurrence of individual genera or fam-

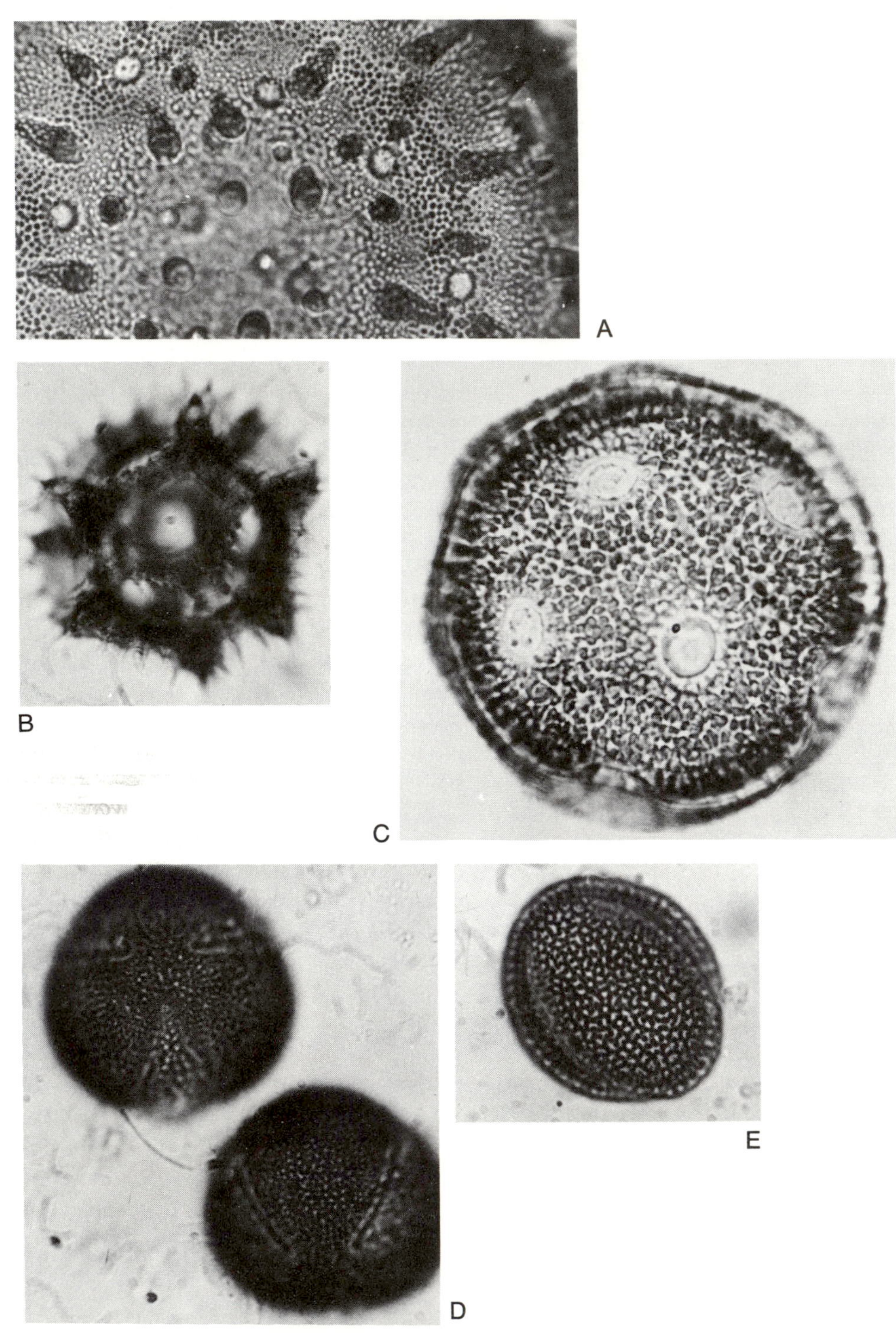

FIGURE 28. Different pollen forms as a basis of pollen analysis. A. musk mallow (*Malva moschata*); B. chicory (*Cichorium intybus*); C. feathered pink (*Dianthus plumarius*); D. marsh marigold (*Caltha palustris*); E. Buckler mustard (*Biscutella laevigata*). Magnification about 1200X.

ilies must suffice. As the ecological requirements of individual species of a given genera or family are in no way the same, this problem often prevents an exact paleoclimatological interpretation of the data obtained. Finally, in every such investigation it must be realized that not only may a vegetational change be the result of climatic fluctuation, but, quite independently of climate, the vegetation may pass through a number of developmental stages. For example, the pollen diagrams from the initial colonizing of a fresh surface (resulting from a landslide) to a complete forest cover would show changes that could overlap with climatic changes.

A good example of such a difficulty is the development of the vegetation on the steep southerly dipping slopes of the Jura in the lower Altmühltal immediately above Kelheim. There, numerous traces of former cultivation have been left behind in the form of small terraces and walls or mounds of loose boulders ("Lesesteine"). Within a horizontal distance of 2.5 kilometers there are to be found rocky summits with the beginnings of colonization by higher plants, open heath without trees or shrubs on the former fields or gardens, shrubs and bushes on the boulder piles, and open thickets and beechwoods, which can be scarcely more than a hundred years old. The spatial distribution and composition of the plant communities show that here there is a nearly complete developmental sequence, a "progressive succession," from exposed rock surfaces which are only a little above float-covered slopes to complete woodland cover.

The flora of every locality, as noted, is composed of members of different floral elements. Table 2 shows the percentage of the individual geographic floral elements in the total flora of the plant community of the region mentioned. In it the following elements can be distinguished: the sub-Mediterranean continental element, which has its center along the northern margin of the Mediterranean and in continental Eurasia; the sub-Atlantic suboceanic floral element, which occurs principally in regions with an oceanic climate, in particular the west coast of Europe; the pre-Alpine and Nordic-Eurasian floral element at home in the colder parts of Europe, preferring the Alps and northern Europe; the temperate continental and sub-Mediterranean floral element, which has its center in the central European deciduous forest zone; and the Eurasian floral element, which is widely

TABLE 2. PERCENTAGES OF THE REPRESENTATIVES OF THE DIFFERENT GEOGRAPHIC FLORAL ELEMENTS IN THE PLANT COVER OF THE ALTMÜHLTAL WHICH BELONG TO A PROGRESSIVE SUCCESSION (SOUTH-FACING SLOPES OF JURA LIMESTONE)

	GEOGRAPIC FLORAL ELEMENTS				
PLANT COMMUNITY	SUB-MEDITERRANEAN CONTINENTAL	SUB-ATLANTIC & SUBOCEANIC	PRE-ALPINE & NORDIC-EURASIAN	TEMPERATE-CONTINENTAL TO TEMP. CONT. & SUB-MED.	EURASIAN
Rock-inhabiting plants	66.7	11.1	3.7	14.8	3.7
Steppe heath ..	53.7	20.9	6.0	10.5	9.0
Rock dumps ...	45.5	21.8	5.4	14.5	12.7
Clearings in beechwoods .	49.1	17.6	1.8	12.3	19.3
Beechwoods ...	27.9	30.2	18.6	16.3	7.0

distributed over the whole continent. The division of the flora of a locality into different geographic floral elements may give indications of the climatic character of that region, as the concentration of species in an area of about the same form and position is determined primarily by the climatic character and physiological requirements of the individual species. Consequently, from the changing proportions of the geographic floral elements on the southern slopes of the lower Altmühltal within the succession described, climatic changes can be read. They progress from a Mediterranean-continental character to a climate tending to essentially oceanic and Nordic whose Mediterranean-continental component is very small. The evidence of the natural succession of plant communities would seem to imply a considerable climatic alteration. Here it is a question only of the climate of the existing plant communities; yet the example has paleoclimatic overtones, insofar as, in a paleoclimatic sense, such a succession can easily be compared to corresponding climatic changes, although these need not have occurred.

Successions of this kind develop rapidly. Thus the danger of errors from this cause in the attempted reconstruction of climate arises only when there is no means of estimating the elapsed time. Yet there is a second sequential process operating over a much greater time period which also leads to changes in vegetation, which in turn can be used as a means of indicating climatic change. This is the change in vegetation caused by soil maturity.

As already noted, in a constant climate and on approximately level surfaces the development of a soil progresses slowly, creating a permanent new plant habitat. One such sequence of soil-development is formed on the loess of central Europe somewhat as follows [448]:

raw loess ⟶ pararendzina ⟶ brown soil ⟶
parabrown soil ⟶ pseudogley ⟶ stagnogley

Parallel to this, the plant world under central European conditions develops from an initial community with a high proportion of different warmth-loving herbaceous plants and some dwarf shrubs, through scrub with woolly viburnum (*Viburnum lantana*), hawthorn (*Crategus oxyacantha* and *C. monogyna*), liguster (*Ligustrum vulgare*), pines, various oaks, and the common maple upon the pararendzina, to beechwood forms upon the brown soil. The same forest type remains upon the parabrown soil, but with the transition to pseudogley the aeration of the soil deteriorates, so that the proportion of hornbeam (*Carpinus betulus*) and, later, alder, birch, willow, and poplar increases, with a contrasting progressive decrease in the significance of oak and beech. Along with these changes, the composition of the herbaceous flora progressively shifts from the woodland herbaceous plants to the less demanding sedges and rushes, until finally only a very poor wood type thrives on the pseudogley.

The development requires more time than the simplified succession described above, which comprises only part of the sequence of soil development as sketched. What is so important in paleoclimatology in this example is that the evolution of the soil alone produces vegetational changes which, without knowledge of the process, might well be interpreted as a climatic change from a dry, perhaps warm, to a warm, moist climate, to dominantly wet-weather conditions. Since many plants belonging to the tree and herbaceous horizons are presently widespread on extremely wet and badly aerated soils in the northern part of western Europe, it might be deduced from the pseudogley and most certainly from the stagnogley that the climate was not only wetter but also cooler, even though the gross climate has indeed not altered. Andersen [13] rendered a great service by his penetrating inquiry into this state of affairs.

4. Conclusions. Study of the methods at present available for the reconstruction of climates of the past has shown that their number is deceptive, for every method is subject to serious and variable defects. Yet this situation is not discouraging, for the probability is that very reliable data can be obtained, whenever indications of climate can be gained, by using a variety of methods. It is certainly true that because the dangers of false interpretation in paleoclimatology are great, any process should be considered which simulates climatic fluctuation, such as the natural succession of plant communities, soil development, and the migrational history of plant and animal species.

2. Subdivisions of the Ice Age

GENERAL PROBLEMS

If the character and intensity of climatic fluctuations in the Ice Age are to be investigated, the basic history of the Pleistocene must be known. Occupying a key position is the question of the number and age of the climatic fluctuations within the various parts of the earth's surface. It is therefore understandable that in unveiling the climatic history of the Ice Age, it is necessary and important to make a distinction between regions with extreme Pleistocene climatic changes and those in which there was a continuous climatic development. Of equal importance is the possibility that climatic fluctuations in different regions were of different age. The views concerning these problems are diverse and must be considered before the climatic history of the Ice Age can be established.

In the Alps the accepted view of four glacials and three interglacials goes all the way back to Penck and Brückner [582]. The ice ages from oldest to youngest were named Günz, Mindel, Riss, and Würm. In contrast to the four Alpine glaciations recorded, it was supposed that there were only three Nordic continental ice sheets—from oldest to youngest, the Elster, Saale, and Weichsel. As Eberl [182] described traces of a glaciation or a cold period which was older than the Günz in foreland of the Schwabian Alps, the apparent contrast between the Alpine and the Nordic glaciations seemed even stronger.

In the meantime, careful geological work has been carried out in the Rhine Estuary and in East Anglia, particularly since World War II. The knowledge obtained impels a reconsideration of former hypotheses, as it can be shown that in the North Sea region six Pleistocene cold periods are recognizable whose intensity equaled that of the already established Alpine glaciations or cold periods. They have been called the Pretiglian, Eburonian, Menapian, Elsterian, Saalian, and Weichselian ice ages or their analogues [775, 776, 815, 816, 818, 788–90, 792]. These cold periods alternate with warm periods, named, from oldest to youngest, the Tiglian, Waalian, Cromerian, Holsteinian, and Eemian.

It would appear today that the number of known glaciations or cold periods was the same in both the Nordic and the Alpine glacial regions. Yet this impression should not conceal the fact that there are still great difficulties concerning the synchroneity of the individual cold or warm periods in the two glaciated regions (see below, p. 58).

In North America only four glaciations are known. They are, from oldest to youngest, the Nebraskan, Kansan, Illinoian, and Wisconsin [cf. reviews in 231, 806]. Furthermore it should be noted that the glacial history of the Wisconsin, at least, was appreciably different from that of the coeval Weichsel-Würm in Europe (see below, p. 233); hence the impression arises that the climatic fluctuations of the Ice Age in North America deviate from those in Europe.

Particular difficulties arise in those regions which were glaciated during the Ice Age but the surface of which was not suitable for the accumulation of glaciogenic sediments or a terminal moraine. This holds for wide regions of eastern Europe and in particular Siberia. Finally, there are regions of the earth never affected by glaciers or melt waters, and here other criteria must be used for evidence of former climatic fluctuations. The difficulties in recognition of evidence of former climatic fluctuations because of the passage of time, and factors operating in that time, are thus further compounded by the surface features of the earth and the climatic character of different regions at the time of glaciation. In this way an explanation can be found for the fact that even today in Russia the existence of the Pleistocene continental glaciations is contested by some, or at best considered as representing a single cold period. Supporters of this view are found among geologists as well as vertebrate paleontologists and paleobotanists. A compromise solution adopted by Quaternary geologists is that there were two glaciations in Siberia of which the older was the greater. It may be correlated with the Saalian. The younger corresponds, however, to the European Weichselian. This younger glaciation in Siberia is split by many workers into two independent glaciations separated from one another by an interglacial that has yet to be recognized in western Europe [see below, p. 226, and references 621, 760, and 277; see also literature in 260].

At the present time in the cool temperate latitudes of the Northern Hemisphere there is a marked contrast between the relatively mild, rainy climates near the oceans and the dry, high continental climate of the continental interiors. This contrast has an analogue in the oceanic regions, where traces of former glaciations are common but are much rarer in the regions of high continental climate, so that there is apparent justification for considering that in Siberia either there were no phases of considerable glacial advance, or these occurred only during the European warm periods. For it should be recognized that the quantity of atmospheric moisture over the continent was lowered as a consequence of the fall in temperature during the European cold periods; hence in Siberia no increase in glaciation occurred during the Pleistocene or at least during the European cold periods. According to this view, the strong glaciation of Siberia was possible only during the most favorable conditions of a European interglacial, for the climate was still cold enough to retard the melting of advancing glaciers. This hypothesis requires the supposition that high continental regions have been subjected to cold periods but not to glaciations, or that the ice ages there were contemporaneous with the European and East Asian interglacials, even though warm and cold periods in continental and oceanic regions were simultaneous [272, 512, 513, 456].

Finally, it has frequently been proposed that local tectonic movements have led to glaciations at different times quite independent of one another in certain mountain ranges or large areas. An example of this is Popov's view [599] that glaciation of the foothills of the west Siberian lowland was due to the tectonic depression of the north and central region of the lowland. The result of the depression was the penetration of polar seas deep into the continent, significantly increasing precipitation in the Urals and the mountainous regions of central Siberia and thereby favoring the development of large ice sheets, which then advanced over the west Siberian lowlands. Popov believes, however, that this glaciation could occur only once during the whole Ice Age.

The foregoing discussion is sufficient to indicate the difficulties encountered in every attempt to establish even the most important stages in the climatic development of the Ice Age. The number and magnitude of the difficulties increase as attempts are made at still finer divisions of the climatic history of individual cold or warm periods. Yet the attempt must be made, for upon the question of the stratigraphic correlation of the Ice Age depends the history of climate in the Pleistocene.

CENTRAL EUROPE

The major sedimentary basins, with their continuous bed-upon-bed deposition, form the best starting point for unraveling the geological development of the earth. One important sedimentary basin is found in the lower reaches of the Rhine, Maas, and Schilt. Here a large number of deep borings have penetrated the thick Pleistocene succession and have been carefully examined from a number of viewpoints. Important datum horizons are formed by marine incursions and by fluviatile beds whose heavy mineral content permits comparison with associated river terraces far to the south outside the depositional area. It is possible, on the basis of such relative dating, and by using techniques of paleobotany, to present a schematic development of the floral history of the region (Fig. 29). Six times during the course of the Ice Age various types of open vegetation replaced a preexisting forest cover in the area of the present-day Netherlands. The sediments that contain the pollen and macroflora indicative of open vegetation often show evidence of strong frost action; furthermore, they sometimes contain local glacial horizons. Such sediments and soil structures are lacking in horizons with woodland pollen and macroflora. From this it is clear that the rapid floral changes indicated reflect the contrast between cold periods or ice ages and the warm periods. The climatic fluctuations which affected the frequent, distinctive floral changes are "first order" fluctuations.

It can be seen from Figure 29 that in addition to the major changes in vegetation, others of a smaller scale are superposed. These are obviously reflections of climatic fluctuation on a smaller scale—that is, second-order changes. The superposition of both orders of climatic fluctuation has the consequence that the oldest warm periods, namely the Tiglian, the Waalian, and the Cromerian, can be divided into two climatic optima separated by a cool phase during which coniferous forests spread far to the south, giving the impression of six rather than only three warm periods. The same may possibly apply to the Holsteinian warm period [134a, 529a], but does not apply to the Eemian according to the data obtained by Dansgaard, Johnsen, Clausen, and Longway [149b] from a deep boring through the Greenland ice cap. Immediately following the Holsteinian and Eemian warm periods, at the beginnings of the succeeding cold periods, there were one or more phases of milder climate which permitted a forest re-advance. The variety of trees in these advance phases was however reduced in such a

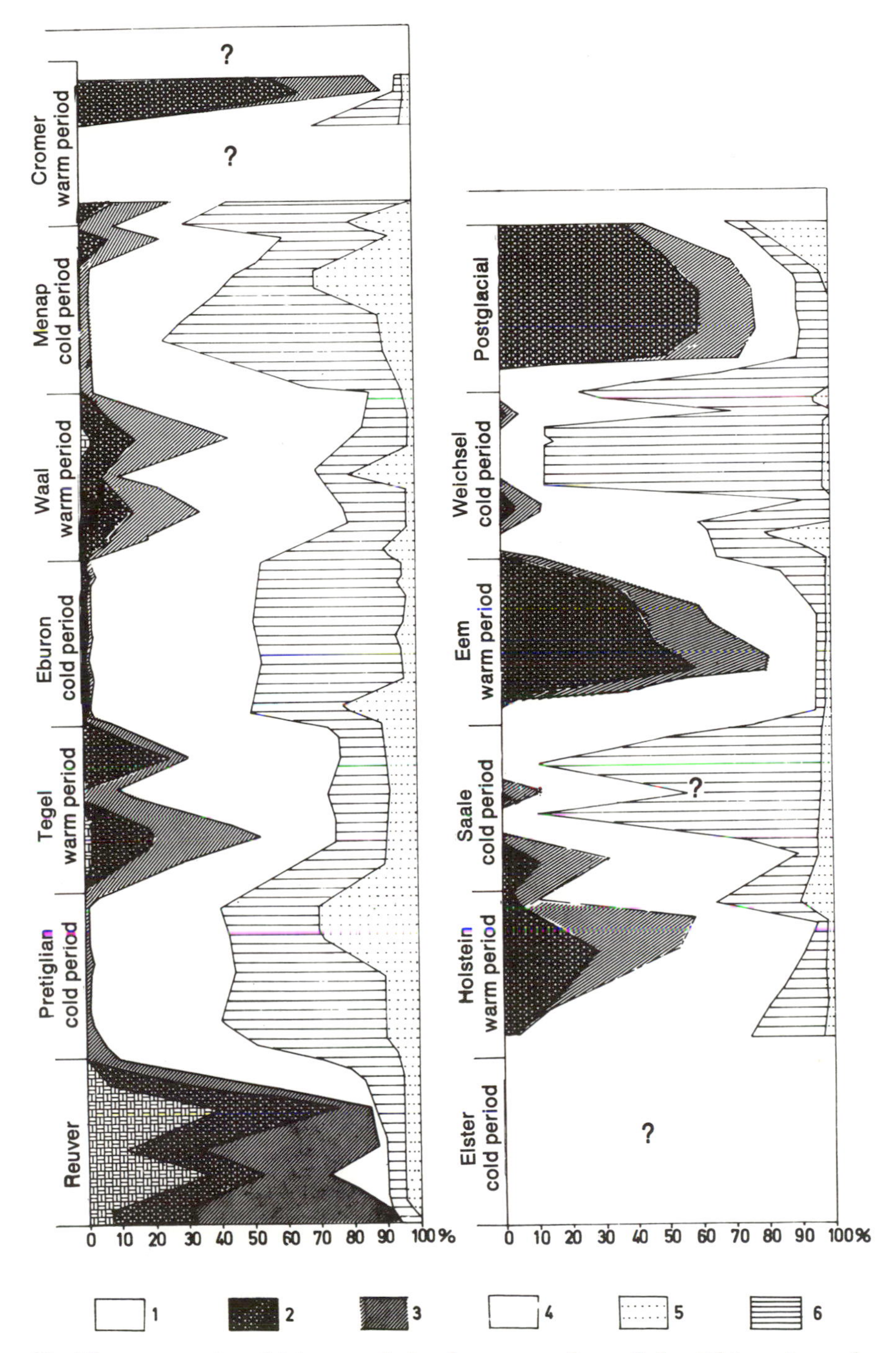

FIGURE 29. The vegetational history of the lower reaches of the Rhine since the end of the Tertiary. The figure depicts a simplified composite pollen diagram in which the total pollen in an individual horizon is normalized to 100%. The right-hand column follows above the left.
1. Tertiary trees, *Sequoia, Taxodium, Sciadopitys, Nyssa, Liquidambar*; 2. thermophilic deciduous species, *Fagus, Quercus, Castanea, Tilia, Carpinus, Corylus, Eucommia, Ulmus, Fraxinus*; 3. woods from damp to wet locations, *Alnus, Carya, Pterocarya, Vitis*; 4. conifers; 5. heath vegetation, *Ericales*; 6. grasses and herbs.

manner that these favorable phases cannot be considered as true warm periods.

The correlation of the succession of ice ages in the Netherlands with that in the glaciated Alpine regions presents considerable difficulties. Some of the uncertainties are shown in Table 3. The problems commence in the Saale-Riss ice age and increase in the still older divisions of the Ice Age. In the region affected by the Scandinavian ice sheet, the period of maximum glaciation—that is the Saale—can be divided into two phases: an earlier, which reached far to the south (Saale advance in a restricted sense, Drenthe glaciation), and a younger, less extensive advance (Warthe-Moscow glaciation). Nothing corresponding to this is known in the Alps with equal clarity, so that it is possible to consider the younger Saale advance in north Germany—that is, the Warthe-Moscow glaciation—as paralleling the Alpine Riss glaciation, and the maximum Saale advance (Drenthe glaciation, Dneprovsk glaciation) as correlating with Alpine Mindel glaciation. As no indubitable interglacial sediments have yet been found between the Warthe and Saale advances [see detailed discussion of this problem in 260], the suggestion does not seem to be justified.

Penck and Brückner [582] observed that the gravels that accumulated during the Mindel glaciation were later weathered to an extraordinary degree prior to the onset of the succeeding cold period. It was assumed from this that the intervening warm period must have been of long duration, and it was consequently referred to as the Great Interglacial. According to Penck and Brückner, this Mindel-Riss interglacial should be equated with the Holsteinian interglacial. Later observations have shown that the sediments formerly regarded as of the Riss ice age contained an important fossil soil that corresponds to, or is even better developed than, today's. Here obviously are traces of a further and until now unknown warm interval. The soils developed are weaker than those formed during the Mindel-Riss interglacial in the terms of Penck and Brückner. Thus the time of the development of the earlier noted massive fossil soil, which Brunnacker [in Graul and Brunnacker, 293] later termed "Riesenböden" must be pushed back in time, and it has to be assumed that the Mindel glaciation is older than it has formerly been considered to be. Brunnacker [94, 95], by his investigation of the developmental history of the middle Main and the Danube near Regensburg, has made it appear probable that the development of the Riesenböden is separated from the present by a long segment of the Ice Age, during which the Main built four terraces (Upper, Middle, Lower Middle, and Lower), and at least three and possibly four layers of loess were deposited. From the count of fossil soils and river terraces and their comparison with the number of glacial phases in other regions, it was established that the Riesenböden must have been formed prior to the Günz cold period. Table 3 shows that it must be considered as formed during the Waalian warm period. This correlation is not without problems, as the floral history of that time [according to Heydenreich, 344], with which the Riesenböden of the middle Main near Marktheidenfeld are probably synchronous, does not correspond to the floral development of the Waalian warm period but shows clear similarities with the Cromerian interglacial. The same is true of the fauna found in sediments of probably the same age near Randersacker on the Main [645]. The dating of the origin of the Riesenböden in Table 3 is thus probably about one warm period too old.

The elucidation of these problems would probably be materially aided by observations on the climate and floral history of the older warm periods. As was previously indicated, the Tiglian and Waalian warm periods in the Netherlands can be divided into two phases separated by a colder period on the basis of pollen analysis. Corresponding data for the Cromerian warm period has not yet been

TABLE 3. THE DIVISIONS OF THE ICE AGE IN THE RHINE ESTUARY, ON THE MIDDLE REACHES OF THE MAIN, ON THE DANUBE NEAR REGENSBURG, IN THE ALPINE FORELAND (BAVARIA AND SWABIA), AND IN CENTRAL GERMANY, AFTER VARIOUS AUTHORS

RHINE ESTUARY		ALPINE FORELAND			MAIN & DANUBE	CENTRAL GERMANY
ZAGWIJN 1963		PENCK AND BRÜCKNER 1909	GRAUL AND BRUNNACKER 1962, SCHÄDEL AND WERNER 1963	MULLER-BECK 1957	BRUNNACKER 1964	CEPEK 1968 (CF. ALSO MANIA 1970)
Warm periods	Cold periods					
	Weichsel	Würm	Würm	Würm	Würm	Weichsel cold period
Eem		Riss-Würm	Riss-Würm	Stuttgart interglacial	1st fossil soil	Eem warm period
	Saale	Riss	Principal Riss terrace Riss	Riss	Riss	Lausitz cold period Rügen warm period Fläming cold period Treene warm period Saale cold period
Holstein		Mindel-Riss Great interglacial	Great interglacial	Steinheim interglacial	2nd fossil soil	Dömnitz warm period Fuhne cold period Holstein warm period
	Elster	Mindel	Old Riss=in part Mindel	N cold period	Mindel	Elster cold period
Cromer		Günz-Mindel	Interglacial	Forest period of Mauer	3rd fossil soil	Voigstedt warm period Helme cold period Artern warm period
	Menap	Günz	Günz	M cold period	Günz	Menap cold period
Waal			Period of Riesen-boden	Warm period	4th and 5th fossil soils in part=forest period of Mauer	Waal warm period
	Eburon	?	Donau	G cold period		Eburon cold period
Tiglian			Warm period	Warm period	?	Tiglian warm period
	Pretiglian	?	Biber	Donau		Brüggen cold period

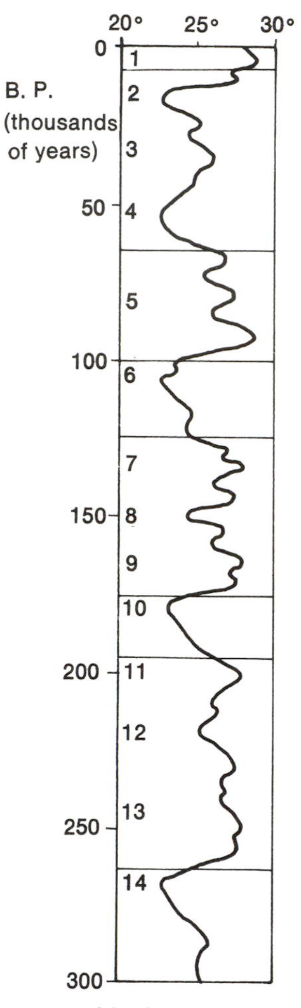

FIGURE 30. Curves of the surface water temperature of the Caribbean Sea during the latter part of the Pleistocene and the Holocene.
1. Post-glacial; 2–4. Weichselian cold period; 5. Eemian warm period; 6. Saalian cold period; 7–9. Holstein warm period; 10. Elsterian cold period; 11–13. Cromerian warm period; 14. Menapian cold period. It may be seen that the warm periods appear to be comprised of very different climatic phases; the Cromerian and Holsteinian warm periods in particular thus seem to have consisted of two warm phases separated by a cold phase. The two youngest warm fluctuations of the Eemian warm period should actually belong to the last cold period. Simplified after Emiliani [199].

observed in the Netherlands material. Near Cromer on the East Anglian coast [178, 789, 792], however, there belong in this warm period the so-called Lower Freshwater Beds, upon which rest successively the Estuarine Beds and the Upper Freshwater Beds. The Lower Freshwater Beds contain a warm temperate flora; the Estuarine Beds harbor marine cold-water fossils; while the Upper Freshwater Beds have the remains of a woodland flora. In the latter, the transition from cold unfavorable conditions through a phase of widespread deciduous forest to a new cold period could be observed. It thus appears that the Cromerian warm period also is composed of three divisions: two warm phases and an intervening cold period. This view is supported by the recent investigation of former surface-water temperatures in the Caribbean Sea [199 and Fig. 30], and on the other by the division of the thick loess deposits of the drier regions of Central Europe, among which those of Czechoslovakia are particularly significant. The oldest Pleistocene warm phases there (including the Cromerian warm period) are composed of two phases of well-developed soil formation under warm conditions separated by a period during which new loess accumulated or surface erosion took place [441, 448 and Fig. 51]. This observation leads to the conclusion that the Riesenböden had their origin not during the Waalian period but during the Cromerian, with several phases.

It seems better to abandon the terms Günz and Mindel cold periods, since their position within the Ice Age has become uncertain. This is directly related to the absence of detailed data or the division of the Donau cold period or even the Biber cold period. It is much more to be recommended that the stratigraphic sequence from the Rhine Estuary form be adopted, especially as it is in broad agreement with the Quaternary glacial history of the Alps.

EXTRA-EUROPEAN REGIONS

In general, as already noted, only four cold periods and three warm periods have been described in the Pleistocene of North America. Clisby [138] on the basis of results from a deep boring in the San Augustin plains in New Mexico noted that the Tertiary passes into the Quaternary and at the end of the Tertiary has established that climatic fluctuations of increasing significance occur toward the Pleistocene boundary. It is probable that in North America, as in Europe, there were further cold periods prior to the initial, Nebraskan, glaciation. This is all the more probable because numerous important climatic fluctuations are known from the changing spectra of the pollen analyses carried out by van der Hammen and Gonzalez [333] on a deep boring in the Sabana de Bogotá in Colombia, which has an elevation of 2,600 meters (Fig. 31). They have shown that, in a region normally forested today, five times within the Pleistocene open vegetation occurred when *Acaena*, a Rosacaea characteristic of the present South American Paramo, was able to spread widely. The correlation by van der Hammen and Gonzalez of many of the individual stages of the Sabana de Bogotá with definite stages in the climatic and floral development of Europe do not appear today to be well founded. Nevertheless, the results in Figure 31 indicate that even in an upland tropical climate the climate changed at least four or five times from warm to cold, the intensity of which can be compared to the first-order climatic fluctuations found in cool, temperate latitudes. One noteworthy fact is that the number of cold periods in the Sabana de Bogotá is greater than that presently known in North America, from which it may be deduced that traces of further cold and warm periods remain to be discovered in the latter region. The correctness of this assumption appears to be confirmed by the stratigraphy of the Pleistocene beds of New Zealand [806, 725, 726, 144 and Fig. 32]. In the southwest

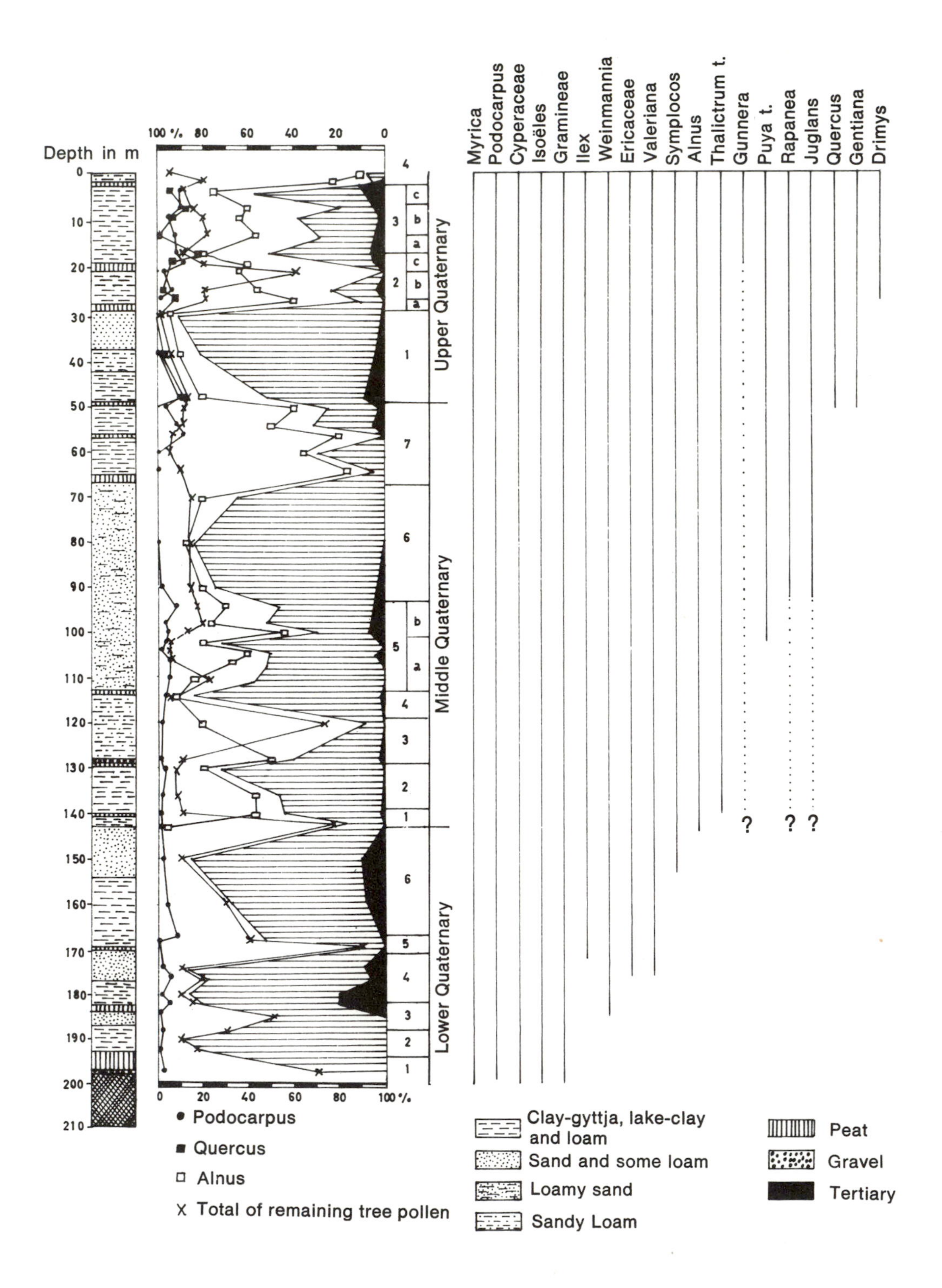

Depth in m
100 % 80 60 40 20 0
0 20 40 60 80 100 %
Upper Quaternary
Middle Quaternary
Lower Quaternary
Myrica
Podocarpus
Cyperaceae
Isoëles
Gramineae
Ilex
Weinmannia
Ericaceae
Valeriana
Symplocos
Alnus
Thalictrum t.
Gunnera
Puya t.
Rapanea
Juglans
Quercus
Gentiana
Drimys
• Podocarpus
■ Quercus
□ Alnus
X Total of remaining tree pollen
Clay-gyttja, lake-clay and loam
Sand and some loam
Loamy sand
Sandy Loam
Peat
Gravel
Tertiary

corner of the North Island in particular, climatic fluctuations can be determined not only from the character of the sediments but also by the fauna and flora they contain. It appears that here, too, the Ice Age is as obvious as in Europe and in the Sabana de Bogotá. Certainly many of the parallels drawn between the Southern Hemisphere and definite stages in the Quaternary climatic development of the Northern Hemisphere must be confirmed or made more precise by subsequent investigation. Nonetheless, the observations in New Zealand and Colombia leave no doubt that the changes recorded basically represent the climatic developments of the Ice Age.

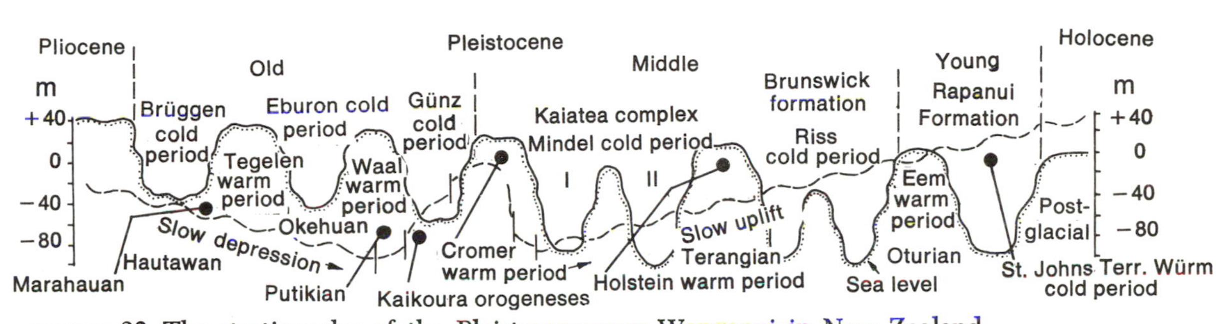

FIGURE 32. The stratigraphy of the Pleistocene near Wanganui in New Zealand. The curves show schematically the movements of sea-level (heavy line) and land areas (dotted line). From Woldstedt [806].

In contrast is the apparently very different division of the Siberian Pleistocene, but the question must be asked whether that impression is correct or whether it is simply a consequence of restricted knowledge.

The northern part of the west Siberian lowland is a large sedimentary basin. In it there is an intertonguing of thick marine, terrestrial, and glaciogenic sediments to a much greater extent than in the lower Rhine. During the warm periods, particularly in the lower reaches of the Ob and Yenissei [568], the tectonically depressed northern part of the plain was flooded by the eustatic rise of sea level caused by the melting of the ice sheets. As a result the sea advanced far to the south, particularly in the lower reaches of the Ob and Yenissei. The last great

FIGURE 31. Floral history of the Sabena de Bogotá 2,600 m. Colombia. The diagram shows the vegetation changes and the migrational history of individual genera. Left hand column = geological profile; second column from left: pollen diagram, total sporomorphs = 100%. At the upper margin of the pollen diagram is the scale for non-tree pollens, and along the lower margin that for tree pollens. In the diagram the sum of the tree and shrub pollens is in white, with grasses shown by horizontal ruling (Gramineae) and Aceana in black. Figures and letters to the right of the pollen diagram: stratigraphic correlation: U. Quaternary: 1. "Riss glaciation"; 2 (a–c). "Eemian"; 3 (probably already 2b). "Wurm glaciation"; 4. "Holocene." M. Quaternary; 1–3. "cf. Cromerian" or the end of the "Tiglian" to the end of the "Waalian"; 4. "Menapian"; 5. cf. "Cromerian"; 6. "Mindel" glaciation or 4–6: "Mindel" glaciation; 7. "Holsteinian."
Lower Quaternary: Either 1. "Reuverian"; 2–6. "Pretiglian and Tiglian," or 1–5 "Tiglian to Waalian," 6. "Menapian." Simplified from van der Hammen and Gonzales [333].

transgression was during the final cold period. This was the Kazancev transgression, which was coeval with the Eemian transgression in the North Sea basin. Then the sea reached as far south as 62.5°S in the lower reaches of the Yenissei. Within the sediments deposited during this transgression is a woodland pollen flora which had developed under warmer conditions than at present. Older than this transgression are moraines that indicate a glacial advance southward as far as the east-west middle section of the Ob. This was the maximum extent of the Samarov glaciation, which probably corresponds to the Saale-Dneprovsk glaciation in northern Europe. The moraines and coeval periglacial sediments of the Samarov glaciation often lie on fluviatile beds in which a rich pollen flora is found at many localities. The pollen indicates a transition in time from a steppe or tundra flora to a spruce-fir woodland and finally to a renewed steppe with birch thickets. The pollen data indicates that the spruce-fir woodland must have developed under a climate similar to or more favorable than the present. It was, therefore, a warm period. This period was preceded by a cold period [20, 21, 22, 466] from which the remains of a dated tundra vegetation near Tomsk are known —that is, in a region which now supports birch-aspen woodland (*Betula-Populus tremula*), grass steppes and spruce-fir-arven woodland (*Picea obovata, Abies sibirica, Pinus sibirica*). Still older are several loess horizons with buried soils that are exposed in the upper reaches of the Ob [629, 630, 631]. They provide information on the frequent alternation of periods of predominantly loess accumulation and renewed soil formation or between phases of very light to a dense steppe vegetative cover. In total there is thus evidence in the west Siberian lowlands of at least three significant cold spells with in part a further outspreading of continental ice sheets and a series of still older significant fluctuations in the thermal and hydrological conditions [260].

These few examples make it clear that even in very strongly contrasted present climatic provinces, traces of marked Pleistocene climatic fluctuations are repeatedly observed. It appears that there were six cold periods and five warm periods, of which the three oldest warm periods can probably be divided into two warm phases separated by a cold phase. The similarity of the climatic history in the given regions leads readily to the hypothesis that the climatic fluctuations, in contrast to older views, were simultaneous events. The correctness of this important surmise must be established in other ways, though the coincidence of marine transgression with warm periods in the Northern and Southern hemispheres already favors the simultaneity of climatic fluctuations.

3. Simultaneity of Climatic Fluctuations

During the Ice Age major climatic fluctuations occurred repeatedly. They were modified by smaller-scale changes, so that the climate within a very short geological time must have altered frequently. Consequently, most standard geological methods of relative and absolute dating are not sufficiently precise to answer the question of the simultaneity of the climatic fluctuations. In view of this state of affairs, the age of the various short-term climatic fluctuations of the last cold period are very significant, for some of them still fall within the range of radiocarbon dating methods. For that reason the absolute chronology of the last cold period will be specified in the following paragraphs; the more so because the climatic history of this cold period must later be discussed in detail.

On the basis of material currently available, the last cold period (Würm-Weichsel-Valdai-Wisconsin) in central Europe can be divided as in Table 4.

TABLE 4. STRATIGRAPHIC DIVISION OF THE LAST COLD PERIOD IN CENTRAL EUROPE

POSTGLACIAL, HOLOCENE
Youngest tundra (=youngest Dryas).
Alleröd interstadial
Older tundra (=older Dryas).
Bølling interstadial
Oldest tundra (=oldest Dryas), including the Pommerian stadial.
Lascaux-Ula interstadial
Main glaciation B, with several interstadial climatic fluctuations.
Paudorf-Stillfried B interstadial
Main glaciation A, with several interstadial climatic fluctuations.
Brørup interstadial (which can be divided into at least two warm phases and one cold phase)
Early glacial cold period.
Amersfoort interstadial
Early glacial cold period.
Last interglacial (Eem, Riss-Würm, Masovian II, Mikulino interglacial)

Only that time interval which begins with the Paudorf-Stillfried B interstadial and encompasses the closing phases of the last cold period can be dated with certainty by radiocarbon in most laboratories. In samples from older time intervals the amount of radiocarbon is very much reduced (cf. above, p. 18), while sources of error such as contamination with younger material together with all instrumental problems augment so that in only a few laboratories can the high precision be met. Even in these cases the results cannot be blindly accepted.

The Bølling interstadial was very short and in all probability very weak, so it is seldom recorded with certainty. The Lascaux-Ula interstadial, which occurred around 16,000–17,000 years B.P., permitted an important migration of woodland flora; but traces of this warm fluctuation have been observed in only a few places in the Northern Hemisphere, so that this interstadial cannot be used in questions of simultaneity of climatic fluctuations. Thus interest is centered on the Stillfried B interstadial and the Alleröd interstadial. In Figure 33 and Tables 5 and 6 is a list of the available radiocarbon dates derived from samples known to belong with certainty to the Alleröd interstadial either by pollen analysis or by macrofossil evidence from the enclosing sediments. The large number of other examples known to belong to one or other of the interstadials on no other grounds than their radiocarbon age cannot be used in this connection.

FIGURE 33. The age of beds formed during the Alleröd interstadial. Half-black circles: beginning of the Alleröd; black points: age of beds formed during the interstadial; open circles: beds from the transition from the Alleröd to the succeeding tundra period. See Appendix 2, p. 245.

TABLE 5. ABSOLUTE AGE DETERMINATIONS (IN YEARS BEFORE A.D. 1950) OF MATERIAL FROM THE PAUDORF-STILLFRIED B INTERSTADIAL OR ITS ANALOGUES (DATING BY RADIOCARBON METHOD)

LOCALITY	ABSOLUTE AGE B.P.
Lower Austria and Moravia:	
Aggsbach, Lower Austria; Gravette cultural horizon	25,540 ± 150
	25,600 ± 100
Stillfried a.d. March; upper surface of the soil[8]	27,990 ± 300
	28,120 ± 200
Paudorf near Krems on the Danube[8]	ca. 30,000
Dolní Věstonice, Moravia; Gravette cultural horizon[1]	25,600 ± 170
Pavlov, Moravia; Gravette cultural horizon[1]	24,800 ± 150
Cave, "pod Hradem," Moravia; Gravette cultural horizon[1]	26,240 ± 300
South Germany and East France:	
Karrestobel, near Ravensburg[9]	27,950 ±500
Arcy sur Cure, Yonne[7]	ca. 31,000 to 30,000
North Germany:	
Geesthacht, interstadial above the Brørup interstadial[1]	26,600 ± 300
Ostrohe interstadial[2]	32,200 to 26,600
South Sweden:	
Göteborg[6]	30,000 to 26,000
European part of U.S.S.R.:	
Molodovo on the middle Dniestr; Gravette cultural layer	23,000 ± 800
in a weak fossil soil immediately above the soil, probably dating from the Stillfried B interstadial[3]	23,700 ± 320
Volga terrace near Rybinsk–Cheremino[5]	28,800 ± 200
	25,900 ± 900
Gormovo on the Belaja, foreland of the central Urals[4]	29,700 ± 1,250
North America:	
Plum Point interstadial, region of Lake Erie[9, 13]	25,500 ± 1,200
	28,200 ± 1,500
	27,500 ± 1,200
	24,600 ± 1,600
Southern High Plains, Llano Estacado, New Mexico[10]	33,500 to 22,500
Searles Lake, Southeast California[11]	32,700 to 24,200
Olympic peninsula, near Vancouver; an interstadial shortly after and one shortly before[12]	27,400 ± 2,000
East Africa:	
Southeast end of Lake Tanganyika, Kalambo Falls[14]	27,000

REFERENCES: [1][315]; [2]Dücker lecture in Cologne, July 9, 1965; [3][360]; [4][364]; [5][271]; [6][87]; [7][463]; [8][219] and Fink [pers. comm.]; [9][313]; [10][325]; [11][723]; [12][342]; [13][806]; [14][855].

Dating the Stillfried B interstadial necessitates the use of the Gravette cultural horizon, which lies generally a little above the fossil soils belonging to this interstadial. The important dates 23,000 to 26,000 years B.P. obtained repeatedly for the cultural horizon certainly indicate a time in which the Stillfried B interstadial was essentially already over. At any rate, the beginning of the cold phase following this interstadial, i.e. of the Main Glacial B of the last glaciation, was interrupted by a relatively short climatic improvement, particularly well observed in the

Molodovo profile on the Dniester in the dry loess province of southwest Ukraine [360]. There follows, above a fossil black earth probably belonging to the Stillfried B interstadial, a thin layer of loess upon which a further but poorly developed soil forms, one which contains the Gravette culture. Its age has been determined between 23,700 to 23,000 years B.P. Similar observations have been made near Dolni Věstonice [407] and near Arcy-sur-Cure [464]. The younger fossil soil with the Gravette culture previously noted, however, can be distinguished from the soil of the true Stillfried B interstadial only in particularly dry regions, as only there, since the end of the Stillfried B soil formation and before the formation of the younger soil horizon, did renewed loess formation permit the separation of the two. In other regions where loess sedimentation does not occur, the two soils merge with each other, and the apparent upper surface of the Stillfried B soil gives too young an age.

In sum it may be considered that the Stillfried B interstadial and its analogues began about 31,000–32,000 years B.P. and ended about 27,000 years B.P. More precise dating does not seem warranted at the present time. Table 5 shows that within this time interval today, in strongly contrasting climatic provinces, former climatic fluctuations can be recognized. The difference in the dating is exceptionally small, and it can be assumed that over the whole earth from 31,000–32,000 to 27,000 years B.P. a climatic fluctuation occurred. The character of this fluctuation is, however, quite different in climatic provinces (cf. below, p. 224). The same is also true of the Alleröd interstadial, which lasted from about 12,000 to 10,800 years B.P. [280, 314]. The data contained in Figure 33 and Table 6 witness the

TABLE 6. ABSOLUTE AGE (IN YEARS BEFORE A.D. 1950) OF MATERIAL WHICH BELONGS TO THE ALLERÖD INTERSTADIAL OR ITS ANALOGUES (CONTAINS DATA NOT SHOWN IN FIG. 33)

LOCALITY	ABSOLUTE AGE B.P.
West Siberia	11,450 ± 250
Mammoth in Mammoth River in the northwest of the Taimyr Peninsula[1]	11,700 ± 300
Palaeolithic station Zabočka near Kokorevo near Krasnojarsk on the Yenessi. This cultural horizon lies immediately under the fossil soil that was probably formed during the Alleröd interstadial.[2]	12,940 ± 270
North America	10,050 ± 270
Alaska, Nome[3]	9,690 ± 400
Moss Lake, Washington[4]	11,900 ± 360
Minnesota before the Two Creeks interstadials[5]	12,650 ± 350
beginning of the Two Creeks interstadials[6]	12,030 ± 200
about the end of the Two Creeks interstadials[5]	11,250 ± 400
White Pine, Michigan[4]	12,600 ± 1,200
Aitkin County, Minnesota[7]	11,700 ± 325
Wood upon fossil soil	11,560 ± 400
Manitowoc, Wisconsin, average of five samples[8] near Richmond, Indiana[9]	11,400 ± 350
	11,700 ± 250
Tappen, Central North Dakota[10] between the Taconic and Green Mountains,	11,840 ± 300
southwest Vermont[11]	10,800 ± 250
St. Pierre, St. Lawrence River[8]	11,050 ± 350
Bermuda Shelf	11,500 ± 700

REFERENCES: [1][401]; [2][134]; [3][350]; [4][313]; [5][376]; [6][262]; [7][212]; [8][312]; [9][392]; [10][812]; [11][795].

correctness of this assumption, with the exception of the data of Hopkins, Macneil, and Leopold [350], of a warm period between 8,500 years and 10,000 years B.P., while the widespread ice of the Taimyr polygons was melted and peat developed in the resulting hollows and grabens in western Alaska. It must remain an open question how far this data will stand detailed criticism, as nothing comparable has yet been observed in subarctic regions.

Flint [233] noted that the sediments resting upon the upper surface of the Two Creeks Forest Bed on Lake Michigan, the North American equivalent of the Alleröd interstadial, are about 400 years older than the better investigated transition from the Alleröd interstadial to the younger tundra period in Europe. This difference is scarcely sufficient to consider the Alleröd interstadial and the Two Creeks interstadial as two differently aged phases of warm climate, as Antevs [19] has recently repeated, for in general the present dating shows that the Alleröd and Two Creeks interstadials are of the same age. From this one would not err in assuming that these climatic fluctuations have causes that affected the whole earth at the same time. It is always well to be aware, however, that different climates simultaneously subjected to the same impulse react in different ways (see below, p. 202). In this connection Mercer [530a] proposed an interesting hypothesis; that the Alleröd interstadial is a purely European climate anomaly. He suggested that the prevailing westerly winds in late glacial times brought about a rapid breakup of the Arctic shelf ice, thus bringing very cold maritime air over the continent during the European Younger Tundra period. Thus, only in western Europe is the oldest part of the true Postglacial warming trend separated by the Younger Tundra period so as to appear as a true interstadial. Only future detailed research will show to what extent this hypothesis can be justified.

It has already been indicated that in Europe between the Stillfried B and Alleröd interstadials there is at least one additional important interstadial, that of Lascaux and Ula. The age of the sediments formed during this fluctuation were found to be 17,000 years B.P. in the Lascaux caves (Montignac sur Vizère, Dordogne [365]); lacustrine sediments in southern Lithuania on the Ula near Zjarvinos contained pollen of a birch-spruce woodland flora which extended to the Baltic and whose age was found to be 16,260 ± 240 years B.P. [42, 271]. Evidently of the same age are two gley horizons near Molodovo on the middle Dniester which contain a cultural horizon whose age was found to be 17,100 ± 180 and 16,750 ± 250 years B.P. [360], and probably also the time of the spread of the pine forests in southern Spain [529]. In North America the data are less certain. There have also been found traces of a relatively significant interstadial lying between the Plum Point (Stillfried B) interstadial and the Two Creeks interstadial. Its age, however, does not agree with the values determined for the Lascaux-Ula interstadial (18,750 ± 300, 19,100 ± 300; Hamilton, Ohio: 20,000 ± 800; central Indiana [286]).

It must be left to future research to bring daylight to the reigning darkness. Perhaps the lack of accord may be solved in other ways, namely that this interstadial too was made up of two warm fluctuations. That possibility is suggested by the occurrence on the Dniester near Molodovo of the already mentioned cultural horizon between two closely following gley horizons within a loess series. It would then be possible to assume that the older determined age of the North American interstadial ("Connersville" interstadial [286]) refers only to the older of the two fluctuations. However, in the present context this question cannot be discussed further. It is much more important that the basic axiom be established, namely that the Stillfried B and Alleröd interstadials are expressions of climatic fluctuations which affected the whole earth simultaneously. The Stillfried B interstadial was longer and more incisive than the Alleröd interstadial.

If, in spite of the difference in magnitude, climatic fluctuations are found simultaneously over the whole earth, then the same may also be assumed for the remaining phases of the last cold period. Corroborating this is the fact that the course of the last cold period, in the furthest corners of the globe, follows the same pattern with the same time scheme. If, however, the climatic fluctuations of the last ice age occurred simultaneously over the entire earth, then it is pertinent to suppose that the same must hold true for the older cold and warm periods. In light of the already numerous factual data, this assumption is substantially more probable than the hypothesis of individual climatic development or contrasting courses of climatic history in different regions. Yet this should be regarded not as an immutable axiom but rather as one that must have its validity tested and proven in each individual case.

The foregoing conception of approximate simultaneity does not require that there be an absolute synchroneity of appearances. In particular, Kac [386, 387] and Heusser [341] have remarked that the postglacial climatic warming, particularly the postglacial climatic optimum of high northern latitudes, set in later and ended much earlier than in low latitudes, since in the north the cooling effect of the residual ice caps and repeated eruptions of cold air must have had a particularly strong effect. Heusser gave an estimated difference of 1,500 years between the beginning of the postglacial warm period in southern British Columbia and Washington on the one hand, and southern Alaska on the other. According to Kac the start of the postglacial warm period in the northern Baltic was delayed around 2,000 years in comparison with south central Europe. These opinions sound convincing, but the basis is insufficiently established. In both cases the time of onset of the postglacial warm period was based upon an investigation of the floral history, and in particular the migration of critical plant species. The method depends not only upon climatic conditions but also upon the velocity of migration, the distance traveled, and the number and vitality of other competing migrating forms. From this it follows that the onset of a particularly favorable climate must of necessity have a later origin in high latitudes than further to the south. A climatic deterioration must equally affect the plant cover in the north earlier than in the south even though it occurs simultaneously, since the cooling at the northern limit of distribution of a species must immediately and decisively affect its vitality. Further to the south a change in the vegetation and flora will first occur when certain fixed threshold values, different for each species, are passed [692]. The solution of the problem raised by Kac and Heusser cannot therefore be successfully investigated in this way, but requires much more exhaustive further study.

4. The Principal Climatic Fluctuations and Changes of the Ice Age

In the preceding chapters an attempt was made to review the basis of paleoclimatic work and the principal sources of error associated with it. The knowledge so acquired will be applied in what follows, where an attempt is made to obtain some glimpses of the climate of the past and the magnitude of the fluctuations.

The critical review of research methods has shown that there is little sense in attempting to give very exact data on past climates. It is not possible today to determine exactly the former temperature and amount of precipitation of various parts of the earth. The values introduced in the following chapters should therefore be regarded as approximations. They provide orientation marks but should in no way be closely compared with present climatological statistics. It is important to keep this fact clearly in view, as all too frequently in the paleoclimatological literature values are calculated to tenths of a degree, simulating an accuracy that does not exist. To the contrary, it may be wondered whether, in view of all the possible sources of error, it is possible to reconstruct past climate with sufficient accuracy. In the present state of knowledge, one has to be satisfied at best with a number of estimates obtained in a variety of ways. Yet these have already contributed to a deeper understanding of the climate of the Ice Age.

There are two aspects of the study of Pleistocene climatic fluctuation which are particularly interesting. One is the qualitative aspect and the alteration of the former climate, and the other is the pattern of atmospheric circulation as either the cause or the consequence of the climatic fluctuation. It is one thing to trace laboriously the principal trends of climatic change as these are indicated by a series of mute witnesses. It is quite another to attempt to learn something about the most important features of atmospheric circulation at various times during the Pleistocene. They have left no direct evidence behind (e.g. wind directions—see below, pp. 230 ff.) but must be inferred from indirect evidence, which may vary considerably in its ability to provide unequivocal data. Many times the evidence is insufficient to provide any compelling argument on the nature of the former atmospheric circulation; as a result, attempts have been made to extrapolate the presumed former atmospheric circulation from a knowledge of present-day extreme weather situations. This path has been trod with increasing frequency in the last few years and has given rise to many interesting hypotheses. Butzer [109] recently very clearly expressed the ground rules for the method when he demanded as a basis for paleoclimatological work on the Ice Age that "All meteorologic implications from the physical evidence must have present-day counterparts in short- or long-term weather patterns. In other words, we cannot accept the existence, in the Pleistocene, of weather patterns unknown today." It appears to me that there are dangers in adopting such a standpoint. Not only is present-day knowledge of atmospheric circulation far too incomplete [239, 653]—for it to be used to provide the basic assumptions about the circulation during past epochs—but there is no compelling reason to believe that the stated assumptions are correct. To use such a basic assumption about the climate of the past severely handicaps from the outset the possibilities of obtaining far-reaching new knowledge.

In the following pages no attempt will be made to introduce new hypotheses or to test and modify existing ones; rather the whole emphasis of the investigation is upon the reconstruction of the actual stages in the development of climate during the Ice Age.

The frequent, contrasting climatic changes during the Pleistocene which find expression in the alternation of warm and cold periods naturally is the center of interest in any study of climatic history. It is understandable that the climatic conditions are more impressive during the cold periods than during the warm ones, although the last-named phases present countless very delicate problems. In the following, the first part of the chapter will be devoted to an exact discussion of the climate not only during the cold periods but also during the warm periods.

The overwhelming impression that results from the study of the flora and fauna whose remains are preserved in the sediments laid down during the warm and cold periods is that a progressive impoverishment of the biosphere took place. It can be assumed from this that the marked, first-order climatic fluctuations were modified by a progressive, slow, climatic change, in such a way that the climate as a whole became progressively unfavorable. This problem is discussed in the second part of the chapter.

Finally, there is another point of general importance which should not go unmentioned: an increase in the cold air masses in the high northern and southern latitudes during the cold periods should lead to a displacement of the polar front toward the equator. From this it is commonly assumed that in the Northern Hemisphere during the glaciations further to the south there were corresponding times of higher rainfall. The latter are termed pluvials. In recent years the question has been raised whether the pluvials of this obvious hypothesis are of the same age on the northern and southern margins of the Sahara. Only by answering this question can some general conceptions about the atmospheric circulation during the cold periods be established. Consequently, this exceedingly difficult problem also must be outlined, without much hope that an explanation will arise from it.

THE CHANGE BETWEEN WARM AND COLD PERIODS

The End of the Tertiary

Figure 3 indicates that climatic fluctuations are recognizable at the end of the Tertiary. Their intensity first increased significantly during the transition to the Ice Age. There are at the present time a number of interesting observations concerning Pliocene climatic fluctuations and the floral changes resulting from them. In this connection work in the following regions is particularly noteworthy: the lower reaches of the Rhine [816], the Kerch Peninsula [521], the middle reaches of the Volga [36, 372, 373, 287], Calabria [200], and the San Augustin Plains in New Mexico [138]. An exact knowledge of the scale and number of climatic fluctuations is, however, still far off.

Immediately prior to the Ice Age, the climate of the present temperature latitudes of the Northern Hemisphere seems to have been particularly favorable for the development of woodland. This was the period of "Reuverian B" in Dutch nomenclature. At that time varied woodlands reached from the Atlantic seaboard in the west to the Sea of Okhotsk and the Sea of Japan in the east [260]. Such a phytogeographical picture has not been repeated since. This creates the impression that at this time, over wide reaches of the present warm and cool temperate latitudes, temperature and rainfall conditions were particularly favorable and resembled those of a subtropical climate. Since this provided the starting point for the development of Pleistocene climatic developments, it is profitable to commence a study of the climatic fluctuations of the Ice Age from the conditions existing at the end of the Pliocene—that is, about Reuverian B.

There is some information in the literature about this questionable interval of time, but unfortunately opinions are strongly divergent. The values determined generally depend upon comparisons of the former and present distribution of various warmth-loving plant species or genera. Van der Vlerk and Florschütz [774] assume that the summer temperature in the Netherlands was only a little higher and winters a little milder than at present. In contrast, Szafer [731] estimated that the July temperatures in Czorstyn in the western Carpathians at approximately the same time was about 29°C (+15° to +16°C at present); the January temperature lying between +4°C and +5°C (0°C at present); the average annual temperature being about +17°C to +18°C (today +6°C to +7°C); with an annual rainfall of about 150 millimeters or twice the present amount. The figures quoted are related to conditions that were deduced from the character of the Kroscienko flora. For the somewhat younger but neighboring Mizerna II flora, the corresponding figures are +24°C average July temperature, +2°C average January temperature, +12°C to +13°C average annual temperature, and a precipitation of about 650 millimeters. In the lower reaches of the Kama (Rybnaja Sloboda flora), Baranov and Jatajkin [36] gave the following estimates: January temperature about 0°C (present day −13°C to −14°C); July temperature about +20°C (about the same as now). The climate in addition was very wet. The snow formerly should have lain a long time. Finally, in the waters around Calabria the minimum secular temperature of the late Pliocene is said to have been lower than the current summer temperature (oxygen isotope analyses [200]). In this sense there is also the suggestion that mountain glaciation was formerly more extensive than at present. Clisby [138] proposed something similar for the high mountains of New Mexico.

With the data at hand it is not easy to select the presumably correct conclusions. On the one hand, the marine animals whose shells are used for oxygen isotope analyses possibly occupied a somewhat different habitat from their present representatives; on the other hand, the plant distribution at the end of the Pliocene was fundamentally different from that of the present day, because of the long time for individual families to occupy fully their available ecological niches (under the given conditions of competition). This today is no longer true, for during the Ice Age and during postglacial times the climatic fluctuations followed one another so rapidly that no stable end point could be reached. Furthermore, there is an additional source of error: because of the climatic fluctuations many of the forms typical of the Pliocene were annihilated in temperate northern latitudes. They were able to maintain themselves only in quite distinct localities from which there were no migration routes back to the areas formerly occupied. Consequently, they were unable to recolonize their original areas, although the climate there today (say in central and western Europe) would favor their spread.

Such difficulties can be overcome only by disposing of as large a number of individual measurements as possible. Accordingly, Figures 34–48 present the current distribution as well as the points where fossil remains are found in sediments of the late Pliocene (approximately Reuverian B) in northern Eurasia. These are localities where beech (*Fagus* sp.), holly (*Ilex* sp.), walnut (*Juglans* sp.), hickory (*Carya* sp.), wingnut (*Pterocarya* sp.), the tulip trees (*Liriodendron tulipifera*), the tupelo (*Nyssa* sp.), the sweet-gum (*Liquidambar*), and the hemlock fir (*Tsuga* sp.) occurred. Most of the genera noted are made up of several species of the present day, and the Pliocene genera were likewise separated into different species, certain of which are identical with present-day forms. Since in much of the fossil material specific identifications have not been carried out and many

FIGURE 34. The distribution of thermophilous plants at the end of the Tertiary in northern Eurasia: beech (*Fagus* sp.). The screened area delimits the present distribution of the species. Large dots indicate the location of fossil remains. Source of data: Appendix 3, p. 246, where data are also given concerning the fossil species.

FIGURE 35. The present distribution of *Fagus grandifolia* in North America.

genera have not been systematically monographed, in the following discussions the determination of climatic figures is based upon the present limits of distribution of the genera, not of the individual species, insofar as such a method seems climatologically meaningful. The necessity for such a limitation can be seen by the former and present distribution of the wingnut (*Pterocarya*), which at the present time is successful in only a few widely separated localities. Clearly this species could occupy a much wider region were there not restrictions to its spread which have little to do with the present climatic conditions. If, then, from the present distribution of the wingnut the former climatic conditions in Europe were to be extrapolated, this process would inevitably lead to false conclusions.

The principal differences between the late Pliocene climate and the present climate are summarized in Table 7. The figures were arrived at by comparison of the present and former distribution of the previously named species. It must again be insisted that these are only estimates, which in the present state of knowledge must necessarily be crude.

FIGURE 36. The distribution of thermophilous plants at the end of the Tertiary in northern Eurasia: holly (*Ilex* sp.). The screened area delimits the present distribution of the various holly species. Large dots indicate the location of fossil remains. Source of data: Appendix 3, p. 246, where data are also given concerning the fossil species.

There are considerable differences in the degree of reliability of the figures quoted. The results from east Siberia seem doubtful. At the end of the Pliocene, the hemlock fir (*Tsuga*) was widespread in woodlands there. Probably this tree can be closely related to the present-day forms, *Tsuga heterophylla* and *Ts. Mertensiana,* which occur in the Rocky Mountains between the northwestern states of the United States and Alaska. The ability of these forms to occupy locations in high mountains makes it difficult to determine the limiting climatic factors, particularly since the number of climatic stations in such regions is far too small.

If heed is taken of the many sources of error, the following estimates can be made. Most of the trees considered could today spread into northwest Germany and Holland, but the July temperature within the Westphalian Basin and in the region of the Lüneberg Heath should be some 3°C higher than at present. Apparently the precipitation was also formerly somewhat higher than at present.

In central and eastern Poland and the bordering regions to the east, the July temperatures must clearly have been some 4°C or 5°C above the present. January

FIGURE 37. The distribution of thermophilous plants at the end of the Tertiary in northern Eurasia: walnut (*Juglans* sp.). The screened area delimits the present extent. Large dots indicate the location of fossil remains. Source of data: Appendix 3, p. 246, where data are also given concerning the fossil species.

seems to have been at least 5°C to 6°C warmer, and the mean annual temperature was 3°C to 4°C above that of today. Also, Table 7 (p. 89) clearly shows an average annual precipitation about 350 to 400 millimeters above the present level. In the middle reaches of the Volga the July temperature was formerly at least 2°C to 3°C above the present level; and in any case the January temperature must have been at least 6°C greater, and possibly even as much as 10°C to 15°C if the occurrence of *Ilex* and *Fagus* were not due to particularly favorable local conditions. As in eastern Poland, the rainfall was some 350 to 400 millimeters higher.

It is notable too that the temperature and amount of rainfall in the regions to the south (lower reaches of the Don and Bulgaria) were probably appreciably above present levels (Table 7), and the same probably holds true for the central parts of eastern Siberia, even if there is still no generally reliable information.

In sum all the data indicate that the climate was considerably more oceanic over appreciably larger areas of northern Eurasia at the end of the Pliocene than it is at present. One is not at liberty to deduce immediately from this that the in-

FIGURE 38. The distribution of thermophilous plants at the end of the Tertiary in northern Eurasia: hickory (*Carya* sp.). The screened area delimits the present extent. Large dots indicate the location of fossil remains. Source of data: Appendix 3, p. 246, where data are also given concerning the fossil species.

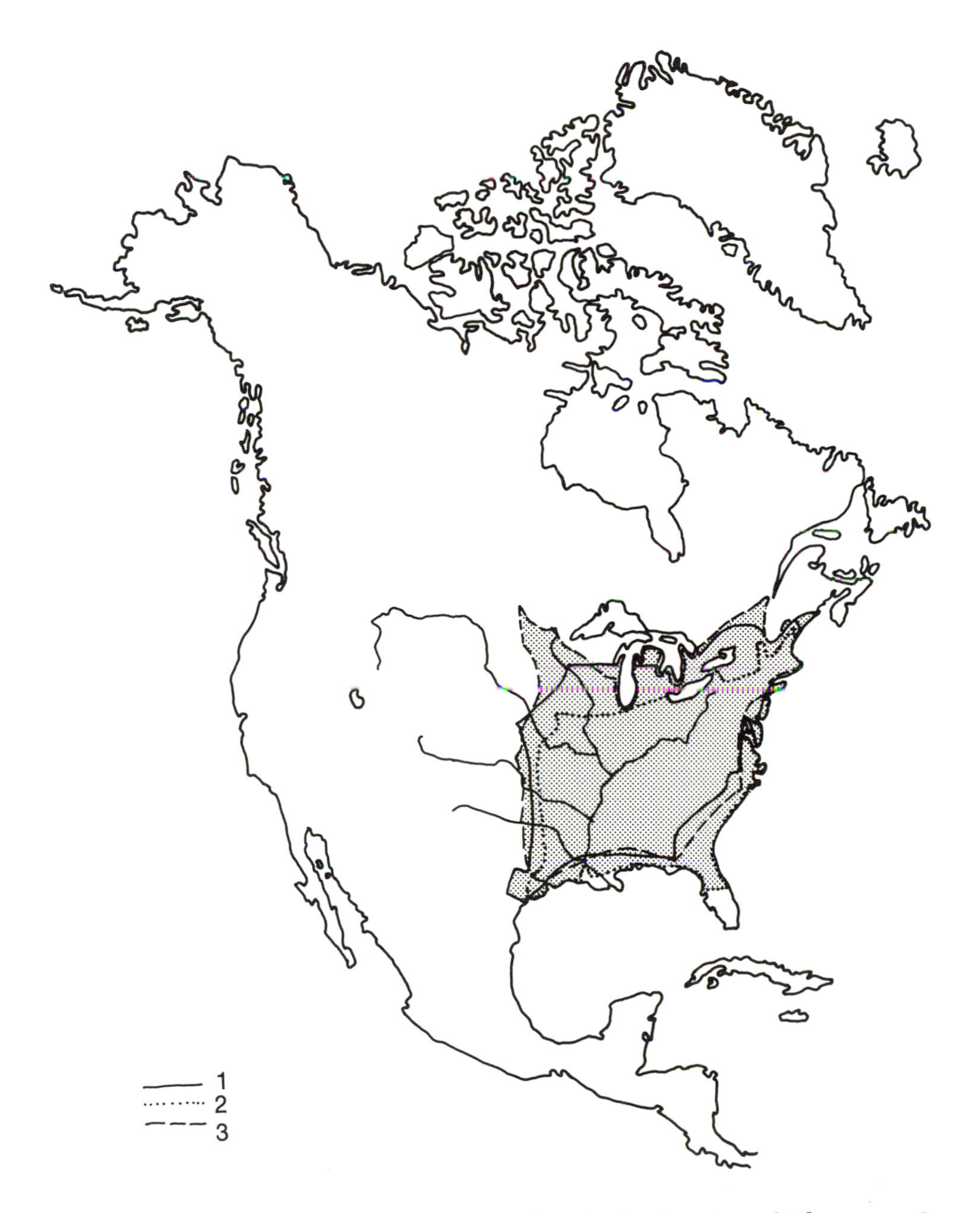

FIGURE 39. The present distribution of the North American hickory species: 1. *Carya ovata;* 2. C. *tomentosa;* 3. C. *cordiformis.* In the figure the distribution of all three, jointly, is indicated by the screened area.

tensity of the atmospheric circulation was increased, for the Mittelgebirge were at that time at least 300 meters lower than at present [260]. In the high mountains and some other regions, strong uplift in part has increased the differences. The mountains formerly were less able to cut off the rain-bearing winds, so that moist conditions penetrated deeper into the continent, strongly ameliorating the temperature extremes. Yet this immediate observation that the July temperatures in the areas of northwestern Germany, more remote from the ocean, were higher than at the present time (and probably the precipitation was also above present levels) suggests that the atmospheric circulation at the end of the Pliocene was actually more intense than at the present time.

Of particular climatic interest is the history of permafrost in Siberia. At the present time it is still unknown whether permanently frozen ground was in existence at the end of the Pliocene. Gakkel and Korotkevich [266] assumed this, but according to Alekseev, Giterman, Kuprina, Medjancev, and Choreva [7], permafrost first appeared in Siberia during the second half of the Pleistocene. The latter

FIGURE 40. The distribution of thermophilous plants at the end of the Tertiary in northern Eurasia: wingnut (*Pterocarya* sp.). The screened area delimits the present area of distribution. Large dots indicate the location of fossil remains. Source of data: Appendix 3, p. 246, where data are also given concerning the fossil species.

view is certainly incorrect, for there are still older indications in Siberia, just as in Europe, of perennially frozen ground. Such evidence has not yet been found at the end of the Pliocene. In this connection it must be noted that definite floral or faunal evidence which would indicate the existence of a tundra zone in northern Eurasia at the end of the Pliocene has not yet been recognized; rather, all present observations indicate that woodlands extended much farther to the north than at present. Yet the actual trend of the coast of northern Eurasia at the end of the Pliocene is equally unknown. Consequently, the possibility of the existence of a former tundra zone somewhat further to the north cannot be excluded, however unlikely it may appear [260].

The absence of all data concerning the former polar limit of tree growth means that July temperatures in high northern latitudes in the Pliocene can be only crudely estimated. At present the July temperature in northern Siberia drops off very rapidly as the Arctic Ocean is approached from the south. At the end of the Pliocene the coastline certainly lay north of its present position. It can be

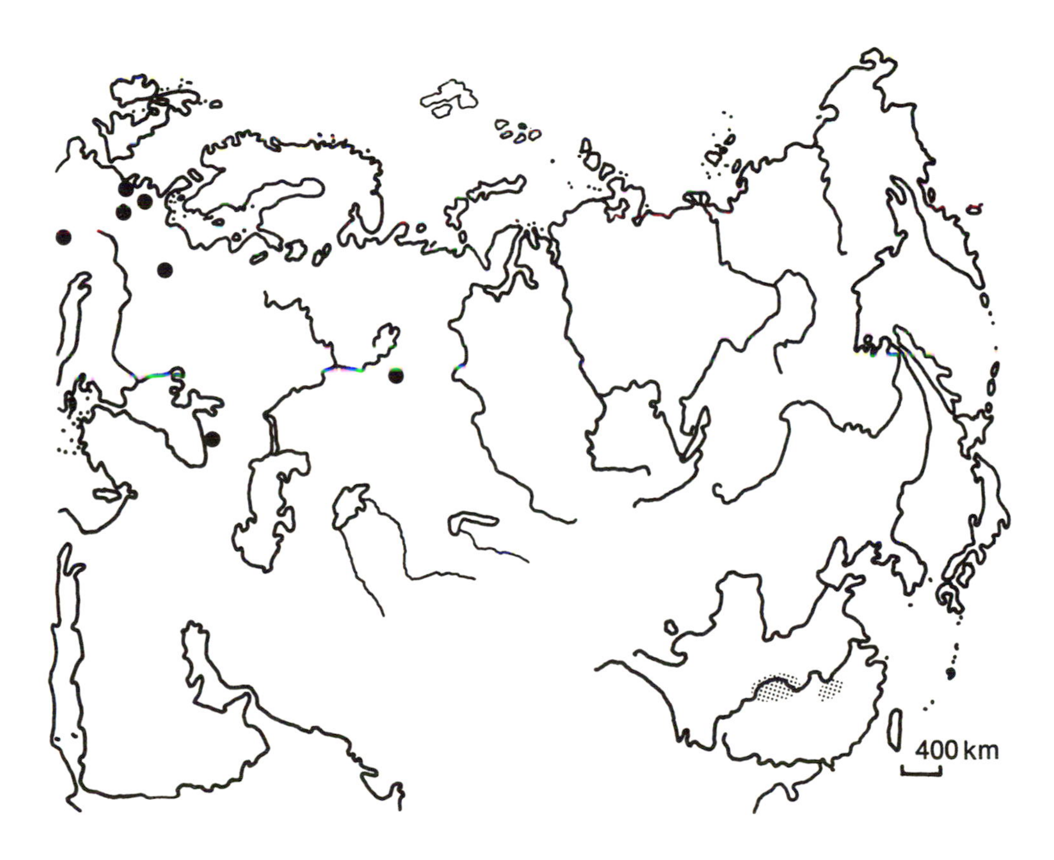

FIGURE 41. The distribution of thermophilous plants at the end of the Tertiary in northern Eurasia: tulip tree (*Liriodendron* sp.). The screened area delimits the present distribution. Large dots indicate the location of fossil remains. Source of data: Appendix 3, p. 246, where data are also given concerning the fossil species.

FIGURE 42. The present distribution of *Liriodendron tulipifera* in North America.

assumed from this that formerly the average July temperature in the present tundra zone dropped more slowly from south to north. If the conjecture that the whole of the present tundra region was under woodland proves to be correct, then it follows from the comparison with the July temperature at the present northern tree limit and the position of the coastline that the average July temperature at the end of the Pliocene must have been at least 5°C higher at the present coast. It should have been in the region of 10°C to 12°C higher. The difference compares favorably with that estimated for the middle reaches of the Volga. This leaves no basis for assessing the former existence of permafrost. Some help may be obtained indirectly. During the last warm period (Eemian) the permafrost area in northern Eurasia was not only appreciably smaller than it is now (Fig. 21) but also less than during the postglacial warm period. In neither of these warm

FIGURE 43. The distribution of thermophilous plants at the end of the Tertiary in northern Eurasia: tupelo (*Nyssa* sp.). The screened area delimits the present distribution. Large dots indicate the location of fossil remains. Source of data: Appendix 3, p. 246, where data are also given concerning the fossil species.

FIGURE 44. The present distribution of *tupelo* in North America: 1. *Nyssa silvatica*; 2. *N. aquatica*. The area of distribution of the genus *Nyssa* is shown.

periods, however, was the climate as favorable as it was during the Reuverian. This leads one to suppose that at the end of the Pliocene, permafrost had not yet got the Siberian soil in its grip, with the exception perhaps of limited regions with locally unfavorable climates or soil conditions. The southern limit of the present permafrost zone is defined by the −2°C to −3°C annual isotherm [366] or the −5°C to −8.5°C annual isotherm [100]. If these numbers are applied to the estimate of the former average annual temperature, it must be assumed that northern Siberia was 8° to 10°C warmer than at present (when −5°C is used as the maximum value of the average annual temperature at the southern permafrost limit). This figure is of the same order as that obtained from the former distribution of *Tsuga*. Of course, the figures are only crude estimates, for the distribution of *Tsuga* immediately permits one to conclude that the former

FIGURE 45. The distribution of thermophilous plants at the end of the Tertiary in northern Eurasia: sweet-gum (*Liquidambar* sp.). The screened area delimits the present distribution. Large dots indicate the location of fossil remains. Source of data: Appendix 3, p. 246, where data are also given concerning the fossil species.

FIGURE 46. The present distribution of sweet-gum *Liquidambar styraciflua*, in North America.

climate of eastern Siberia was considerably wetter than at present, so that even the former existence of a thicker snow blanket alone reduces considerably the possible areas where permafrost could develop compared with the present time. Thus the distribution of permanently frozen ground cannot provide a true indication of former annual temperatures at the end of the Pliocene.

The climatic conditions at the end of the Pliocene in North America are still largely unknown. As has already been noted, only four cold periods are recognized. It is therefore possible that many formations that are placed by American geologists in the Pliocene would be regarded as Pleistocene in the European sense. In all probability, conditions in America cannot have differed fundamentally from contemporary events in northern Eurasia.

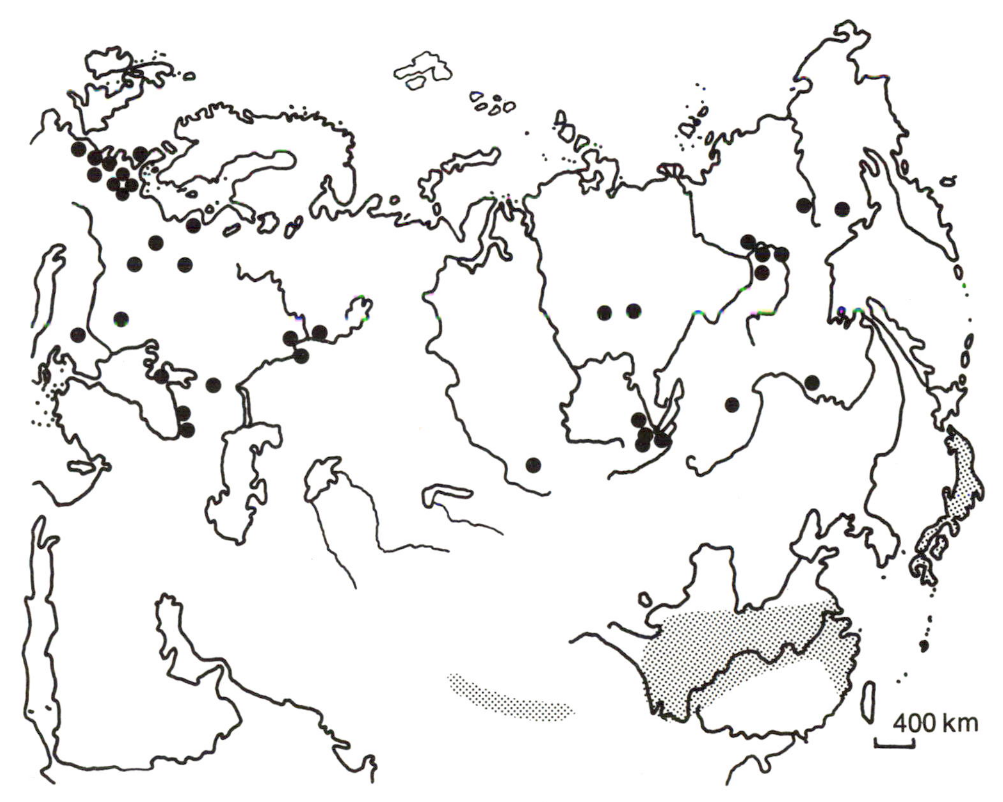

FIGURE 47. The distribution of thermophilous plants at the end of the Tertiary in northern Eurasia: hemlock-fir (*Tsuga* sp.). The screened area delimits the present distribution. Large dots indicate the location of fossil remains. Source of data: Appendix 3, p. 246, where data are also given concerning the fossil species.

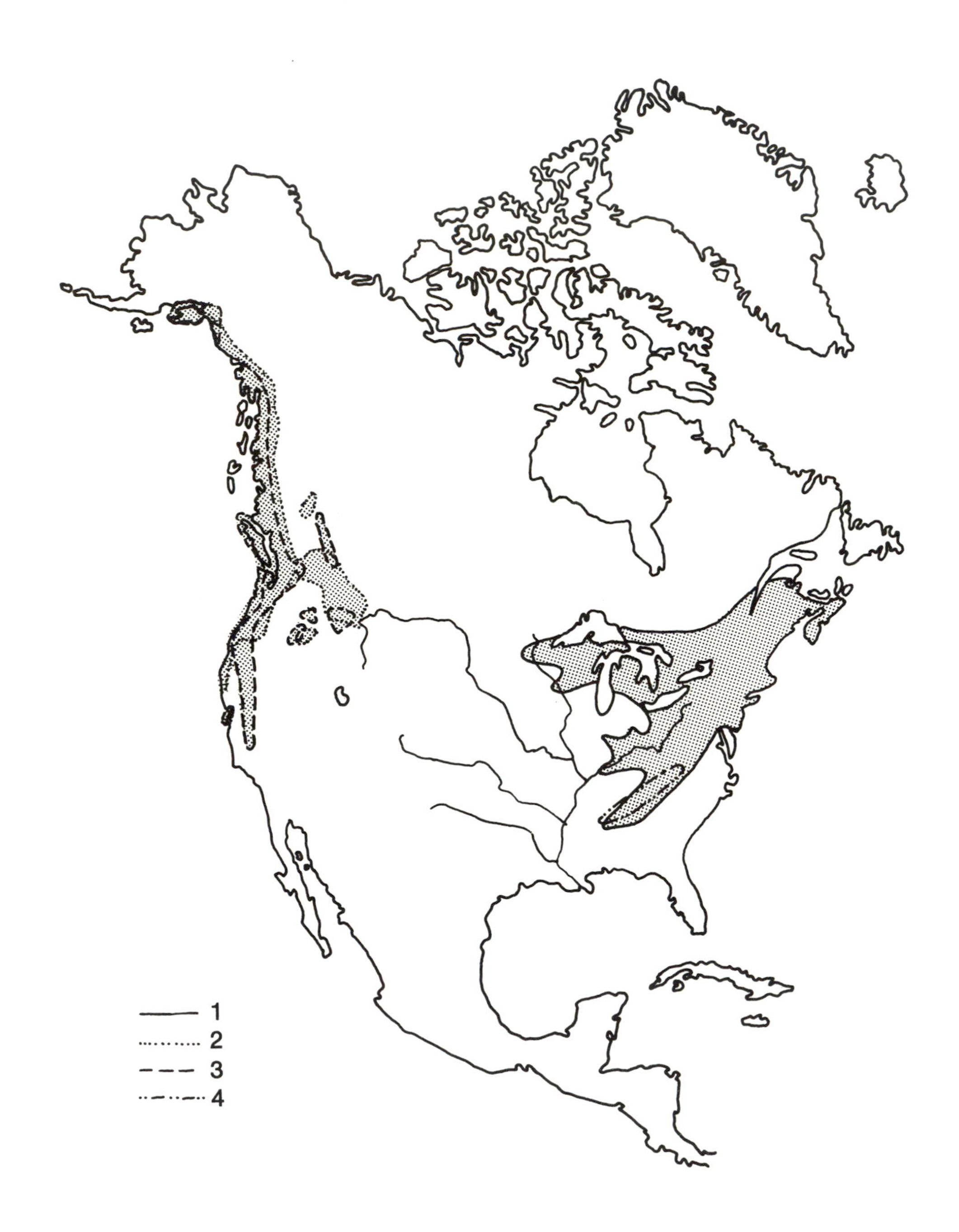

FIGURE 48. The present distribution of the hemlock-fir in North America. 1. *Tsuga canadensis;* 2. *Ts. heterophylla;* 3. *Ts. Mertensiana;* 4. *Ts. caroliniana.*

TABLE 7. DIFFERENCES IN SOME OF THE CLIMATIC ELEMENTS BETWEEN THE END OF THE PLIOCENE (REUVERIAN B) AND THE PRESENT TIME (ESTIMATED FROM CLIMATIC VALUES BASED UPON PRESENT-DAY DISTRIBUTION OF CERTAIN GENERA OR SPECIES)

REGION AND GENUS	MEAN TEMPERATURE			ANNUAL PRECIPITATION (MM.)
	JULY IN °C	JANUARY IN °C	ANNUAL °C	
Northwestern Germany, Netherlands				
Liquidambar	+3 to 4	*	= or +1 to 2	*
Nyssa near coastline	*	*	*	*
Edge of the Mittelgebirge	+3	*	*	*
Tsuga (*canadensis*)	+2	*	*	+300
Carya and *Liriodendron*				
near coastline	*	*	*	*
edge of Mittelgebirge	+3	*	*	*
Ilex	*	*	*	*
Fagus	*	*	*	*
Central and eastern Poland, Lithuania, and White Russia				
Liquidambar	+5	+5	+3 to 4	+350
Nyssa	+4 to 5	+1	+3 to 4	+250
Tsuga canadensis	+2 to 3	+3	+1	+400
Carya	+3	*	*	*
Ilex	*	+5 to 6	*	+150
Fagus	*	+4	+3	*
Central reaches of the Volga				
Tsuga (*canadensis*)	+2 to 3	+3	+1	400
Carya (and *Liriodendron*?)	+3	+6	+5	350
Ilex	*	+15	*	250
Fagus	*	+12	+5 to 6	200
Lower reaches of the Don				
Tsuga (*canadensis*)	*	*	*	+350
Fagus	*	+7	+3	+200
Bulgaria central lowlands				
Liquidambar	+3	+1	+1 to 2	+300
Tsuga (*canadensis*)	*	*	*	+150
Eastern Siberia				
Tsuga (*heterophylla* and *mertensiana*)				
Anadyr	+3	*	*	*
Yakutsk-Irkutsk	*	+10 to 15	+10	+200

*Signifies that the particular plant could occur under presently existing climatic conditions in that particular region.

The Pretiglian Cold Period

In sharp contrast to the very favorable climatic and phytogeographical conditions of the Reuverian B are those of the first cold period, which in the Netherlands is called the Pretiglian. The full extent of the change unwinds slowly.

The vertebrate remains found in numerous localities seem to indicate that at the beginning of the Ice Age, savanna conditions covered wide areas of the broad plains of Eurasia, for the closest relatives of these animals today occupy tropical and subtropical wooded plains and grasslands. From this some have deduced that the climate was seasonally dry and appreciably warmer than at present. A whole sequence of recent observations has led to increasing doubts concerning the correctness of this hypothesis. Today the opinion is gaining ground that the climate during the Pretiglian was drier but also perceptibly colder than at present. The study of the climate during the Pretiglian, in contrast to that of the late Pliocene, may thus cast light on the decisive climatic changes at the boundary between the Pliocene and Quaternary.

Fleming [228] concluded from the character of the fauna of the Hautawan Series (Fig. 32), which is probably of Pretiglian age, that the former sea temperature in this part of New Zealand was about 5°C lower than at present. This temperature fall is also shown by the coeval pollen flora. In North Island, near Frankton, in the region formerly occupied by thermophilous species of the southern beech (*Nothofagus matauräensis, N. cranwellae,* etc.), there occurred woods made up of *Metrosideras* sp. and many *Nothofagus menziesii* and *N. cliffortiöides,* which at the present time form the highest forest levels in the mountains. The thermophilous trees had thus retreated considerably [144]. If the Hautawan truly corresponds to the Pretiglian, then the cooling must also have been felt strongly below latitude 38°S. This would seem to be confirmed by the pollen analyses of van der Hammen and Gonzalez [333] on the 2,600-meter high Colombian Sabana de Bogotá. The present tree limit is at about 3,200 meters, about 600 meters above the sampling points. The current average annual temperature in the Sabana de Bogotá is around 13°C to 14°C; in the Páramos, characterized by *Acaena,* the temperature at 3,600 meters is about 6.5°C [322]. Figure 30 shows that at the beginning of the Ice Age, open vegetation, in which *Acaena* plays an important role, occupied an area which up until that time had been blanketed by woodland. Thus the tree line during this first recognizable cold period was at least 600 meters lower, and, if wind-transported pollen is taken into account, probably 1,000 meters below its present position. As the former woodland was replaced by a plant community characterized by *Acaena,* the average annual temperature during the presumed Pretiglian interval in the Sabana de Bogotá must have been at least 6° to 7°C lower than at present, a figure which matches that in New Zealand. This is important, for it may be seen that there was a very sharp temperature decrease even within the low tropical latitudes. Comparable climatic conditions seem to have occurred in high latitudes. In this connection particular note should be made of the former observations in northern Eurasia. Figure 49 shows the general distribution of phytological trends of the principal northern Eurasian vegetation types, to the extent that they may be reliably determined for the Pretiglian. The detailed information on which

FIGURE 49. Sketch map of the vegetation belts during the Pretiglian cold period. 1. projected northern coastline of the continent; 2. large basins of internal drainage, in particular the Akčagyl Sea in the Caspian Depression; 3. possible extent of former glaciation; 4. localities in which traces of "glaciation" have been found; 5. floral, faunal, and loess localities; 6. pine-spruce-larch taiga; 7. subarctic parkland; 8. wooded steppes, with loess in places; 9. steppe and desert; 10. fringing forests; 11. pine-spruce and fir forest; 12. fir, spruce, and beech woods and the mixed forest of Colchis.

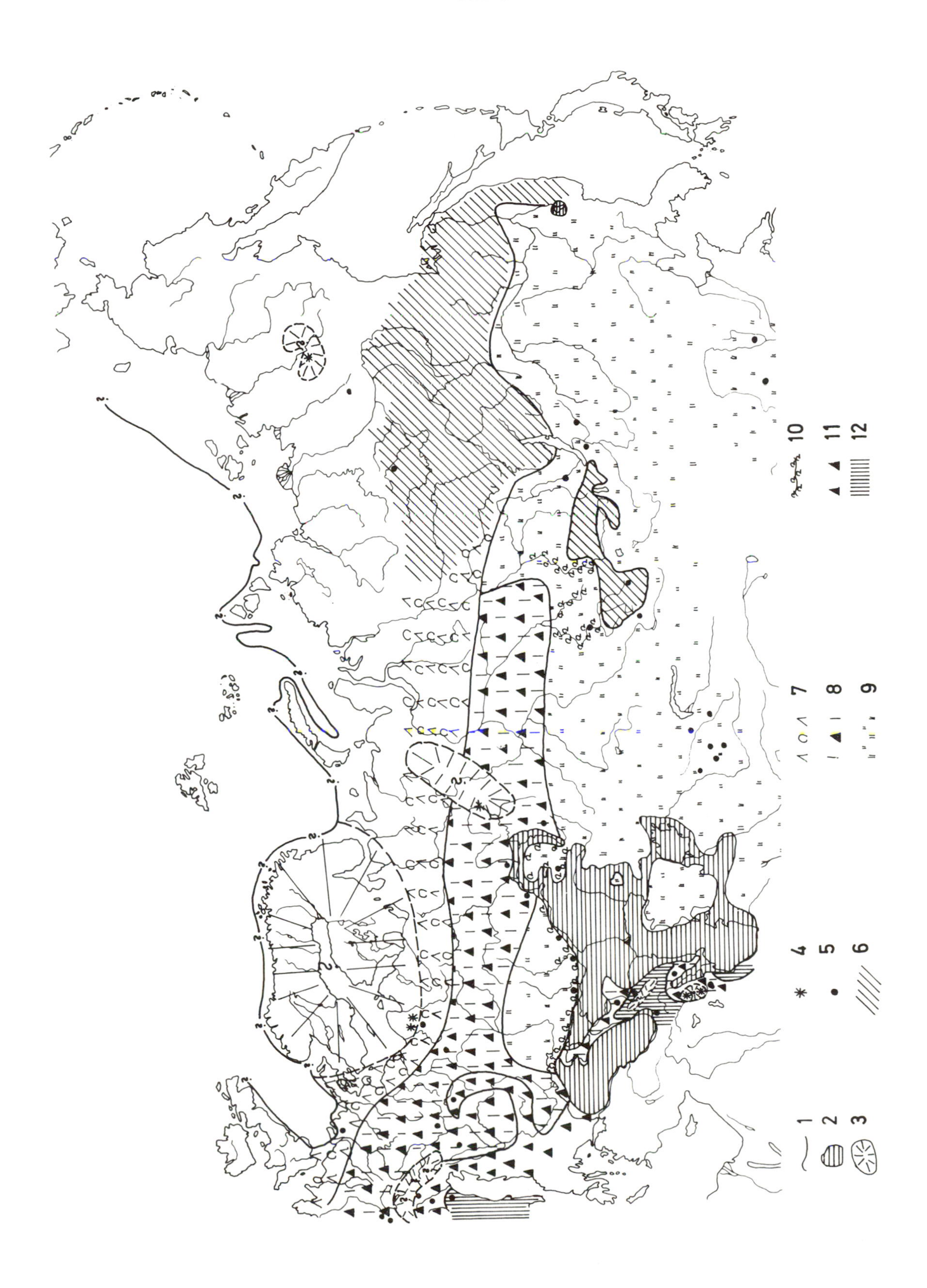
1
2
3
4
5
6
7
8
9
10
11
12

this reconstruction is based will be considered in a separate work [260]. It will be remembered, however, that during the final phases of the Pliocene the forests of northern Eurasia were made up of a large number of different trees, broad-leafed and conifers, with a present-day center of distribution in the region of mild climate in North America, eastern Asia, and the Caucasus. With the commencement of the Pretiglian the beginning of a complete change in flora and climate can clearly be seen. There occurred steppes and wooded steppes with interspersed poorly developed stands of coniferous forest over wide regions that were formerly under forest, and appeared much like the present-day northernmost limits of the coniferous forest belt. In some regions, for instance southern France and Moravia, loess sedimentation seems to have occurred far beyond its present extent, and frost had intensively penetrated the soil.

In spite of this remarkable change in the vegetation and in the character of the landscape, our inexact knowledge makes it difficult to estimate the extent of the former climate. A starting point may be furnished by the following observations. As early as 1949 and 1954, Wolters had described soil structures in the lower Rhine region which appeared to indicate the presence of ground frost in the period preceding the Tiglian warm period. In beds of the same age, large, poorly rounded quartzite blocks and pebbles of a very soft phyllite in a predominantly fine-grained matrix were repeatedly found. This heterogeneity of the sediment and the character of the large blocks is evidence of a very efficient fluviatile ice transport. Zagwijn [816] and Ahorner and Kaiser [4] subsequently noted that the "frost" structures described by Wolters could have other interpretations. Nonetheless, nothing alters the reliability of the picture presented by Wolters, for Ahorner and Kaiser were also able to demonstrate in the lower Rhine evidence of ground frost in sediments of Pretiglian age. They demonstrated that the frost intensity was approximately equivalent to that of the younger tundra during late glacial times in the last cold period in the same region. If now from the similarity in form and intensity of the ground frost during the Pretiglian cold period and the younger tundra period it can be supposed that the climates were also similar, it must be assumed that the average annual temperature during the Pretiglian was 9° to 10°C lower than at present. Such temperatures, however, are incompatible with the existence of a true savanna in central and eastern Europe. The figures must therefore be put to the test, to see how far they can be confirmed by other observations.

In Lithuania, near Rudamina and Kalwarÿa, moraine-like sediments were observed [320] below sediments which contained a flora of a warm period, probably the Tiglian. Should this observation prove correct, an appreciable drop in temperature must be considered, for at the same time in the Netherlands open vegetation had replaced the woodlands that were dominant at the end of the Tertiary. In the former only isolated clumps of birch or pine are found. It follows from this that the climate must have been very dry. A further conclusion may be drawn, that the proposed glacial advance in Lithuania (Fig. 29) was caused not by an increased precipitation but through increasing aridity, with an appreciable fall in temperature. Notwithstanding, it is still open to question whether the material near Rudamina and Kalwariÿa can be regarded as a true moraine. Even if the moraine-like sediment proves to be only a pseudomoraine—that is, solifluction material formed as the result of intense frost—it would be an indication of average annual temperatures at least 8°C lower than at present if the possibility is accepted that Lithuania was then approximately at the southern limit of permafrost. However, the observations are still uncertain and no great weight may be attached to them.

In 1959 and 1960 Moskvitin described traces of ground frost in Akchagyl sediments—that is, in the oldest Pleistocene—near Rostov on the Don, on the Sal River some 310 kilometers east of Rostov, and on the Belaya River in the western foothills of the Urals. On the Belaya the soil formerly thawed to a depth of only 75 centimeters, but on the lower reaches of the Don it extended to a depth of about 2 meters. These observations of Moskvitin have been confirmed and augmented by Prjachin [608, 609] as well as by Vostrjakov, Mizinov, Moskvitin, and Chigurjaeva [778]. Goreckÿ [288] and Shancer [646, 647] nonetheless have questioned the interpretation of the observations, since the character of the fauna found in the sediments on the lower Don was formerly regarded as evidence of a warm climate. It seems doubtful, however, that the character of early Pleistocene faunas are actually a reliable means of estimating former temperatures, at least in the case of the faunas of dry regions as in this example, for such animals even today can support not only dryness but also very low winter and night temperatures. Furthermore, it seems inconceivable that Moskvitin, to whom so much is owed for his work not only upon the division of the Pleistocene but also in relation to the recognition and significance of ground frost, would be in error in this case. I therefore assume that the observations are correct, in which case it follows that the lower reaches of the Don during Akchagyl times—that is, during the Pretiglian cold period—lay along the southern margin of the permafrost zone and must have had an average annual temperature some 11°C below present levels. In the Belaya Basin a minimum temperature fall of about 4°C should be considered. The same figure of an annual temperature four or five degrees lower can also be derived from the former distribution of forest types on the lower Kama. It is important to note in this connection that the temperatures given represent minimum values, which can easily be exceeded. They are thus not contrary to the temperature fall for the lower Don; rather, the latter is confirmed by observations made in western Georgia (Fig. 50 [649]). There, near the sea in the oldest Pleistocene Kujal'nic horizon, are remains of widespread forests of spruce, fir, and pine where today stands particularly rich deciduous woodland, which on account of its richness in species was many times regarded as the refuge of the Tertiary forest types. The pollen diagram shows that the former lower limit of the subalpine coniferous forest lay 1,200 meters below present-day levels. This depression of the treeline leads one to suppose that temperatures in Colchis were some 8° to 10°C lower than they now are, a figure that agrees closely with those already detailed. The same is also true in northern Italy. Near Castell'Arquato near Piacenza [477] there were formerly widespread forests of pine (*Pinus*), spruce (*Picea*), fir (*Abies*), and hemlock fir (*Tsuga*). Under natural conditions at the present time there are instead thermophilous woods rich in deciduous types. The strong contrast between the former and present-day natural vegetation suggests a depression of the altitude limits of the forest types of some 1,500 meters, which corresponds to a lowering of the mean annual temperature of about 10°C.

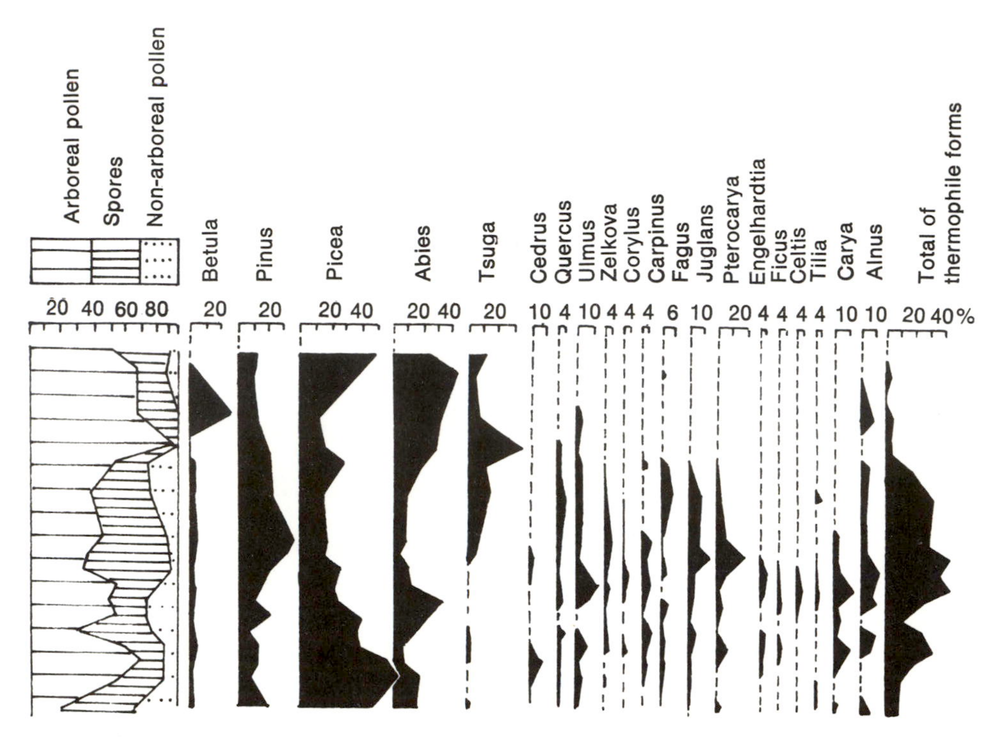

FIGURE 50. The effect of the Pretiglian cold period on the vegetation of Colchis: pollen diagram of the Kujal'nic beds, which probably belong to the oldest Pleistocene; Cichis-Perdi, western Georgia. Instead of the present thermophilous deciduous woodland, former (lowermost 4 horizons) subalpine coniferous forests occurred. Somewhat simplified after Shatilova [649].

The observations given are based upon material of uncertain reliability, obtained from very different regions. As they nevertheless give values for the depression of the mean annual temperature comparable to those of the present day, it may be supposed that these figures are not too far from the truth. The difference between the mean annual temperatures of the Reuverian B and the maximum of the Pretiglian cold period in Europe must then have been between 13°C and 15°C. It cannot be concluded from this that the temperature change was everywhere equally great, but tentatively no more exact figures are possible.

Figure 49 shows that during the Pretiglian, steppelike formations constituted a particularly characteristic element of the former vegetation. In these steppes of central and eastern Europe there were, nonetheless, relatively numerous stands of conifers. Probably the remarkable "loess à bancs durcis" of southern France is of the same age; the loess of Brno also seems to be coeval [441 and Fig. 51]. It follows from this that both the temperature and the total rainfall were considerably less. At the present time the forested steppes of southern Siberia with an average annual temperature of −1°C receive total precipitation of about 350 to 400 millimeters. If it can be assumed that this type of vegetation can be compared with wooded steppe and steppes of the Pretiglian cold period in central and eastern Europe, it must be concluded that the Netherlands and Moravia formerly received from 150 to 200 millimeters less precipitation than at present. In White Russia the calculated differential rises to 250 to 300 millimeters, and even in the

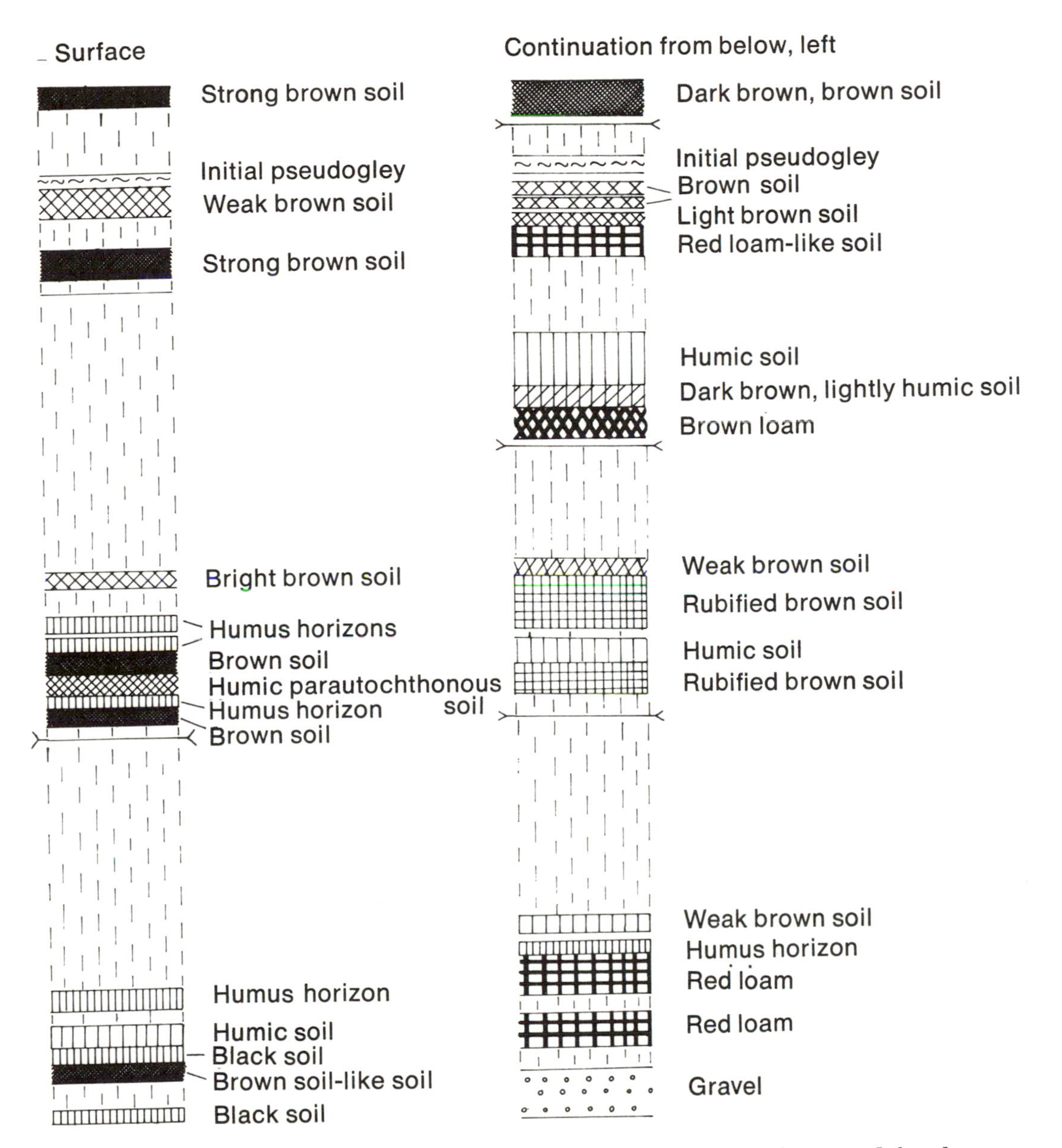

FIGURE 51. Simplified stratigraphy of the loess (vertical broken ruling) and fossil soils in the Červeny kopec claypit near Brno. Several fossils lie upon one another, so that one can speak of a soil- or pedocomplex (PK). Because of this, the impression created is that the more significant warm fluctuations of the Ice Age were made up of more than one phase, such that even two climatically equivalent warm phases may be separated by a short phase of unfavorable climate. The oldest pedocomplex, which is shown in the right-hand column as a double red loam above a gravel, appears to be formed from loesslike material and be divided by the accumulation of this material. The right-hand column should be placed below the left horizontal lines, with arrows indicating where the profile has been shortened in order to represent it in this diagram. Simplified after Kukla and Ložek [441].

upper reaches of the Ob near Barnaul, where formerly loess steppe appears to have been widespread [629, 630], there must have been approximately 100 to 150 millimeters less rainfall annually than occurs today.

Should these calculations prove generally correct, the climate during the Pretiglian in central Europe must have been about 13° to 15°C colder and with 500 to 600 millimeters less rainfall than during the preceding Reuverian B, and the Pretiglian cold period must therefore mark the distinctive boundary of the Pleistocene and can be compared climatically to the subsequent cold periods of the Ice Age. Yet up to the present time there are no completely reliable observations on the former spread of glaciers. As was noted earlier, Gudelis [320] described traces of glaciation of this age in Lithuania, and reports of comparable material have come from Solikamsk [545], the Caucasus [265, 364], and northeast Yakutsk [765, 583, 351], where moraine-like material is reported from the Pinakul Series. Thoral and Gauthier [744] thought there was a coeval Alpine glaciation. These data are shown in Figure 49, in which further paleogeographic consequences can be seen. How correct the data are must remain an open question. Nonetheless, there is another means of tackling the problem, for if water is abstracted from the oceans in large amounts, the sea level falls eustatically. With the melting of the ice, the sea level rises again, and if more ice is melted than exists at the present time, the level will be above present mean sea level. Figure 32 shows that the sea level during the formation of the Hautawan Series in North Island, New Zealand, which probably corresponds to the Pretiglian, lay considerably lower than it did either before or afterward. The same result can be obtained from Suggate [725]. On the Chukch Peninsula in the northeastern corner of Yakutsk, the sea level between the development of the late Pliocene Kojnatchun Series and the Pinakul Series, which is probably of Pretiglian age, was notably lower. If it can be assumed that this marine regression in both areas was not of tectonic origin, it would appear from the reduction in sea level that a considerable volume of water must have been locked up as ice. This could also be inferred from the reigning very low temperatures. It is therefore probably correct to regard the Pretiglian cold period as a true ice age, even if the scale of the glaciation remains unknown.

The climatic development between the previously described conditions of the Reuverian B and the Pretiglian are still largely uninvestigated, for the amount of available data is still too small. It seems as if, in central Europe, the transition from Reuverian B to the Pretiglian was characterized by increased humidity, which particularly favored the development of peat. Whether it can be concluded from this that the total precipitation increased in comparison with conditions in the preceding Reuverian B is questionable, for the same effect can be produced by the fall in temperature already established, through a reduction in evaporation and plant transpiration. A similar pluvial phase can be observed at the end of subsequent warm periods; they will be studied in detail in the following section.

The Warm Periods

1. The problem. The Pretiglian cold period was finally replaced by an important warm period, during which in northern Eurasia the condition of the vegetation resembled in many respects that of the late Tertiary. This was the

Tiglian warm period [819]. As has already been noted (Fig. 29), it was composed of two long phases of warm climate separated by an intervening phase during which unfavorable weather conditions prevailed. The climate during this latter phase does not appear to have been as severe as that during the true cold periods (ice ages) of the Ice Age. At the present time there is unfortunately very little information available concerning regional differences in climate during the three phases of the Tiglian. This is a result of the great geological-stratigraphical problems which make themselves felt, for just as it is difficult to date with some certainty sediments of the Tiglian warm period outside the Netherlands, so it seems hopeless to recognize unequivocally formations belonging to one of the three phases. The same probably holds true for the three divisions of the Waalian warm period and is likely to be true for the Cromerian warm period as well.

Consequently, only the climatic changes during the last two warm periods—that is, the Holsteinian and the Eemian—will be considered in the following, and particular emphasis will be placed upon data from northern Eurasia. So much evidence is available that an attempt can be made to analyze exactly the climatic history of these two periods. In other parts of the earth the material for such attempts is much less satisfactory; even so, there is a considerable fund of data, particularly in North America. Yet the number of pollen analyses that are of great importance is markedly lower there, and in these regions no exact analysis of climatic change during the warm periods is yet possible. The following considerations are therefore largely restricted to northern Eurasia—that is to say, the regions lying to the north of the Pyrenees, the Alps, the Caucasus, and the central Asiatic and southern Siberian mountains. Occasional reviews of other regions will show to what extent the climatic conditions there resembled or deviated from those in northern Eurasia.

In the Northern Hemisphere in that part of the cool temperate latitudes now occupied by deciduous forest, the postglacial floral migration pursued to a very great extent the following course: over the steppes or tundra existing at the end of the preceding cold period spread pine and birch woods in which the elements of true deciduous woodlands—that is, the central European mixed oak forest—became increasingly important. Finally, the thermophilous forest was established, and soon in its turn a woodland indicative of a moist climate developed. To this belong the beech woods of central Europe. Simultaneously there developed raised bogs in many areas, which today have been destroyed to a large extent by human activity.

From this succession of definable forest types, which Rudolph [640] called the "historische Grund-Succession" in central Europe but which Firbas [223] referred to as the "mitteleuropaische Grundfolge der Waldentwicklung," it can be deduced that at first the climate was cold and dry, for tundra and steppe reigned in those regions now occupied by woodland in temperate latitudes. During the spread of pine and birch woods, the climate must have remained dry, but the indication from plant remains in sediments of this period are that the climate was already quite warm. With the development of the central European mixed oak forest, the climatic optimum seems to have been reached. It was warm and humid. Finally there set in the cooler humid climate in which we are still living today.

The history of the postglacial vegetation shows many similarities to the Eemian

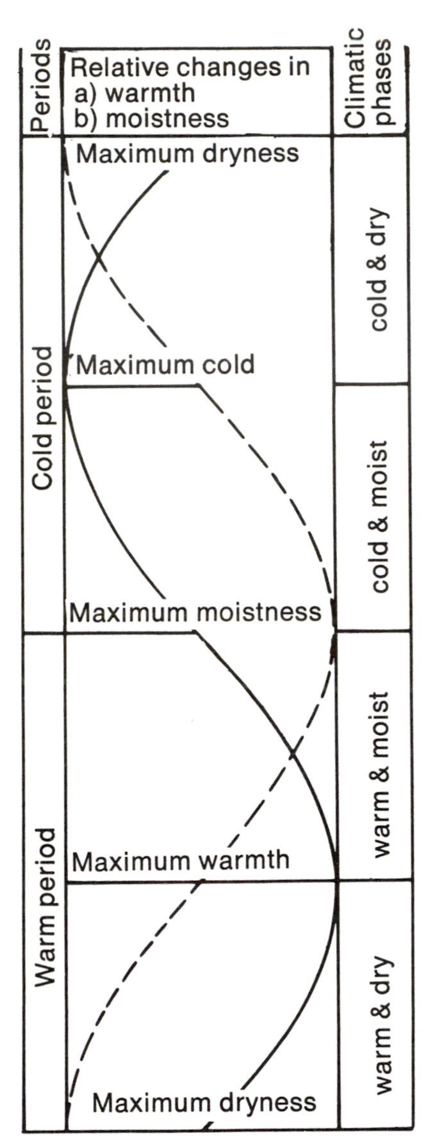

FIGURE 52. Schematic representation of the variation of important climatic factors within a warm period and the succeeding cold period, after Grichuk [303, 305]. The continuous curve represents temperature change, the broken line precipitation changes. A curve bending to the right indicates an increasing value of that factor, an inflection to the left a decrease.

and probably also to the Cromerian warm periods. This is the basis of Grichuk's proposed scheme of climatic fluctuation during a cold and a warm period (Fig. 52). According to this the climate of a warm period was first cold and dry, only later becoming wetter, with a simultaneous decrease in temperature. In favor of this scheme is the fact that not only in postglacial times but also during the Eemian warm period indications of increasing wetness appear at the end of the actual warm period. During the first phase of the succeeding cold period the climate, according to Grichuk and Butzer [105–7, 109], was initially very wet

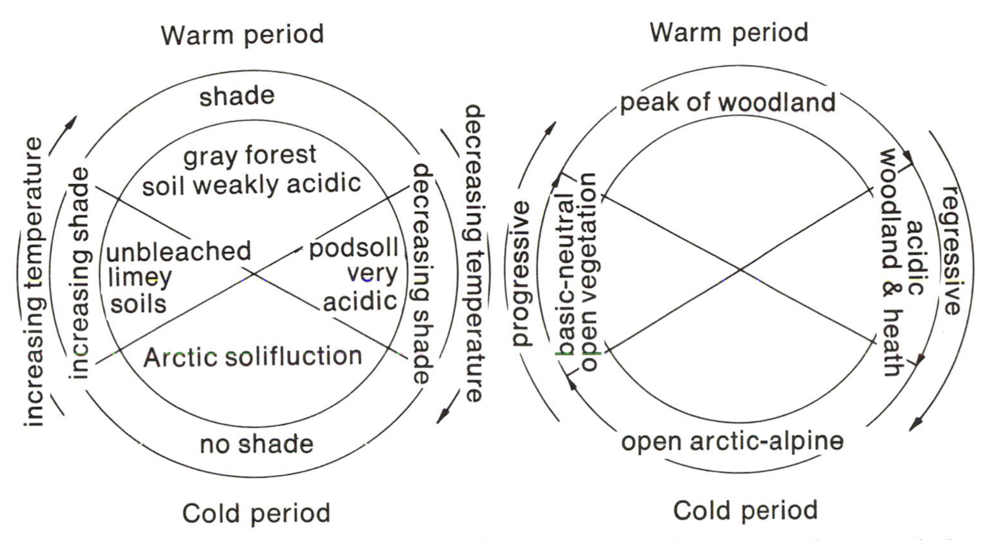

FIGURE 53. Schematic representation of changes in soil, existing climate of the principal plant communities, and vegetation changes during the transition from a cold period to the succeeding warm period, and finally back to the following cold period. From Iversen [363], somewhat simplified.

and cold, until finally it passed into a cold dry phase, with the greatest dryness at the transition into the following warm period. The warm periods were thus, according to Grichuk, warm and dry at first, later becoming cooler and moist. Parallel to this was the development of vegetation and soils, which for their part also influenced the development of vegetation (cf. above, pp. 51 ff.). Iversen [363] diagrammatically represented the processes at work (Fig. 53) showing that accompanying the rising temperature during the first part of the warm period was an increasing, protective plant cover. The soil was still basic to neutral, and leaching played no role. Later under the optimum temperature conditions and with a thick protective cover of forest trees, brown, somewhat acid forest soils developed, which became more strongly acid as the temperature fell, so that raised peat bogs finally became widespread. Figures 52 and 53 illustrate the basic steps in the process, which must be taken into consideration in any reconstruction of the climate of warm periods. Yet there are in detail many modifications necessary to the scheme reported, and these will be discussed in the following paragraphs.

The course of development of the vegetation during the Holsteinian warm period differed significantly from that during the Eemian and Cromerian and probably also from the still older warm periods. A distinguishing contrast lies in the fact that over wide areas of Eurasia during the Holsteinian warm period there was a coniferous forest cover in which spruce (*Picea*) and fir (*Abies*) were particularly important. During the other warm periods, in place of this the "Grund-Succession" established for the postglacial is valid over wide limits. In it apparently the warmest section of the climatic history permitted a far-reaching increase in the area of thermophilous deciduous forest (Table 1).

In contrast to the rapid succession of forest types during the warm periods in Europe (with the exception of the Holsteinian warm period), the contemporaneous floras of Siberia changed very little. Almost without exception coniferous forests reigned in which stands of spruce (*Picea obovata*), fir (*Abies sibirica*), and Arolla pine (*Pinus sibirica*) were particularly noteworthy in the western Siberian lowlands and in southern Siberia, while in the northern part of central Siberia and in northeastern Siberia the forests were formed from various species of larch (*Larix* sp.), *Pinus silvestris*, and *Picea obovata* [260]. It thus seems that there are certain similarities in the development of vegetation during the Holsteinian warm period of Europe and all the Siberian warm periods. This would indicate a similarity of climate in the sense that the Holsteinian warm period in Europe was quite cool and continental; and therefore, in Europe at least, a distinction must be made between "warm" and "cool" warm periods. Moskvitin [547, 549] has gone into this question in detail. Selle [672] and Środoń [709], however, have drawn attention to the possibility that the unusual vegetative history of the Holsteinian warm period may only be the result of differences in the history of plant migration and the greater number of competing forms, since in the beds of Holsteinian age in Europe there occur countless indications of the

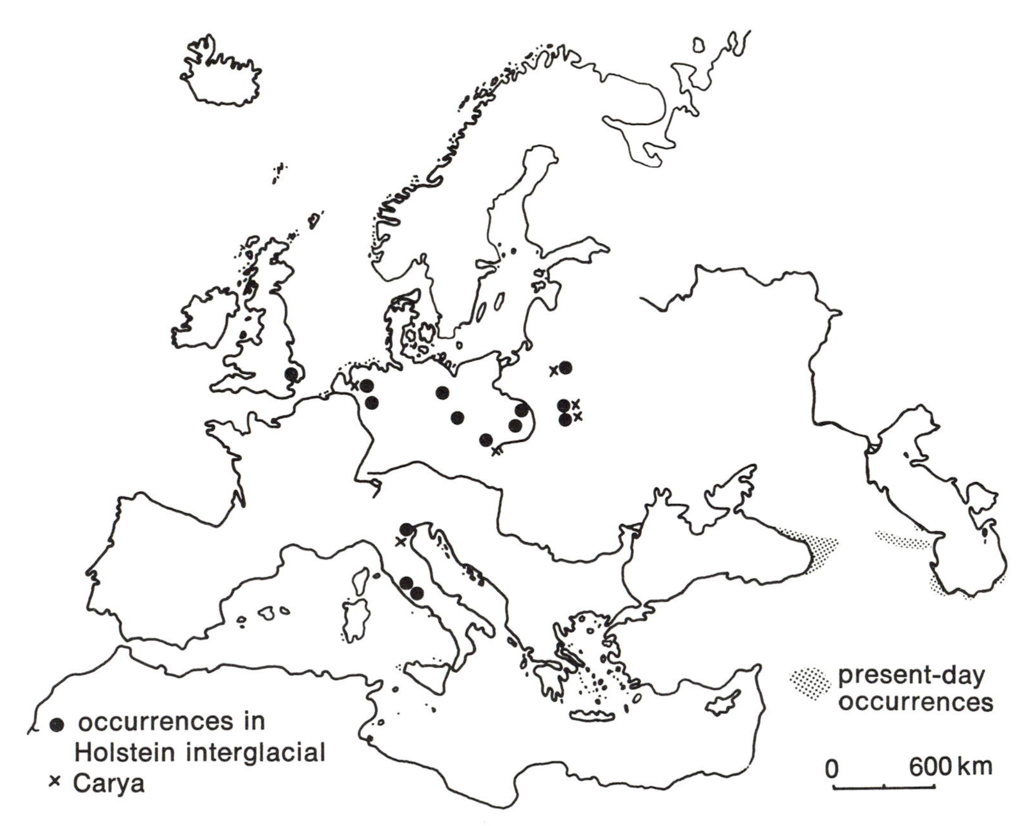

FIGURE 54. Holsteinian warm period: the distribution of localities where wingnut (*Pterocarya* sp.) and hickory (*Carya* sp.) have been found. Data source: [260] and Appendix 4, p. 248.

former widespread distribution of plant types which are today regarded as thermophilous. The distribution of the wingnut (*Pterocarya*) and the grape (*Vitis silvestris*) can serve as example of this (Figs. 54, 55). The climate, therefore, cannot have been very unfavorable. On the other hand, we cannot overlook the fact that the spruce (*Picea abies*) formerly spread far beyond its present western limits and grew in profusion in the British Isles, which seems to indicate that the winters there were colder than at present, for Enquist [201] has shown that the western limit of the spruce is reached where the temperature sinks below freezing on fewer than 120 days during winter (Fig. 25). According to this the western limit of the spruce in Europe is determined by mild winter conditions. This example clearly illustrates how difficult it is to reconstruct climate on the basis of the comparison of the present and former limits of distribution of individual species only. The difference between the former and present-day distribution of some species has sometimes suggested too that certain plant species that are at present widely distributed in favorable climates had formerly different ecologic requirements—e.g. *Azolla filiculoïdes* [697]. It is always instructive to make clear that it is such methodological difficulties which make our knowledge of the climatic character of the Holsteinian warm period appear unreliable.

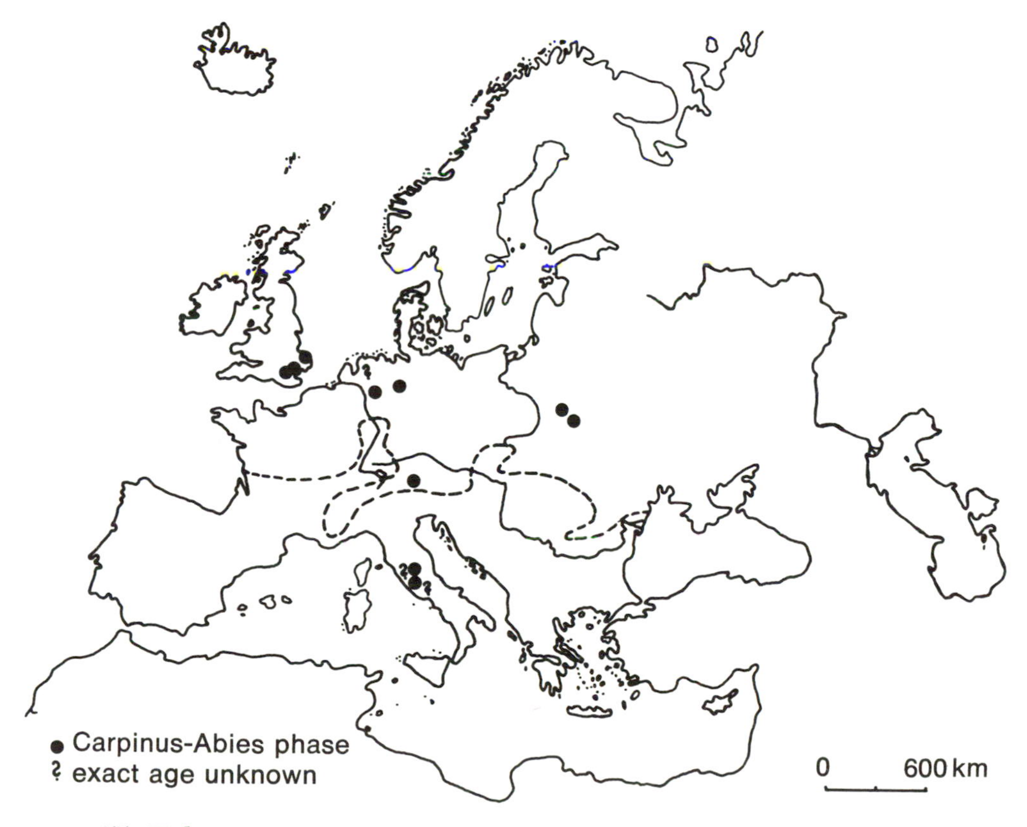

FIGURE 55. Holsteinian warm period: the distribution of localities where the vine (*Vitis sylvestris*) occurs. The broken line indicates its probable northern limit under natural conditions. Data source: Appendix 4, p. 248.

Zagwijn [818] and Stevens [720] assumed that the mean July temperature during the Holsteinian climatic optimum in the Netherlands and England was approximately the same as at present. Kac [384] presented an alternative opinion —that the climate during the thermal optimum of this warm period was characterized by cool summers but with a relatively high humidity, and that this extended from the Netherlands to central Russia. The majority of workers believe that the climate during the Holsteinian warm period was more favorable than at present [for central Europe: 47, 202, 371, 414, 415, 484, 488, 560, 696, 707, 709, 650, 755; for Siberia: 6, 9, 834].

Even with respect to the climatic development of the Holsteinian warm period, opinions differ. It is generally assumed that the climate during the time when hornbeam and fir were widespread in central Europe, and when the wingnut also occurred, must have been oceanic [177, 202, 371, 381, 382, 388, 414, 415, 484, 488, 670, 673, 697, 753, 782]. Yet in the preceding phase of development of the vegetation, oak and elm dominated in the oceanic region and spruce to the east of the Elbe. From this it is sometimes deduced, in accordance with Grichuk's assumptions, that the climate was relatively warm and dry [177, 202, 381, 753]. In contrast, Kneblová [414, 415], from the former wide distribution of conifers, postulated a cooler climate, and Jarón considered the climate to be cool and moist. These contradictory opinions are invariably explained according to the former distribution of certain plant species or even whole plant communities, a persuasive argument for the danger of such interpretations. Regrettably, the marine fauna of the North Sea also fails to provide clear indications of former climates, although giving unique data for the last warm period [337].

Unlike these conflicting results, the observations on the climate of the last warm period (Eemian-Mikulino-Kazancev) are in much better agreement. Today it can scarcely be doubted that the temperatures then were higher than at present. Archipov, Koreneva, and Lavrushin [23] even talked in this context of the first truly warm interglacial in Siberia. A similar view was expressed by Zubakov [833], who nevertheless mistakenly placed this warm period between the maximum advance of the Saalian glaciation and that of the Warthe. Aljavdin [10] found that this former climate was warmer than that during the preceding warm periods, and Aleshinskaja [9] and Lavrova [453] determined on the basis of the distribution of certain marine mollusks that the waters of the shelf sea from the Murman coast in the West to the Kara Sea in the East was some 3° to 4°C warmer than at present. The water temperatures of the North Sea were also above present levels [103], for the proportion of "lusitanian" faunal elements—forms now widely distributed in southern sea areas—was markedly higher in the eastern North Sea during the Eemian warm period than today (Oldenbüttel in Holstein: 18 percent, compared with 5.5 percent today [336]). The same thing is indicated by the types of soil developed on the continent (central Europe [96], Czechoslovakia [490, 499, 610, 611], northern Italy [507], Montenegro [514]); and the validity of these observations is further confirmed by paleobotanical work (Denmark [13, 14, 381], central Europe [384, 590, 614, 618, 624, 670, 676, 706, 729, 662, 717, 754, 772, 817], eastern Europe [384, 728] northern Italy [570]). It is generally assumed that the mean July temperatures were then 2° to 3°C higher than at present.

Netherlands (*Zagwijn* 1961 [817]) Estimated mean July temperature	N. Germany (*Selle,* 1941 [670], 1962 [676])		N. E. Poland (*Środoń* 1950/51 [706])	
	Climate	*Vegetation Phase*	*Climate*	*Vegetation Phase*
0° 10° 20°C	subarctic	pine phase	subarctic (?)	immigration of tundra
	cool atlantic	pine-spruce-fir phase	severe climate as in the taiga of N. Russia	spruce-pine phase
	montane	spruce phase		
	temperate atlantic	hornbeam-spruce phase	generally colder and more continental—	hornbeam-alder spruce phase
	warm atlantic	hornbeam phase	warm and moist much warmer than today	
	warm continental	hazel-linden phase hazel-(mixed oak forest) phase pine-mixed oak forest-hazel phase	much warmer than at present moist	linden-alder phase
			somewhat warmer than today moist	oak-elm phase and hazel phase
	warm continental	pine-mixed oak forest phase pine birch phase	climate becoming evidently warmer and moister, possibly similar to present conditions	pine-birch phase
	moist, cool subarctic	birch phase subarctic steppe	cold but somewhat moister	pine-spruce phase
	arctic	tundra	cold, severe climate	tundra

FIGURE 56. Schematic illustration of the climatic changes during the Eemian warm period in the Netherlands (temperature curve), in northern Germany and northeastern Poland. After Zagwijn [817], Selle [670, 676], and Środoń [706].

It thus appears relatively easy to obtain reliable data on thermal conditions during the climatic optimum of the last warm period in Europe. The same is far from true, however, for conditions of rainfall. During that interval of the last warm period, when a warm climate is indicated by the floral and faunal evidence, oak and elm forests dominated at first but were replaced after a relatively short time, during which widespread and conceivably very dense thickets of hazel bushes (*Corylus avellana*) prevailed. They were replaced once more by forests made up almost exclusively of various species of linden, namely the broad-leafed linden (*Tilia platyphyllos*), the small-leafed linden (*T. cordata*), and a southern European linden (*T. tomentosa*). It has generally been assumed that the climate of the oak-elm phase was warm and dry and warm and moist—that is, oceanic—during the closing linden phase [384, 590, 618, 624, 706, 662, 754, 728]. On the other hand, it has sometimes been assumed that the climate was oceanic as early as the oak-elm phase and that it can scarcely be distinguished from the linden phase [381, 614, 670, 676]. This question will be considered subsequently. Figure 56 summarizes the present state of knowledge of the climatic fluctuations during the last warm period (from the data of Zagwijn [817], Selle [670, 676], and Środoń [706]). From it, an increase in the wetness of the climate appears to occur toward the end of the warm period, so that in many regions of northern, eastern, and central Europe, *Sphagnum* peat bogs began to form, which soon became a characteristic element in the vegetation.

In conclusion it must be remarked that our knowledge of the details of the climatic history of both the warm periods considered is very limited. In the following sections an attempt will be made to consider more exactly the questions raised here.

2. The Holsteinian warm period. In Figures 54, 55, and 57–63 the distribution of fossil localities of various plants is shown by means of which the flora of the Holsteinian warm period can be clearly distinguished from the postglacial. Without doubt, the most striking is the occurrence of the wingnut (*Pterocarya*), whose fossil remains show similarities to the present-day species found in the Caucasus and to those in eastern Asia [260]. Regrettably, the former distribution of the wingnut provides no climatic data, for the distribution of *Pterocarya* in both the Caucasus and the Far East is very severely restricted locally by mountainous regions and unfavorable events in the history of the development of the areas in which the tree is found. Nevertheless, fossils of *Pterocarya* have been found

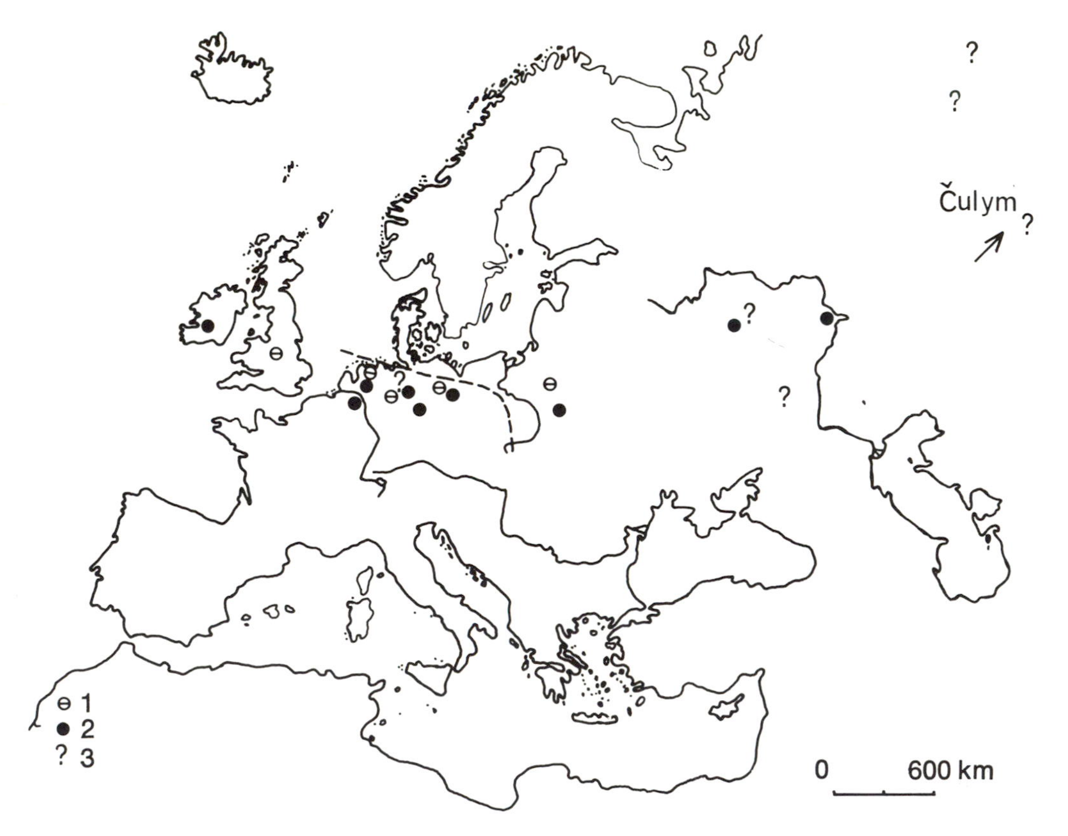

FIGURE 57. Holsteinian warm period: The distribution of *Azolla* sp.: 1. first part of the warm period (spruce-alder phase); 2. second part of the warm period (hornbeam-fir phase); 3. occurrences of unknown age. The broken line approximates northern distribution limits at the present day. Data source: Appendix 4, p. 248.

only in sediments of the hornbeam-fir phase. In them the remains of other plants requiring warmth are also found, so that it may correctly be assumed that the climate was particularly favorable. On the other hand, the data assembled in Figures 61–63 also shows that the climate over long stretches of the preceding spruce-alder phase cannot have been unfavorable, for the occurrence of the broad-leafed linden (*Tilia platyphyllos*) is of some significance. At least these finds show that the climate cannot be regarded as cold. Rather, the thermal conditions of both phases appear to have been equally favorable, if the very beginning and end of the warm period can be excluded. From the distribution data in the figures of the various thermophilous plant species and genera the results contained in Table 8 have been calculated for both phases in the development of the vegetation. The values in this table, as well as those in Tables 7 and 10, are obtained from the comparison of the climate prevailing at the present limit of distribution with that of the fossil locality.

It must be noted that these data represent minimum values. Differences in the tabulated data which arise out of differences in the distribution of various plant species found in the same area are, above all else, the consequence of the

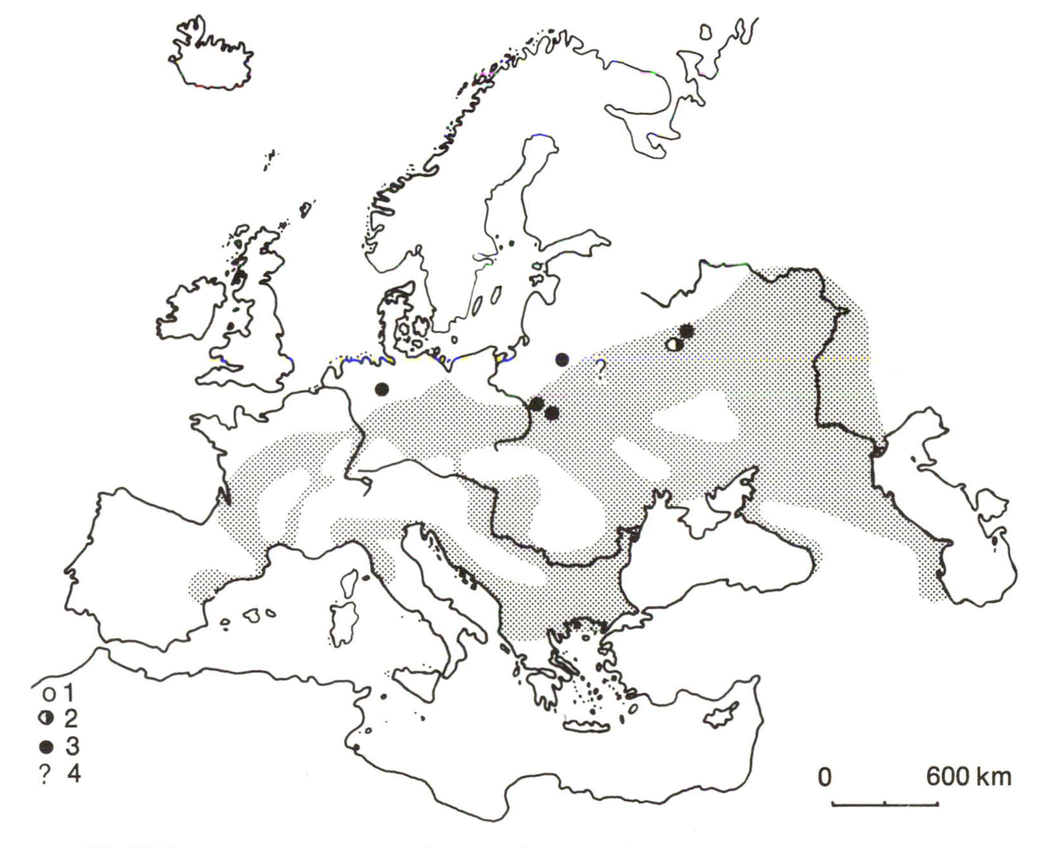

FIGURE 58. Holsteinian warm period: Distribution of waternut localities (*Trapa* sp.): 1. first part of the warm period (spruce-alder phase); 2. first and second parts of the warm period; 3. second part of the warm period (hornbeam-fir phase); 4. localities of unknown age. The screened area represents the present distribution. Data source: Appendix 4, p. 248.

different demands of the plants examined on the climatic character of the environment. It is quite understandable that the values obtained from different species will differ from one another, for their present distribution may be limited by either the lowest summer temperatures or the minimum winter temperature, or in other cases by the minimum rainfall. In sum, the data from Table 8 permit those of Table 9 to be recognized, with some reservations.

The data concerning thermal conditions in the western Siberian lowlands are uncertain, for they depend entirely upon the assumption that the remains of *Azolla interglacialica* (probably identical with *A. filiculoïdes*) found at various localities are actually *in situ* deposits, not transported by rivers from regions far to the south nor derived from the reworking of older beds. The data on mean annual temperature on the middle reaches of the Volga are beset with even greater uncertainties. These are based upon the former occurrence of *Azolla* and holly. The holly pollen could be redeposited, and the *Azolla* may have been transported far to the east and introduced into locally favorable water conditions by aquatic birds. Yet the trend in these values indicative of former climatic differences agrees well with the pattern derived from much richer material to the west. Furthermore, the absence of permanently frozen ground in the lower reaches of the Vilyui River in central Siberia (Alekseev [6] was unable to detect there traces of syngenetic

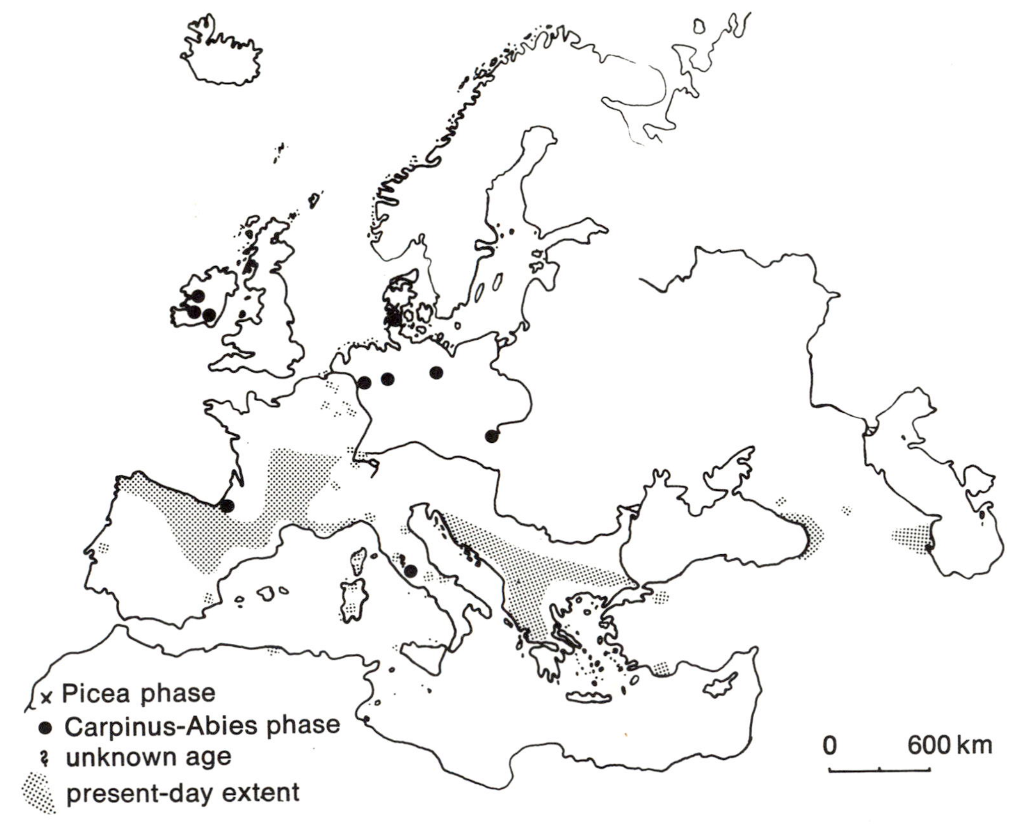

FIGURE 59. Holsteinian warm period: Distribution of the box tree localities (*Buxus sempervirens*). Data source: Appendix 4, p. 248.

freezing in sediments of the Holsteinian interglacial) shows that the data derived from the occurrence of *Azolla* in the western Siberian lowlands cannot be entirely false.

In short the finds indicate that during the spruce phase, just as in the hornbeam-fir phase of the Holsteinian warm period in central and eastern Europe, the mean winter, summer, and annual temperatures were higher than today, and of such a type that it must be concluded that the climate was more oceanic, with a relative intensity increasing eastward more strongly than in the vicinity of the ocean, by comparison with present-day conditions. Supporting this view is the change in amount of precipitation, for the minimum difference between the former amounts of rainfall and those of the present day is less in the east.

Probably the rainfall of central Europe was also higher than at present. This does not show up very clearly in Table 8; but as previously noted, the wingnut (*Pterocarya*) occurs in various places in central Europe during the hornbeam-fir phase. At the same time the *Tsuga* sp. appear to have been sparsely distributed in central Europe. Their fossilized remains have been described from Essen-Vogelheim [435], Kostenthal near Cosel (Goscięcyn [709]), Olszewice near Łódź [470], Raków on the south side of Łysa Góra [433, 434], Sciejowice near Kraków [501], and from the area of the Moravian gate [500]. In the regions where

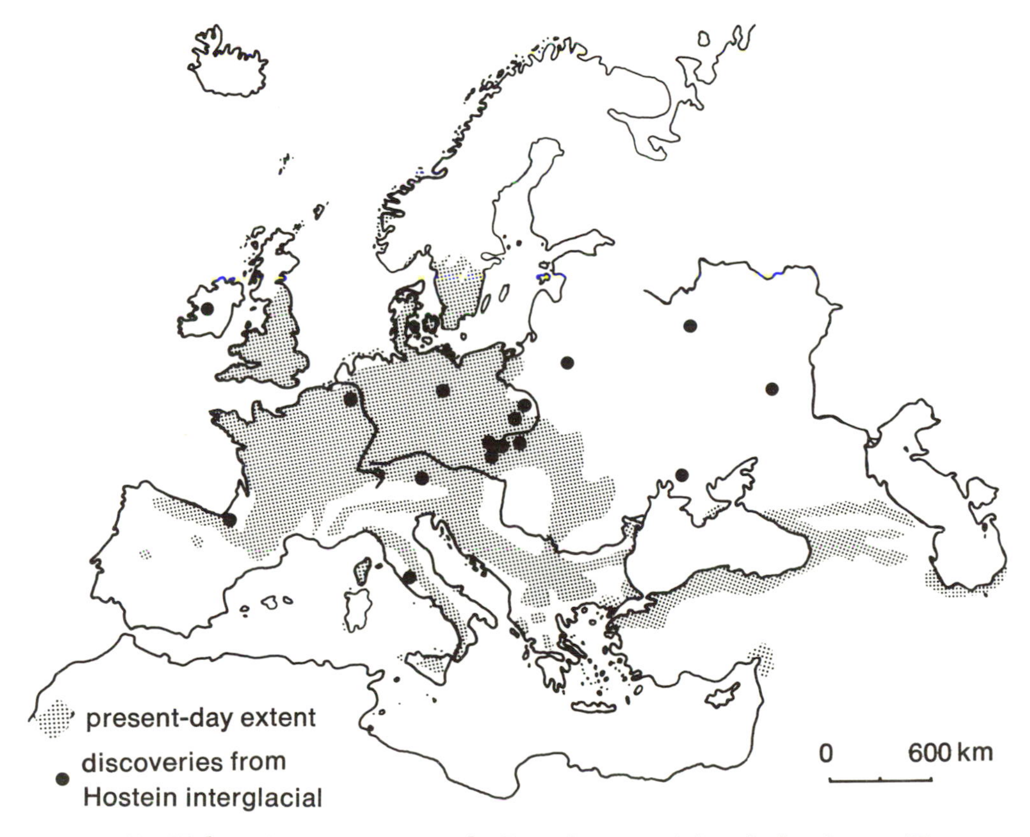

FIGURE 60. Holsteinian warm period: Distribution of beech localities (*Fagus silvatica*). Data source: Appendix 4, p. 248.

Pterocarya caucasica is presently distributed, the annual precipitation seldom falls below 1,200 millimeters; *Tsuga canadensis* flourishes generally in regions that receive more than 800 millimeters rainfall. It may be deduced from this that the total precipitation in north Germany and Poland during the hornbeam-fir phase of the Holsteinian warm period was at least 200 millimeters greater than the present, if in these calculations the minimum annual precipitation in the present-day distributional region of *Tsuga canadensis* can be taken, since established *Pterocarya* flourishes in moist locations, and the high moisture requirements could formerly have been met from the groundwater level.

Thus the climate of the Holsteinian warm period over wide regions of northern Eurasia was in general warmer and more oceanic than today. It shows many notable similarities with the outgoing Tertiary (p. 89).

These conditions are valid, not only in the regions discussed but also in the northern Mediterranean. Follieri [247–52] and Paganelli [576] described a very rich woodland flora of the Holsteinian warm period from the vicinity of Rome and from the Po Delta, which resembled the present lowland forests of Colchis. The flora also strongly resembled the flora in the Po Delta sediments of the last warm period discovered by Marchesoni [570]. As a direct consequence, the climatic

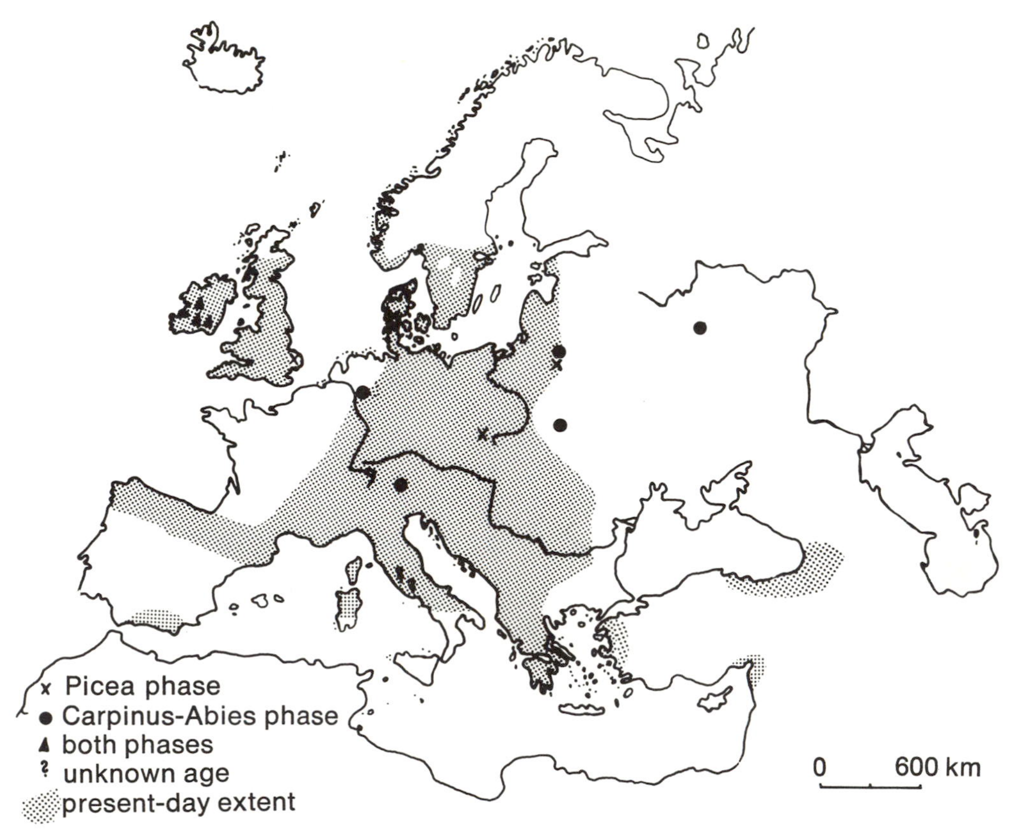

FIGURE 61. Holsteinian warm period: Distribution of yew localities (*Taxus baccata*). Data source: Appendix 4, p. 248.

figures determined by him for the last warm period can be applied to the penultimate warm period, the Holsteinian. Marchesoni assumed that the former mean July temperature in the Po Delta was probably 2°C above present levels. The January temperature would have been some 3°C to 4°C warmer than at present, with a total annual precipitation of about 1,800 millimeters instead of 740 millimeters, as at present. It might be questioned whether the rainfall was actually so high or whether the rainfall was differently distributed during the year, since in one of the fossil lakes near Rome dating from the Holsteinian warm period, annual kieselguhr bands lead one to conclude that the summer and autumn precipitation was higher than that in the winter and spring [251]. Today, however, Rome lies in the zone of winter rainfall of an Etesian climate.

The difference in the seasonal precipitation is shown by the fact that the beds formed during the summer season (p. 15) were appreciably thicker than those formed during winter; yet the greater summer thickness does not imply a heavy influx of sediment due to sudden rainstorms, such as occur today as a result of the destruction of the vegetation by human activity. Furthermore, since the winter temperatures were also higher during the Holsteinian warm period, it can scarcely be assumed that the variation in thickness of the individual beds was due to con-

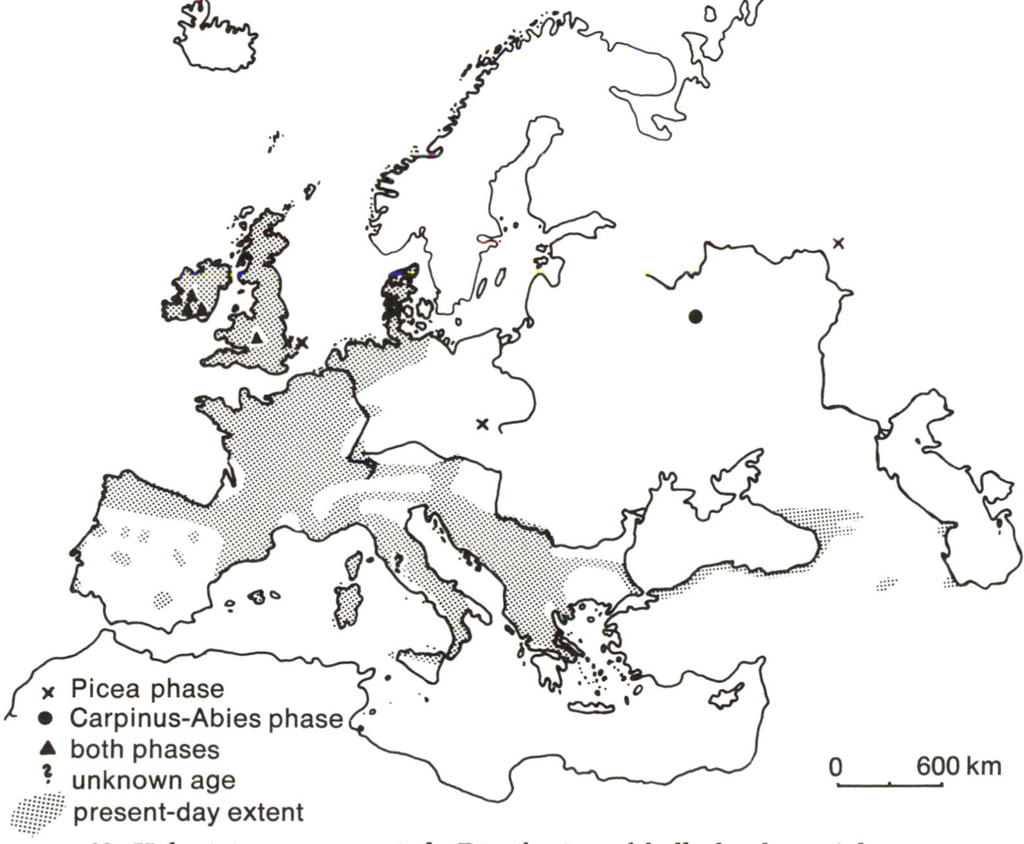

FIGURE 62. Holsteinian warm period: Distribution of holly localities (*Ilex aquifolium*). Data source: Appendix 4, p. 248.

siderable differences in the production of organic matter as a consequence of the winter to summer change. It seems much more likely that a regular summer rainfall was the cause; moreover it seems clear that a warm moist climate extended into the north Mediterranean area and there was linked with a change in the annual distribution pattern of rainfall when compared to present-day conditions.

The greater part of the discoveries that permit conclusions about a higher rainfall come from the hornbeam-fir phase. Yet it should not be concluded that the climate during the preceding spruce-alder phase was very warm and dry. The former distribution of holly (*Ilex aquifolium*) and the broad-leaved linden (*Tilia platyphyllos*) within this phase allows one to conclude that not only temperature but also rainfall was higher—that is, in general, the former climate was much more oceanic than the present. If this is true, the difference in the floral character between the two phases is more readily explained in terms of differences in the expansion of these plants from the refuges occupied during the preceding cold period than as a climatic difference, though it will also be conceded that the oceanic influence on climate during the hornbeam-fir phase was greater. Yet with the material presently available, this cannot be determined with certainty. At any rate the difference in the climatic character of the two phases cannot have been very great.

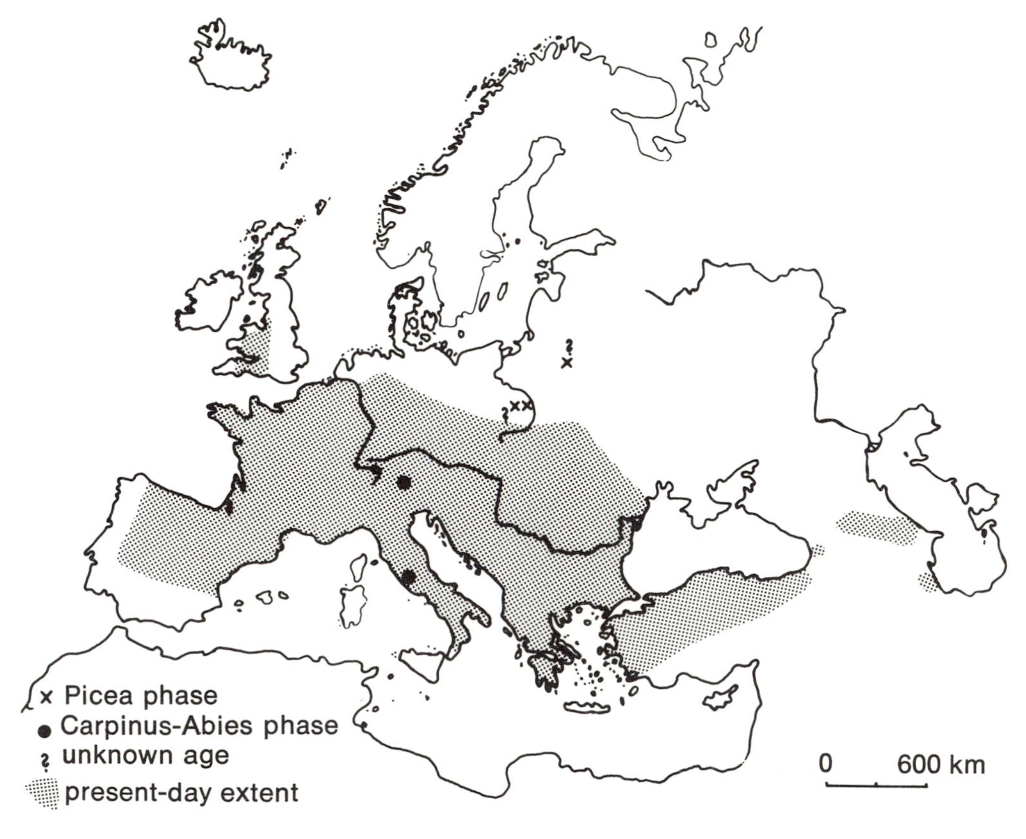

FIGURE 63. Holsteinian warm period: Distribution of broad-leaved linden localities (*Tilia platyphyllos*). Data source: Appendix 4, p. 248.

TABLE 8. DIFFERENCES IN CERTAIN CLIMATIC FACTORS BETWEEN THE HOLSTEINIAN WARM PERIOD AND THE PRESENT DAY (DETERMINED FROM THE DISTRIBUTION OF CERTAIN PLANT SPECIES AND GENERA; FIGS. 55, 57–63)

LOCALITY AND PLANT SPECIES	MEAN TEMPERATURE			ANNUAL PRECIPITATION (MM.)
	JANUARY °C	JULY °C	ANNUAL °C	
British Isles and Ireland				
Vitis sylvestris	=	+2	+1	=
Azolla filiculoïdes	=	+2	=	?
Denmark				
Vitis sylvestris	+1 to 2	+3	+2 to 3	=
Buxus sempervirens	+1 to 2	+1	+1 to 2	=
Trapa natans	?	+2	?	?
Northwestern Germany and the Netherlands				
Vitis sylvestris	=	+2	+1	=
Staphylea pinnata	=	+2	=	=
Poland				
Vitis sylvestris	+2	=	+1 to 2	=
Azolla filiculoïdes	+3	=	+1	?
Fagus silvatica	+3	=	=	=
Tilia platyphyllos	+1 to 2	=	+1	+50
Ilex aquifolium	+3 to 4	=	=	=
Southern Lithuania				
Tilia platyphyllos	+5 to 6	+3	+4	=
T. tomentosa	+6	+5	+7	=
Carya sp.	=	+3	+1	+150
Central Russia				
Azolla filiculoïdes	+8	=	+5	?
Ilex aquifolium	+9 to 10	=	+3	=
Tilia platyphyllos	+6 to 9	+2	+4 to 6	+100
Fagus silvatica	+10	=	+5 to 6	=
Trapa natans	?	=	?	?
Taxus baccata	+3 to 4	=	+3	=
Middle reaches of the Volga				
Azolla filiculoïdes	+11	=	+3 to 4	=
Ilex aquifolium	+14	=	+6 ?	+300
Lower reaches of the Irtys				
Azolla filiculoïdes	+24	+1	+12	?
Surrounds of the Tomsk Romosk				
Azolla filiculoïdes	+20 to 22	+1	+11	?
Central Siberia				
Distribution of permafrost	?	?	+6	?

TABLE 9. MINIMUM VALUES OF THE CLIMATIC CHANGE DURING THE HOLSTEINIAN WARM PERIOD COMPARED TO PRESENT-DAY CONDITIONS IN VARIOUS REGIONS OF NORTHERN EURASIA

REGION	MEAN TEMPERATURE			ANNUAL PRECIPITATION
	JANUARY °C	JULY °C	ANNUAL °C	(MM.)
British Isles and Ireland	=	+2	+1	=
Denmark	+1 to 2	+3	+2 to 3	=
Northwestern Germany and the Netherlands	=	+2	+1	=
Poland	+3 (to 4)	=	+1 to 2	+50
Southern Lithuania	+5 to 6	+5	+7	+150
Central Russia	+8 to 10	+2	+5 to 6	+100
Middle reaches of the Volga	+12 ?	=	+6 ?	+300
Western Siberia	+20 to 22	+1	+11 to 12	?
Central Siberia	?	?	+6	?

3. The Eemian warm period. In common with earlier cold periods, the climatic conditions of the Saalian cold period reduced the number of plant species in central Europe. As a consequence, during the Eemian period many of the former plant species so characteristic even of the Holsteinian warm period no longer occur, so that the impoverishment of the flora does not have to be explained by a particularly unfavorable climate during the Eemian warm period as compared with the Holsteinian warm period.

(a) The climatic optimum must first be considered. In Table 10, using already familiar methods, the calculated minimum value for the differences of the mean annual temperature of the coldest and warmest months and of the rainfall of the Eemian period with respect to the present day are listed. An exact analysis of the history of the vegetation quickly shows that climatic conditions most favorable for woodland were often attained during the oak-elm phase and the following linden phase. Very unfavorable conditions persisted during the first part of the hornbeam-fir phase, which followed the *Tilia* phase. Consequently, the figures derived from the distribution of the beech (*Fagus silvatica*) are given in Table 10, although it must be noted that the beech first appeared during the hornbeam phase in central and eastern Europe. It is generally assumed that the species during this and the preceding warm period no longer occurred north of the Alps. Yet it happens that most of the discoveries of beech described occur only in the hornbeam-fir phase of the Holsteinian warm period or from the hornbeam phase of the Eemian warm period. This distribution can no longer be explained by redeposition of older material or long-distance transport. Rather, it must be concluded that, in contrast to the commonly held view, the beech occurred during both the warm periods named in central and eastern Europe even if it was not one of the quantitatively important types of trees, perhaps with the exception of the immediate vicinity of the Carpathians and the Alps [260]. Quite independently of whether or not *Fagus* actually occurred during the Eemian warm period in central and eastern Europe, the data in Table 10 show that the climate in the area under investigation must have been very agreeable.

In this case, too, the same conditions are valid as for the estimates given in Tables 7–9: the various minimum-difference values determined for the increased temperature, as deduced from the distribution pattern of individual genera or species, are mostly due to the fact that for some species minimum summer temperatures are decisive whereas in others minimum annual or maximum winter temperatures are regarded as the limiting factors. With regard to the value estimated for the western Siberian lowlands, it should be noted that they are lower than in the Holsteinian warm period; *Azolla interglacialica* is no longer to be found in the sediments of the last warm period in Siberia, so that the minimum temperature difference in comparison with the present day can be determined only from the spread of particularly exacting forest types, namely the spruce-fir-arolla-pine forest. Its former northern limit is not everywhere known with certainty, for the polar seas simultaneously spread far to the south over the continent. It is therefore possible that the forest never reached its maximum possible northern limit. The estimated values are thus in all probability lower than they actually were. Thanks to this favorable climate, some plants were able to extend far to the north over the western Siberian lowlands. Thus the lower reaches of the Ob below

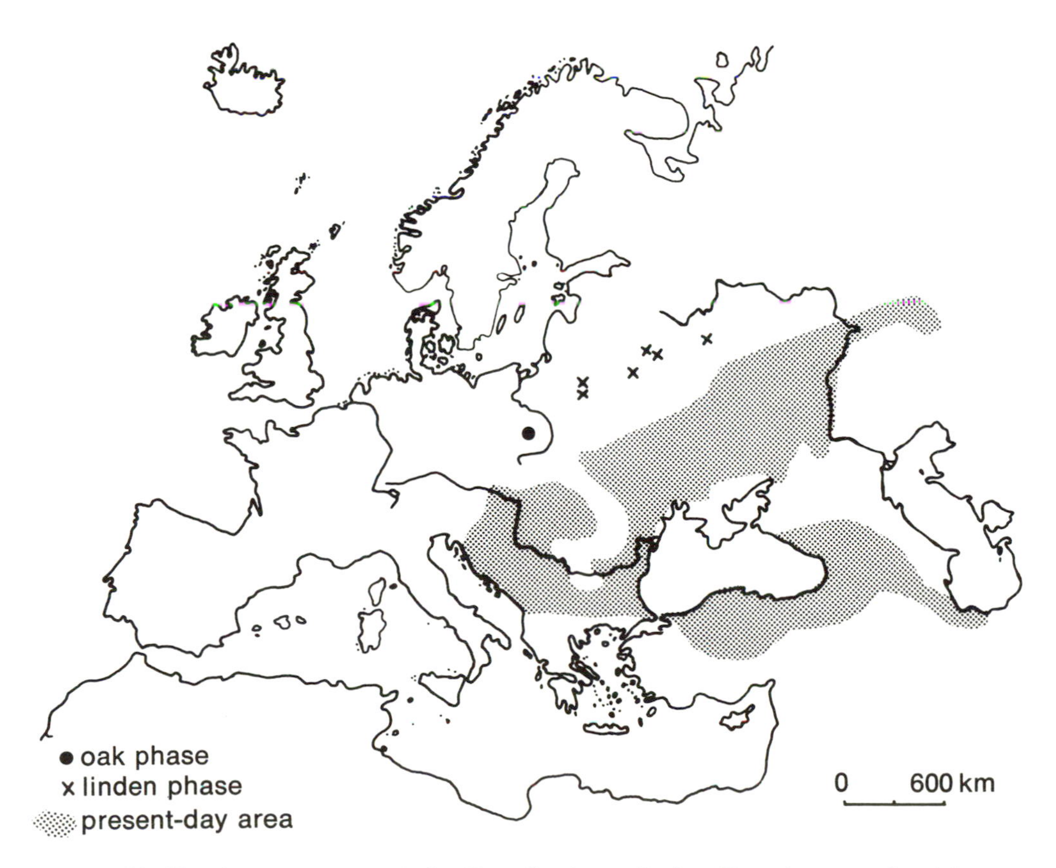

FIGURE 64. Eemian warm period: Distribution of the Tatarian maple (*Acer tataricum*). Data source: Appendix 5, p. 249.

about 62°N near Karymkary was reached by the quillwort (*Isoëtes echinospora*), the foliated pondweed (*Potamogeton obtusifolius*), Najadaceae (*Najas marina*), the frogbit (*Hydrocharis Morsus-ranae*), the water lily (*Nymphaea candida*), and the wild strawberry (*Fragaria viridis*). The present northern limit of these species lies some 3° to 4° further south. However, the winters there were also very cold, for the sediments containing this flora show clear signs of contemporaneous (syngenetic) forms of seasonal freezing [520]. The absence of *Azolla* has little paleoclimatic significance, as in this context the particularly severe glaciation of western Siberia during the preceding cold period has to be considered (Fig. 74).

In sum it can probably be assumed that the climate over considerable areas during the oak-elm, linden, and hornbeam phases of the last warm period were approximately warmer and moister than the present day by the amounts shown in Table 11.

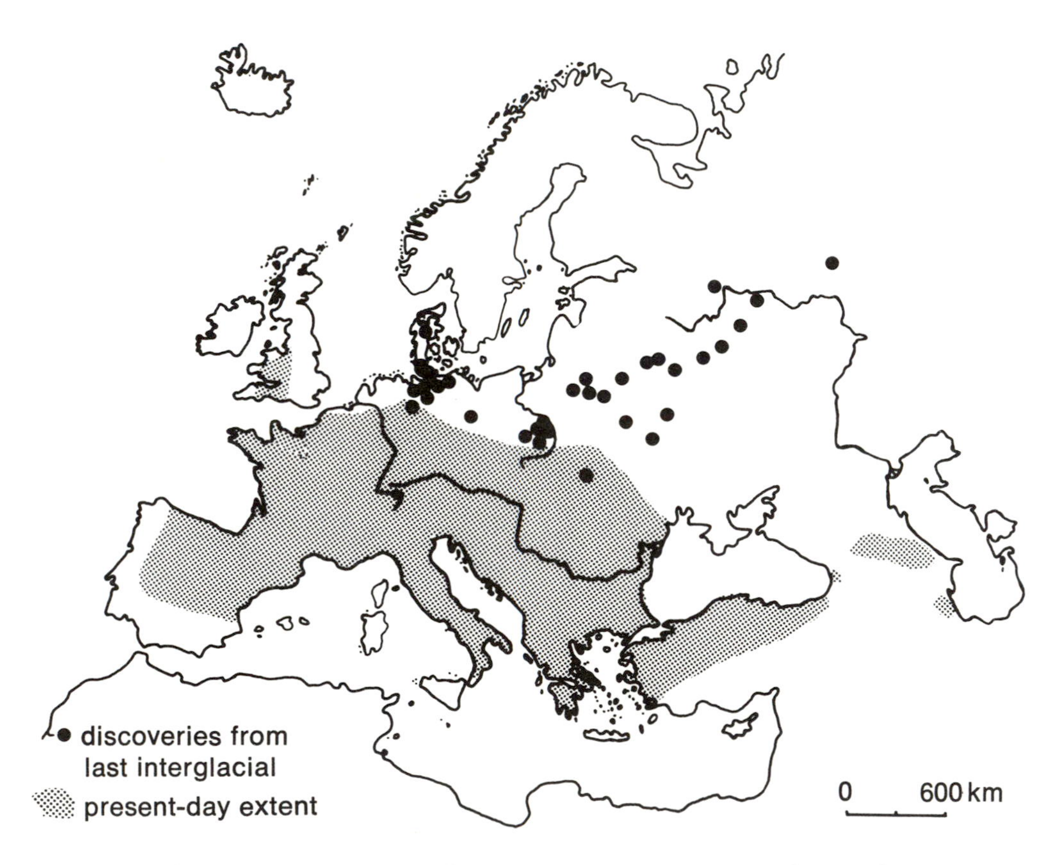

FIGURE 65. Eemian warm period: Distribution of the broad-leaved linden (*Tilia platyphyllos*). The particularly obvious errors in the occurrence of localities within the present-day range of the species is not explained by an assumption that the broad-leaved linden was then restricted to areas north of its present region, but rather lies in the favorable conditions for preservation of fossils in the former "young-moraine" regions of the preceding Saale cold period, in contrast to the less favorable conditions for fossilization outside the former glaciated region. Data source: Appendix 5, p. 249.

These values agree remarkably well with those determined for the last but one warm period. They show that the climate during the Eemian warm period was also warmer and more oceanic than at present. This holds true, as has been noted above, for the Mediterranean area also, for which region Marchesoni [510] has indicated January temperatures 3° to 4° higher, July temperatures about 2° higher, and a perceptibly higher total rainfall. It is, nevertheless, not clear whether, as in the preceding warm period, the principal rainfall occurred during the summer months. However, the great similarity of the two warm periods and their contrast to present conditions would make the latter seem more than likely.

Tables 10 and 11 also show that the water supply for the plant community during the Eemian climatic optimum appears to have been more favorable than the present one. Schönhals [654, 655], Macoun [499], and Moskvitin [544] have nonetheless described thick steppe black earths in Rhine-Hessen and from central Russia, or soils of a warm but somewhat dry climate [499], as particularly characteristic of this warm period. Blagoveshchensky [64], on the basis of his palynological research in central Russia, required an oak woodland steppe for the first phase of the last warm period. These observations stand in apparent contradiction to the preceding. However, Schönhals has recently revised his former opinion,

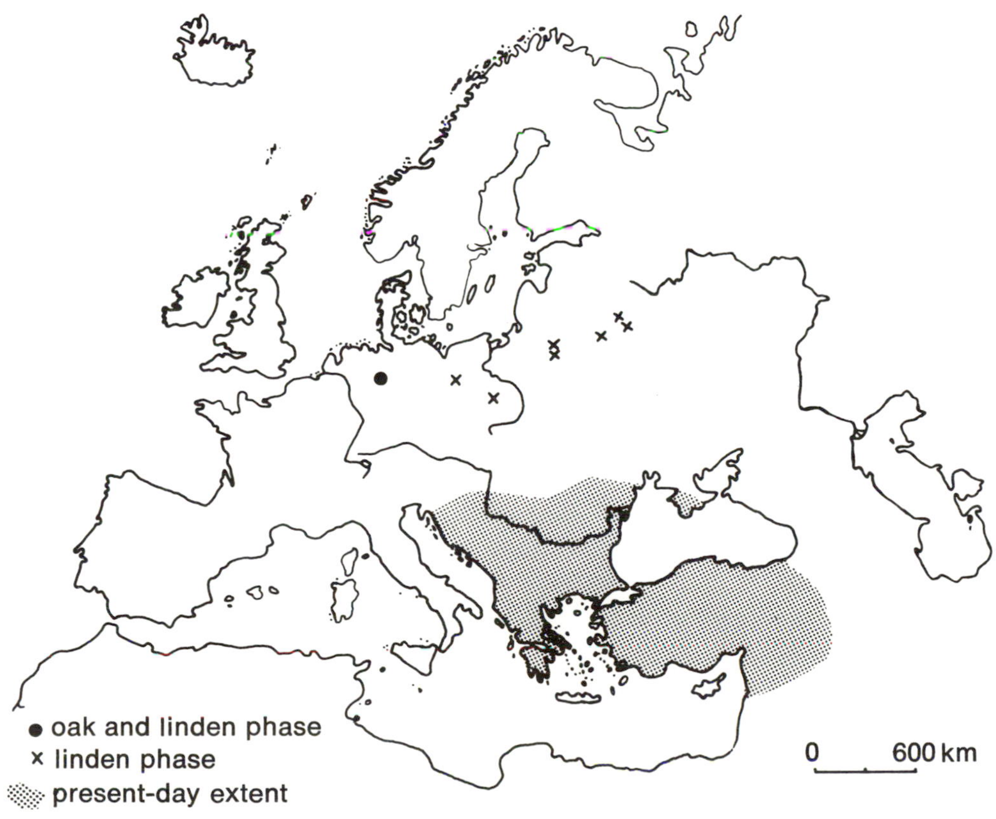

FIGURE 66. Eemian warm period: Distribution of the silver linden (*Tilia tomentosa*). Data source: Appendix 5, p. 249.

since the soils are frequently confused with those of the first warm fluctuation of the last cold period [659], and the black earths of central Russia too seem to date from the oldest known warm fluctuation of the last cold period and not from the last warm period. Finally, the results of all the most recent pollen analyses oppose Blagoveshchensky's assumption of an oak-woodland steppe phase early in the last warm period. These corrections, recently made possible, are very important, for they show that there is no longer room for the view that steppe or forest-steppe covered wide areas of the present forest zone of central and eastern Europe during the appreciably warm Eemian warm period. This makes the hypothesis of the natural occurrence of open plant communities in the vegetation of central Europe during the postglacial warm period seem improbable, and all the more so because it was somewhat cooler than during the last warm period.

Up to the present time the question whether the climate of the oak-elm phase was drier than that of the linden phase was left open. The former distribution of the box tree (*Buxus sempervirens*, Fig. 69) and the holly (*Ilex aquifolium*, Fig. 70), that is, of trees which at present favor an oceanic climate, clearly speaks against the assumption of a markedly drier climate during the oak-elm phase in

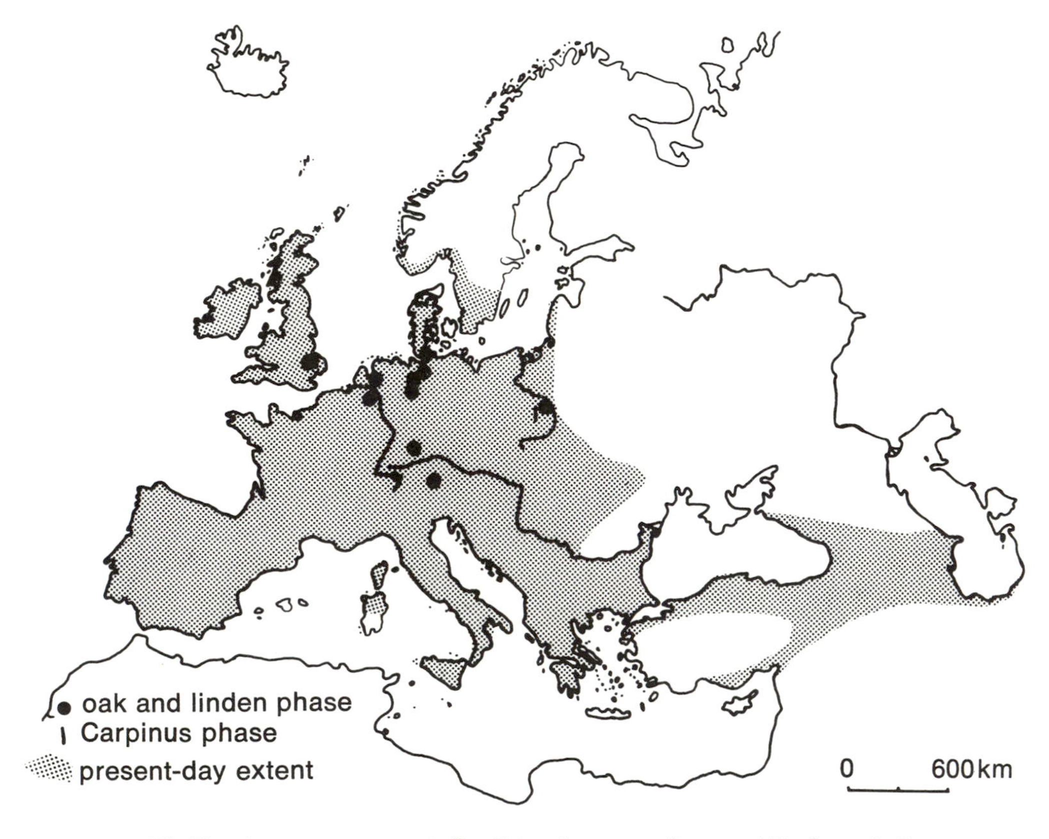

FIGURE 67. Eemian warm period: Distribution of ivy (*Hedera helix*). Data source: Appendix 5, p. 249.

comparison with the linden phase. Probably the climate of these two phases was similar. Yet the very limited occurrence in time of the broad-leaved linden (*Tilia platyphyllos*) only within the linden phase of this warm period seems to indicate that the climate was particularly wet at that time only, for the demands of the broad-leaved linden on the water budget at its site are particularly high. The number and distribution of the fossil localities of *Buxus sempervirens, Ilex aquifolium,* and the yew (*Taxus baccata,* Fig. 68) which, during much of the Eemian warm period, were widely distributed in central Europe, does not permit such a conclusion. It seems much more likely that the very oceanic climate during the linden phase and perhaps also during the hornbeam phase was spurious, resulting from variations in the migrational history of various plant types. Namely, it appears [260] that the oak and elm, which determined the character of the woodlands of the oak-elm phase, are light-loving trees—that is, species which during their younger stages in particular require a high intensity of light for their development. In contrast, the later-occurring and rapidly spreading hazel and linden require shade. Their young growth develops better in shady woodland, and in consequence they can replace an oak-elm forest without a climatic change if only

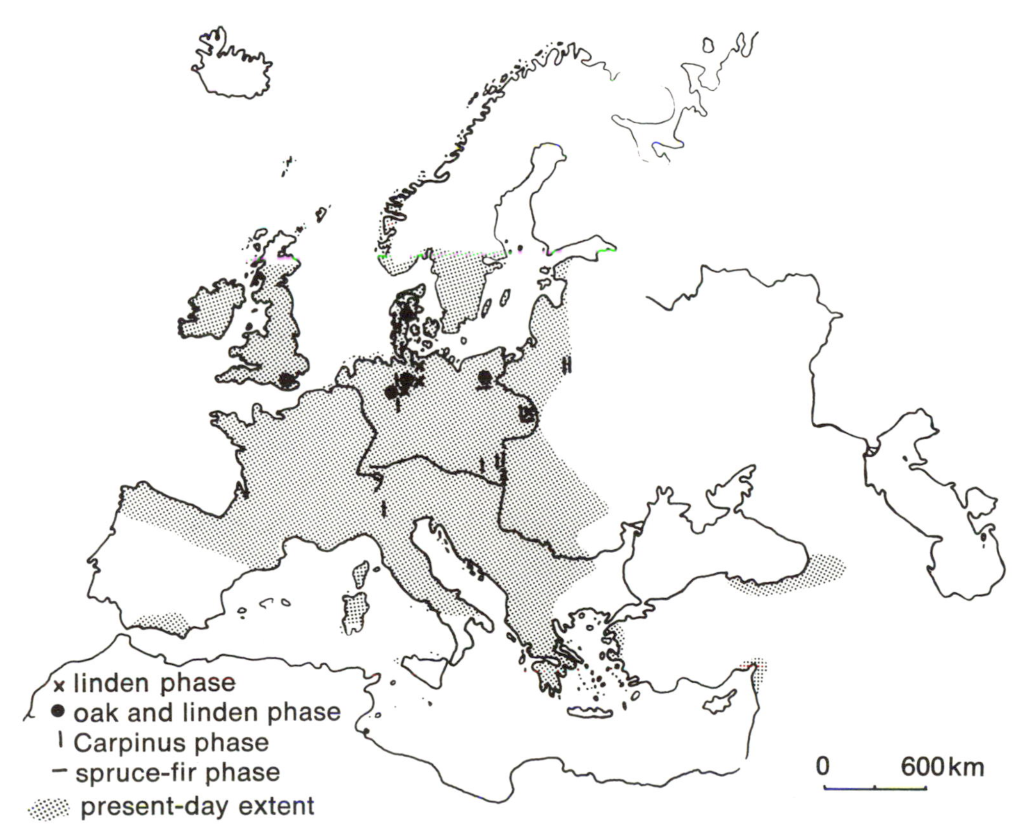

FIGURE 68. Eemian warm period: Distribution of yew (*Taxus baccata*). Data source: Appendix 5, p. 249.

the existing climatic conditions favor such an advance. Such a favorable climate without doubt occurred during the phases of the last warm period being examined, so that it should be assumed that the remarkable change in woodland composition, from oak-elm to a widespread development of hazel and linden, was the consequence of a natural succession. This was modified only in that the linden, which first began to immigrate during the second part of the oak-elm phase in central Europe, thereafter was able to replace oak and elm rapidly [260] as the result of the favorable climate. The later transition to a predominantly hornbeam forest appears to be a consequence of an intervening further impoverishment of the soil.

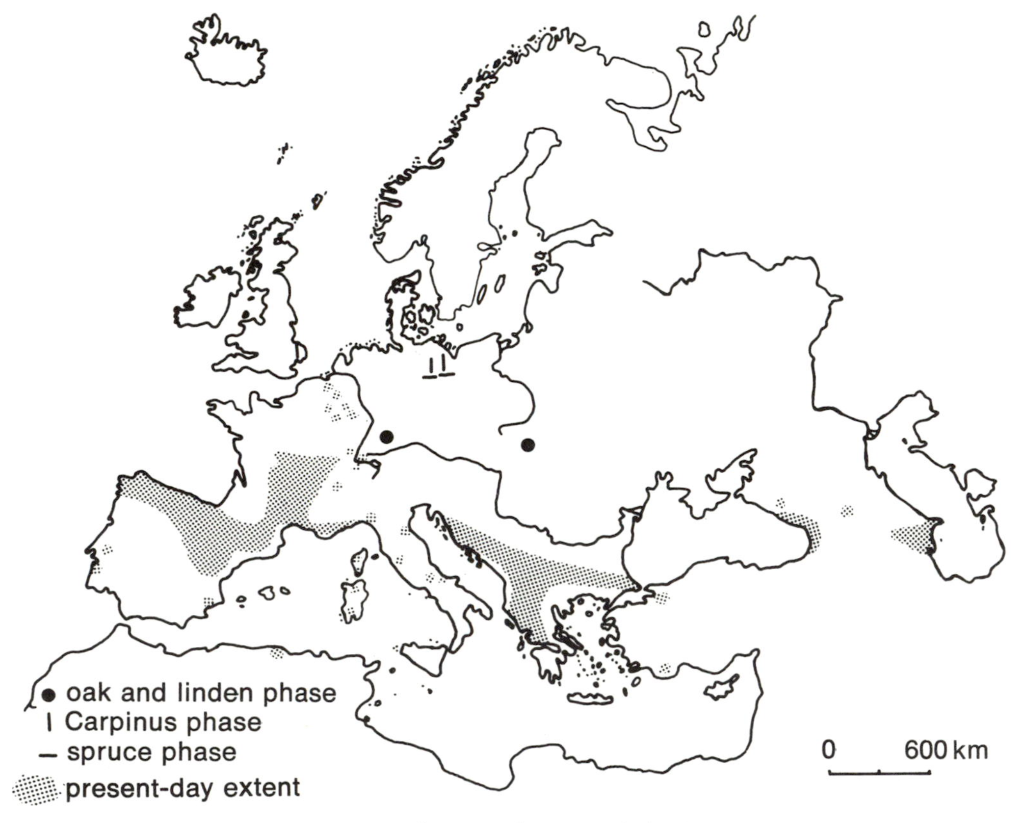

FIGURE 69. Eemian warm period: Distribution of the box (*Buxus sempervirens*). Data source: Appendix 5, p. 249.

(b) The problem of a wet phase at the end of the last warm period in central and eastern Europe must also be considered. It can be deduced from the data presently available that no certain evidence exists that the climate during the last warm period from the oak-elm phase to the linden phase and possibly into the hornbeam phase became more moist. Yet there exist many observations in various parts of northern Eurasia of increasing wetness toward the end of the last warm period.

In the present woodland climate of northern Eurasia and North America bodies of standing water or sluggishly flowing water are very rapidly silted up. In this process there generally occurs a definite succession of plant communities which

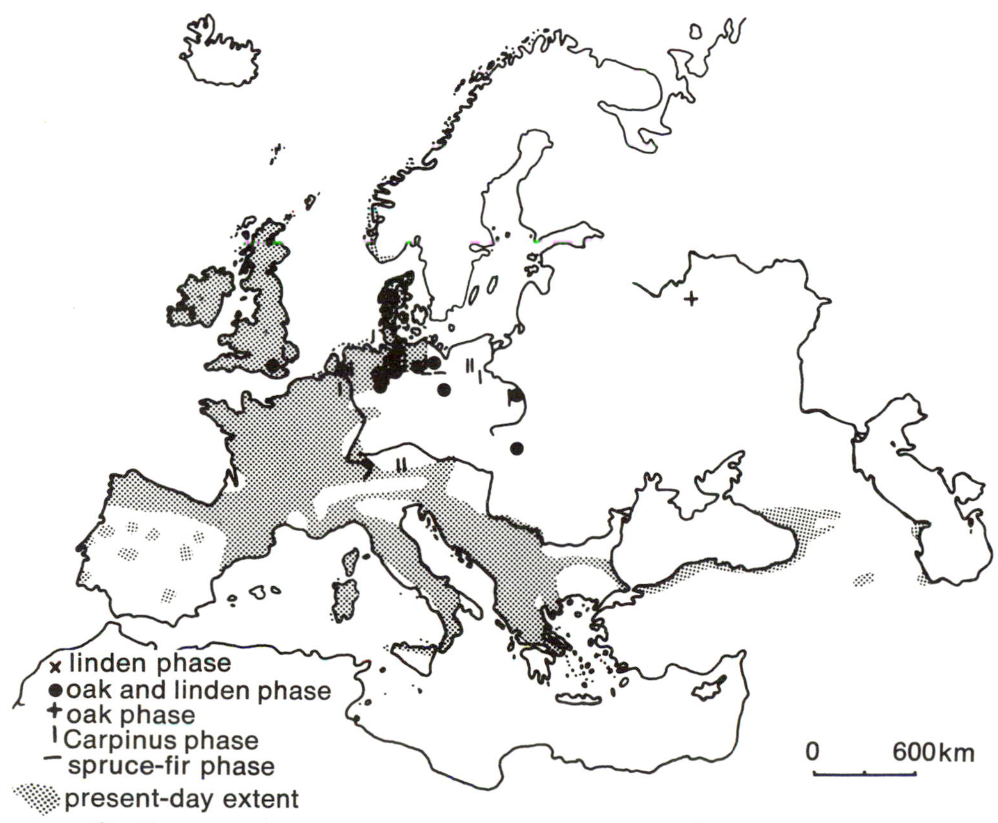

FIGURE 70. Eemian warm period: Distribution of holly (*Ilex aquifolium*). Data source: Appendix 5, p. 249.

prepare the way for one another (Fig. 75). After the outermost girdle of pondweed (*Potamogeton* sp.), which is rooted in the lake bottom and whose blooms many times extend above the water surface, there follows a girdle of floating plants in which flourish water lilies (*Nymphaea alba* and *N. candida*), pond lily (*Nuphar luteum*), and other plants with well developed floating leaves. The plants of both zones with their submerged stem and root systems retain the rain of organic and mineral matter so that the level of the lake bed rises. Finally, when the water becomes sufficiently shallow, it is colonized by bulrush (*Scirpus lacustris*), reeds (*Phragmites communis*), cat tails (*Typha latifolia*), and burr-reed (species of *Sparganium*). Their underground and submerged organs also help

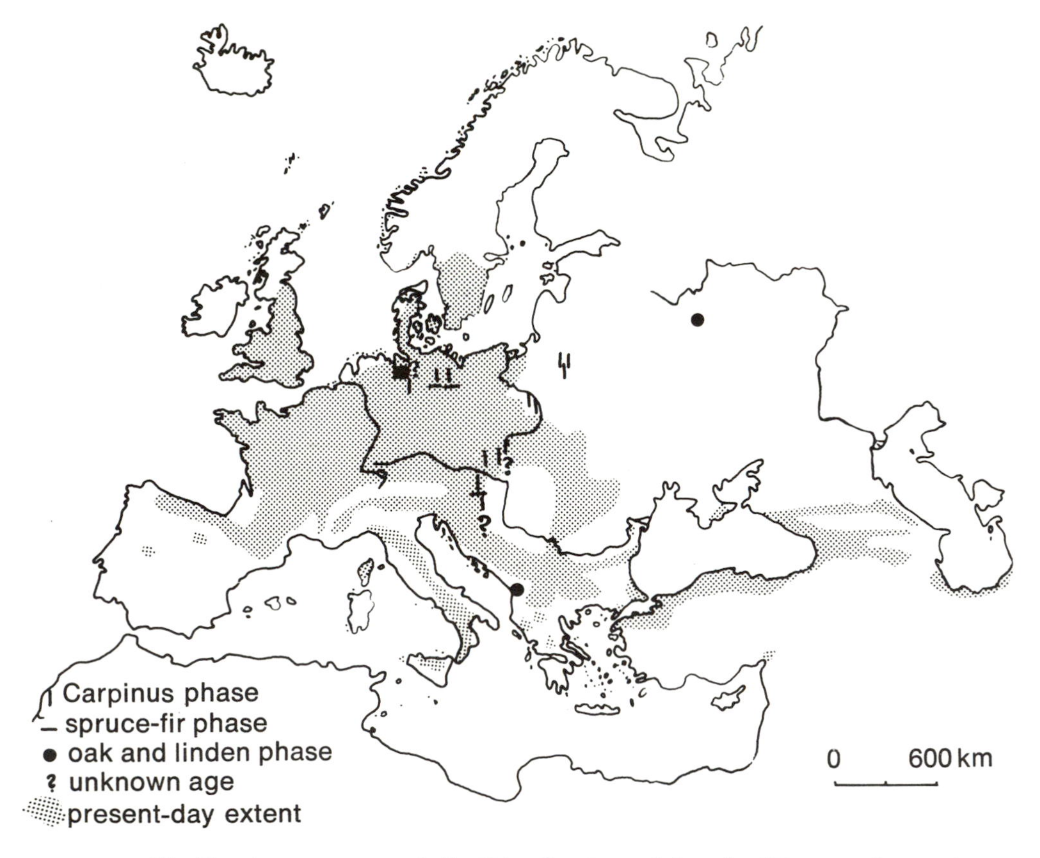

FIGURE 71. Eemian warm period: Distribution of beech (*Fagus silvatica*). For a discussion of the problem of the beech during the last two warm periods, see [260]. Data source: Appendix 5, p. 249.

retain the rich fall of organic debris, so that the lake becomes still shallower, until it is finally colonized by tall sedges upon which in a humid climate with cold winters peat moss (*Sphagnum*) can spread. The silted-up lake is thus gradually transformed into a bog. At first it remains in contact with ground water rich in nutritive compounds. It is then spoken of as a carr (or reed-swamp). Gradually its level rises because of the rapid growth of *Sphagnum*, and the water budget of the upper part of the peat becomes independent of the ground water. It has thus become a raised bog whose surface arches upward. Once a certain height is reached, the *Sphagnum* growth rate diminishes as rain water drains away too rapidly from the elevated portion, so that the moss plants remain dry for long

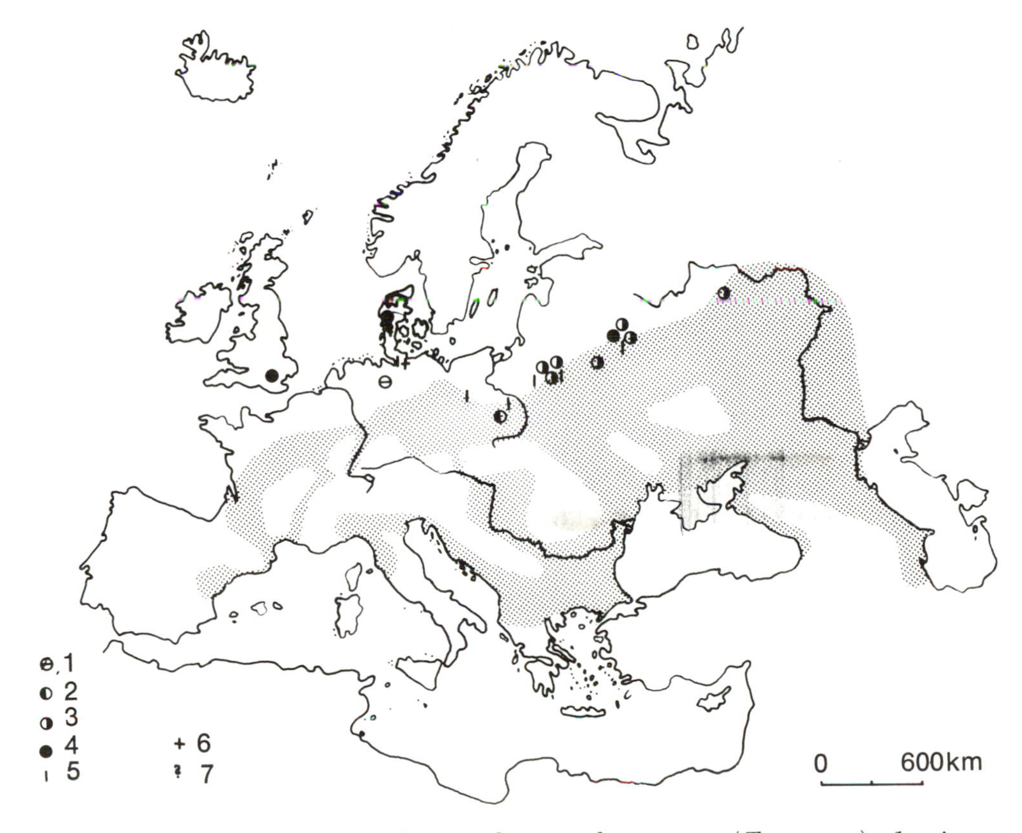

FIGURE 72. Eemian warm period: Distribution of waternut (*Trapa* sp.): 1. pine-oak mixed forest phase; 2. oak-elm phase; 3. linden phase; 4. oak-elm and linden phases: 5. hornbeam phase; 6. spruce-fit phase; 7. exact age unknown. The screened area indicates present distribution. Data source: Appendix 5, p. 249.

parts of the year. Under these conditions, dwarf shrubs, mostly *Ericaceae* (heath vegetation) and trees can spread over the surface of the bog. Earlier, their young growth was inhibited by the more rapid growth of moss; now they are able to replace them, and thereby the raised ombrogenous bog (caused by rain) changes to a heath—or wooded moorland. When this stage occurs, the deterioration in the water budget of the bog depends upon the gross climatic conditions on the one hand and upon existing local climate and the form of the relief on the other. The initiation of the low bog, as also the growth of the raised bog, are equally functions of climate on the one hand, but linked on the other to the supply of nutrients and the aeration and depth of the basin being

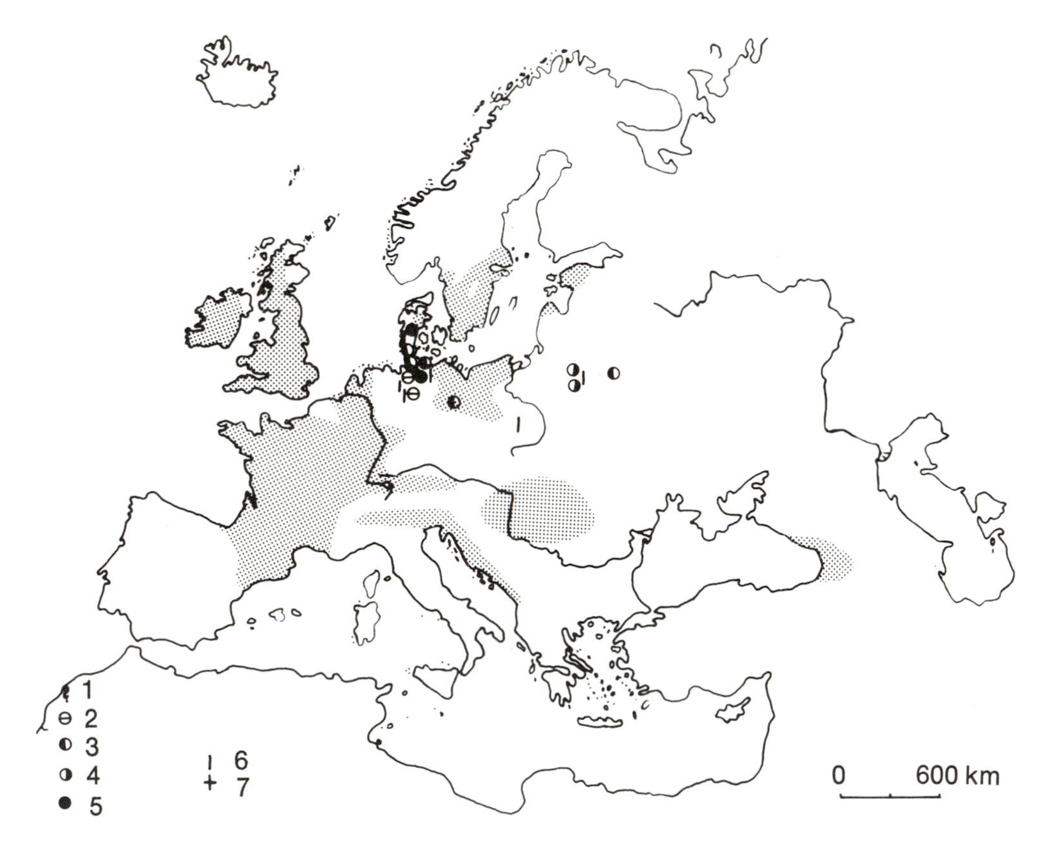

FIGURE 73. Eemian warm period: Distribution of cut-sedge (*Cladium Mariscus*): 1. exact age unknown; 2. pine mixed with oak forest phase; 3. oak-elm phase; 4. linden phase; 5. oak-elm and linden phases; 6. hornbeam phase; 7. spruce-fir phase. The screened area indicates present distribution. Data source: Appendix 5, p. 249.

TABLE 10. MINIMUM VALUES FOR THE DIFFERENCE BETWEEN MEAN ANNUAL TEMPERATURE, JULY AND JANUARY TEMPERATURES, AND THE ANNUAL PRECIPITATION DURING THE OAK-ELM, LINDEN, AND IN PART THE HORNBEAM PHASE OF THE EEMIAN WARM PERIOD IN COMPARISON WITH THE PRESENT DAY (FIGS. 64–73)

REGION AND PLANT SPECIES OR GENUS	MEAN TEMPERATURE			ANNUAL PRECIPITATION (MM.)
	JANUARY °C	JULY °C	ANNUAL °C	
Denmark				
Tilia platyphyllos	+2	+1 to 2	+1 to 2	=
Trapa natans	=	2	?	?
Northern and central Germany				
Tilia tomentosa	=	+3	+2 to 3	=
Buxus sempervirens	+1 to 2	+1	+1 to 2	=
Central Poland				
Fagus silvatica	+3	=	=	=
Tilia tomentosa	=	+3	+3	=
T. platyphyllos	+1 to 2	=	+1	+50
Acer tataricum	=	+1 to 2	=	=
Ilex aquifolium	+3 to 4	=	=	=
White Russia				
Acer tataricum	=	+3	=	=
Tilia platyphyllos	+5 to 6	+3	+4	=
T. tomentosa	+6	+5	+7	=
Trapa sp.	?	+1	?	?
Central Russia				
Fagus silvatica	+10	=	+5 to 6	+
Tilia platyphyllos	+6 to 9	+2	+4 to 6	+100
Acer tataricum	=	+2	=	=
Ilex aquifolium	+9 to 10	=	+3	+100
Trapa natans	?	=	?	?
Northwest Ukraine				
Tilia platyphyllos	+2 to 3	=	+1	+50
Western Siberia				
Distribution of spruce-fir-arolla pine forest	+4	+3	+3	+100
Central Siberia				
Distribution of permafrost	?	?	+6	?

TABLE 11. MINIMUM VALUES OF THE CHANGE IN CLIMATE DURING THE EEMIAN WARM PERIOD IN COMPARISON WITH PRESENT-DAY CONDITIONS IN VARIOUS PARTS OF NORTHERN EURASIA

REGION	MEAN TEMPERATURE			ANNUAL PRECIPITATION (MM.)
	JANUARY °C	JULY °C	ANNUAL °C	
Denmark	+2	+1 to 2	+1 to 3	=
Northern and central Germany	+1 to 2	+3	+2 to 3	=
Central Poland	+3 to 4	+3	+3	+50
White Russia	+5 to 6	+5	+7	=
Central Russia	+9 to 10	+2	+4 to 6	+100
Northwestern Ukraine	+2 to 3	=	+1	+50
Western Siberia	+4	+3	+3	+100 ?
Central Siberia	?	?	+6	?

FIGURE 74. Map showing the maximum spread of continental and mountain glaciation of northern Eurasia during the last three cold periods. 1. Elsterian cold period; 2. Saalian-Dneprovsk-Samarov cold period; 3. Weichselian-Valdai-Zyrjanka cold period.

FIGURE 75. Silting up of a body of standing water. Grosses Heiliges Meer near Rheine, Westphalia. The silting up of open water (to be seen in the upper left of photo) passes from the pond weed girdle (right foreground) directly to the zone of thick tall sedges (right foreground, middle of photo) and then to older brush peat (background). The floating girdle, represented by water lilies, is absent here, and the reed belt is only poorly developed.

silted up. It must be further realized that silting up can be accomplished in other ways too, perhaps by a floating bog, which floats upon water, and by a succession which instead of leading to a raised bog usually ends with the formation of alder-meadow woodland (Fig. 75). All these processes result in the lake finally becoming more or less a land area. During the last warm period this organic sedimentation and infilling process took place at many localities, filling up old oxbows of the still untamed rivers, and old lakes formed either by the melting of dead ice or excavated by glaciers of the Saalian ice age. In general the silting up began in the linden phase, and very many lakes were completely filled up by the hornbeam phase, whether in the form of forest peat swamps or various types of bog.

Toward the end of this warm period a reverse tendency became apparent in many localities. Bogs and forest swamps were once more flooded and drowned. Such a regressive succession under constant environmental conditions is unthinkable. As this remarkable wetness in addition left traces over a wide area, which reached from the Netherlands in the west to central Russia in the east, local variations in water budget can be excluded as a means of explanation, and it has been presumed as a result that a climatic fluctuation is indicated.

The presently available data on the onset of flooding of fifty-six bogs is shown in Figure 76. It appears that toward the end of the warm period there was an appreciable increase in the number of bogs flooded. During the linden phase 3.6 percent were flooded, and a further 7.2 percent during the hornbeam phase.

During the first part of the following spruce-fir time 16.1 percent of the bogs were flooded; in the second part 35.7 percent, and in the spruce-pine period, which belongs to the last cold period, a further 37.5 percent. This increasing wetness appears to be a clear response to a progressively wetter climate. However, Figure 76 shows that the wet phase must have set in earlier in the east than in the west. This follows from the data if represented in the following manner. The region being investigated was divided into an eastern and western section by a line running from the estuary of the Memel to the middle Weichsel. In each section the number of bogs was normalized to 100 percent and the percentage of bogs flooded during each time interval was calculated.

	Western Section	*Eastern Section*
Hornbeam phase	4	9
Spruce-fir phase	28	58
Spruce-pine phase	68	31

FIGURE 76. The time of drowning of certain bogs at the end of the last warm period in Europe: 1. linden phase; 2. hornbeam phase; 3. first part of the spruce-fir phase; 4. second part of the spruce-fir phase; 5. pine-spruce phase and the beginning of the last cold period. Source: Appendix 6, p. 250.

This state of affairs is in sharp contrast to the assumption that the flooding of the bogs must be a result of increased precipitation. That should have affected the more oceanic regions most strongly, rather than the relatively continental central Russia. It must also be added that the flooding in the east led to the covering of the former bog or forest swamp with layers of sand or loam [central Russia: 25, 26, 129, 147, 306, 531; White Russia and the Baltic: 125, 165, 297, 320, 687; Poland: 590, 699, 712, 748; northern Germany and the Netherlands: 246, 614]. In many areas in central Russia and Poland *Artemisia* (wormwood) was already flourishing. At present this plant is found in dry locations. In the west, however, immediately after flooding, new Sphagnum peats formed in which heath plants (*Ericaceae*) rapidly became important. This shows that the heath moorland developmental stage soon reached a stage in which water requirements were less than those of true raised bogs [northern Germany: 28, 41, 82, 309, 424, 587, 675; Netherlands: 246, 817; Poland: 63, 75; White and central Russia: 123, 297]. These observations throw doubt on the commonly held belief that the flooding was a consequence of increased precipitation. It is particularly significant in this context that the flooding set in earlier in the east than in the west. It suggests that in this we may be glimpsing the effects of lower temperatures—that is, a climate with very cold winters was already established. Fortunately, the assumption may be tested by the former distribution of elm, oak, and linden. There is one difficulty, however: as a consequence of the relief of the central Russian uplands and its division into hills and valleys, local climatic conditions might have favored the more demanding plants, although the gross climate was highly unfavorable. It is possible as a result that climatic values determined from the spread of specific trees may be somewhat too mild in the region of the central Russian uplands (Table 12).

TABLE 12. MEAN JANUARY AND JULY TEMPERATURES DURING THE FIRST HALF OF THE SPRUCE-FIR PHASE OF THE LAST WARM PERIOD FOR VARIOUS REGIONS OF CENTRAL AND EAST EUROPE (FIGURES DERIVED BY COMPARISON OF THE FORMER [260] AND PRESENT DISTRIBUTION OF *QUERCUS*, *ULMUS*, AND *TILIA*)

REGION AND TREE	MEAN JANUARY TEMPERATURE		MEAN JULY TEMPERATURE	
	DIFFERENCE FROM P.D. °C	FORMER TEMPERATURE °C	DIFFERENCE FROM P.D. °C	FORMER TEMPERATURE °C
Northeastern Poland				
Ulmus sp.	−4.5	−9.0	=	+18.0
Quercus sp.	−4.0	−9.2	−1.5	+17.5
Tilia sp.	−4.5	−9.0	−1.5	+17.0
Lower Oka				
Quercus sp.	−4.5	−14.5	=	+19.0
Tilia sp.	−8.5	−19.0	−1.5	+17.0
Bug				
Ulmus sp.	−11.0	−14.5	=	+19.0
Hamburg district				
Ulmus sp.	=	+1.0	=	+17.0
Quercus sp.	=	+1.0	=	+17.0
Tilia sp.	=	+1.0	=	+17.0

It has already been pointed out that the values determined for central Russia in particular may be in error. Yet the table serves to show that the mean January temperatures in central Russia and in White Russia even at the beginning of the spruce-fir phase—that is, simultaneously with the flooding of the bogs—had dropped considerably, whereas the mean July temperatures were still about up to present levels. In the east, therefore, in contrast to the conditions prevailing during the hornbeam phase, the climate must have become increasingly more continental, but this had not yet been able to establish itself in the west. Possibly under these conditions a very thick soil layer froze each winter in the east, and the frost remained in it for a long part of each year. Without doubt this must have favored the flooding of the bogs, as the annual freezing blocked the soil drainage of rain and melt waters and encroached upon the vegetation, so that with less protection, increasing quantities of inorganics were washed away. The occurrence of wormwood in the east and the formation of heath-raised bogs in the west indicate that the precipitation was not high, at least no higher than during the preceding hornbeam phase. It may be concluded that the flooding of the bog at the end of the last warm period was a result not of higher precipitation but of increasing frost action in the ground. With this the principal argument for the assumption that the precipitation increased toward the end of the warm period is removed; rather in contrast it must be considered to have decreased during the spruce-fir phase. Quite apart from results from the floral history mentioned above, the formation of loess in many places in Europe during the following first cold phase of the Weichselian cold period prior to the Amersfoort interstadial (cf. below, pp. 212 ff.) also favors this conclusion. This first cold climate phase represents only the younger part of the pine-birch phase, which previously was shown as the time of greatest flooding of the bogs, or at least followed immediately upon it. This will be referred to again in considering the climatic conditions of the last cold period. However, it must be remembered that loess can be formed only under cold dry climatic conditions, and not those under which existing bogs are flooded by increased precipitation.

If it can be assumed that a considerable drop in the average winter temperatures during the spruce-fir phase—and especially during the following spruce-pine-birch phase—caused the flooding of the bogs at the end of the last warm period, then the temperatures during the preceding hornbeam phase become particularly interesting. The former distribution of the more demanding plants rapidly indicates that conditions cannot have been unfavorable. During this phase the yew (*Taxus baccata*) reached its maximum easterly extent, far beyond the region it was able to occupy during the preceding linden phase and during the still older oak-elm phase (Fig. 68). The box tree (*Buxus sempervirens*) simultaneously reached its maximum northeasterly extent (Fig. 69). This is also true for the beech, and the area occupied by the holly (*Ilex aquifolium*) was not less than previously, although this plant is particularly sensitive to cold winters (Figs. 70, 71). The cut-sedge (*Cladium mariscus*) cannot support frosts more severe than $-2°$ to $-5°C$ [605, 606]. Yet it flourished near Grodno (Fig. 73) during the hornbeam phase of the last warm period. The present northern limit of the waternut (*Trapa natans*) coincides approximately with the 18° July isotherm [381]. It was profusely developed during the hornbeam phase in the waters from

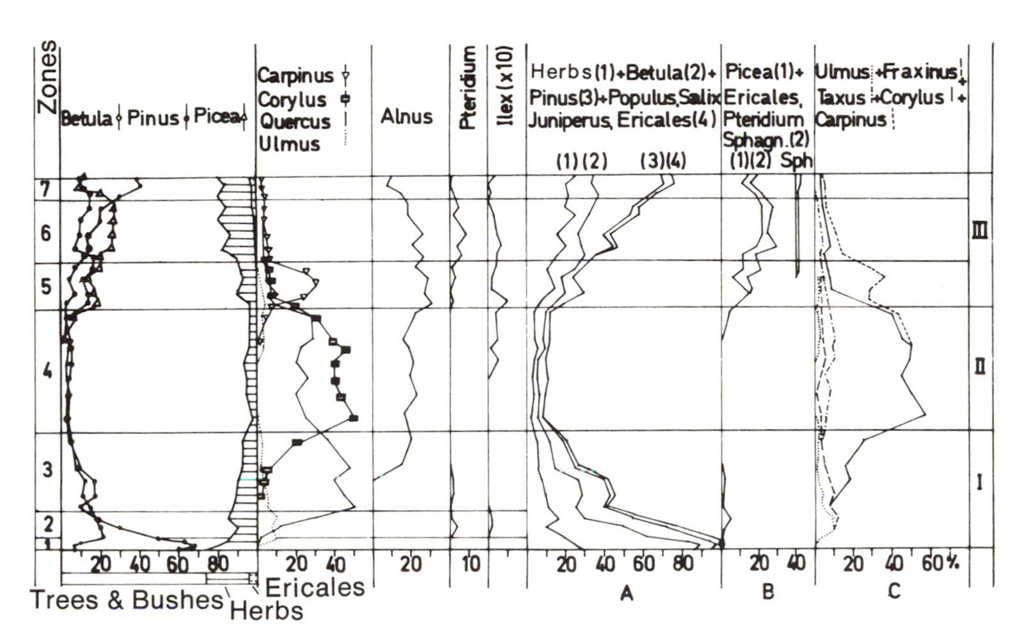

FIGURE 77. Vegetation history of the last warm period, pollen diagram from Herning in Denmark. The diagram shows the changes in the significance of certain ecological groups of plants. A. weak competitors and plants which do not support shade protection; B. plants growing on an acidic substrate; C. plants requiring a mild nutritive humus soil. I. "protocratic phase," time of forest in migration; II. "mesocratic phase," period of dominant deciduous forest; III. "oligocratic phase," period of changes in the woodland cover by progressive soil maturity and finally including climatic deterioration. Arabic numerals at the left indicate the forest zones of the diagram. From Anderson [13].

the Elbe Estuary in the west to the region of Smolensk in the east (Fig. 72). This shows that the July temperatures in the region considered must have been at least 1°C warmer than at present. The distribution of the remaining plants either indicates the same or that July and January were still warmer. Yet in contrast to this, the distribution of the elm [260] suggests that the January temperatures between Moscow and Leningrad were 3°C lower, and the July and mean annual temperatures 1°C lower than today. This anomalous result, like the sudden advance of the hornbeam, is probably not to be explained directly in terms of climate, but is associated with soil development. The effects of soil maturity on vegetation can be studied in Figures 77 and 78. These present pollen diagrams from Herning in Denmark and from Bedlno in central Poland as representatives of very different climatic regions of today recalculated according to a suggestion of Andersen's [13]. In them, the pollen sums of plants with similar ecological requirements are plotted together. As a result changes in the degree of protection, acidity, and nutrient content of the soil can be rapidly reviewed. With the transition from the linden to the hornbeam phase, the acidity of the soil in both regions gradually increases. As has been noted, this is part of the natural process of soil development, for it occurs on horizontal surfaces without the accumulation of new sediments or changes in the water or thermal budgets. The European linden and elm, with the exception of the mountain elm (*Ulmus scabra*), today

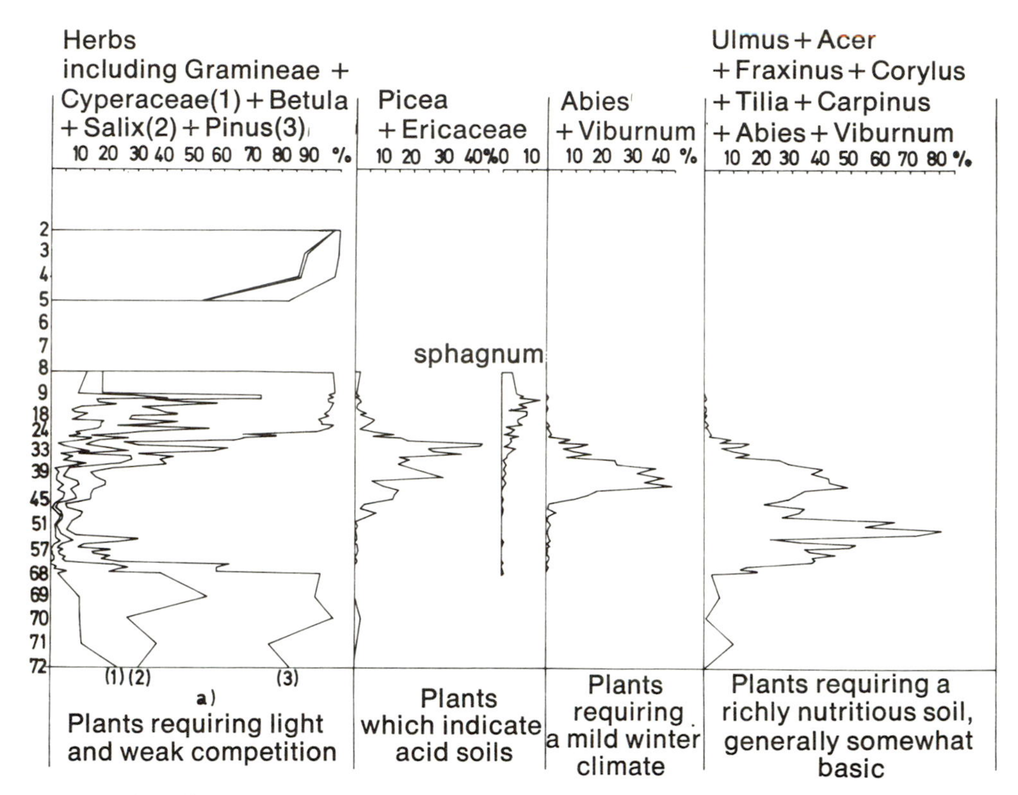

FIGURE 78. The vegetation history of the last warm period, pollen diagram from Bedlno in central Poland. This figure, like Figure 77, shows the changing significance of certain ecological groups of former plants. It is, however, taken from a different climatic zone to permit a comparison between the vegetation and soil development in an oceanic and continental region. It can be seen that the vegetation changes are in the same sense in both regions. Recalculated from Środoń and Golsbowa [712].

prosper particularly well on mild to moderately acid soils; the hornbeam instead tolerates appreciably greater acidity and more impoverished soils. Furthermore, the hornbeam, like the linden, is a shade tree and because of this was able to replace the elm relatively easily. Now, as there is no indication for a notable climatic deterioration during the hornbeam phase as compared to the linden phase, the rapid spread of the hornbeam and the simultaneous displacement of the linden and the elm can be seen as the result of progressive soil development. This conclusion is the more likely because there was no intense flooding of the bogs during the hornbeam phase. The few known examples of dating from this time occur only at the end of the phase, immediately prior to the transition to the spruce-fir phase, when a climatic deterioration slowly becomes perceptible.

From observations made along the coastline too, the onset of a strong cooling can be recognized only at the end of the hornbeam phase or at the beginning of the spruce-fir phase. In every warm period as a result of glacier melting a marine transgression occurs. That of the last warm period is referred to as the Eemian transgression, Boreal transgression, or Kazancev transgression in northern Eurasia.

In the North Sea region it lasted until about the beginning of the spruce-fir phase [80, 793]. The same holds true for the Boreal transgression on the northern coast of Eurasia [454, 155–59, 298]. Local insignificant differences in time at the beginning of the regression had a tectonic cause which modified the course of the regression brought about by the deterioration of the climate (as water was again locked up as ice). A strong cooling is thus recognized in the spruce-fir phase, but not during the earlier hornbeam phase.

(c) The observations from northern Eurasia presented so far indicate a contrast with the commonly held view that the climate in eastern and central Europe from the oak-elm phase until almost the end of the hornbeam phase was warmer and moister than at present. During this long interval of time no changes in climatic character are recognized. The contemporaneous development of the vegetation from oak-elm woods through a hazelnut phase to linden and finally hornbeam woods is, in the first instance, the result of a natural succession of plant communities influenced by progressive soil development and the relatively late influx of the linden. The latter factor is probably not so much related to the warm period climate as to that of the preceding cold period. The oak-elm phase in central Europe was preceded by the pine-birch phase, during which the migration of oak and elm began. This phase very closely resembles the history of the pine-birch phase of the postglacial. During the latter half of the pine-birch phase of the last warm period, the oak occupied approximately its present area in Europe. The elm, however, had not progressed as far as the oak, although theoretically the elm as a species spread by wind (anemochore) can spread more rapidly than one spread by animals (zoochore), like the oak. The elm and the oak appeared at the end of the last glaciation but one, approximately simultaneously in the European lowlands north of the Alps. It may then be surmised that the later spread of the elm was retarded because of a relatively unfavorable climate, so that the limiting climatic factors may be deduced from its former extent [260]. According to this the January temperatures in central Russia and the Baltic during the latter half of the pine-birch phase was still some 3° to 4°C below present levels. The mean July temperature, however, corresponded approximately to now, with the mean annual temperature about 2°C lower than the present. The amount of precipitation may have corresponded to present-day values or have been a little less. Also during the second half of the spruce-birch period the waternut (Fig. 72) occurs in central Europe in the Luneberg Heath, and in northwest Germany the cut-sedge (Fig. 73) flourished. From this it seems that the July temperature was a little above present values, and possibly the same was true in central Russia. The winter temperatures in northwest Germany were in general never below −2° to −5°C. Nevertheless, they may have been appreciably below present-day figures. Up to the present time there is no detailed data on the amount of precipitation in central Europe.

Table 13 summarizes what has been said of the climatic development of the last warm period. In it, values are given for northeast Poland—because this region lies centrally in the area considered and because of favorable conditions of

TABLE 13. REVIEW OF THE CLIMATIC DEVELOPMENT OF THE LAST WARM PERIOD IN NORTHEAST POLAND IN COMPARISON WITH PRESENT-DAY CLIMATE

VEGETATION PHASE	MEAN TEMPERATURE JANUARY °C	JULY °C	ANNUAL °C	ANNUAL PRECIPITATION (MM.)
Pine Spruce-birch phase	−3 to −4	same or a little higher	−2	same or lower
Oak-elm phase Linden phase	+3 to +5	+3 to +5	+5	+50
Hornbeam phase Beginning of the pine-fir phase	−4 to −5	−1 to −2	−2	? perhaps as today
Spruce				

preservation and because the indefatigable activity of Professor W. Szafer and Professor A. Środoń on well preserved material has had such far-reaching influence on our knowledge of the vegetation and climatic development of central Europe.

4. General paleoclimatic problems of the Holsteinian and Eemian warm periods. It has been established that the climate of both warm periods at their optima in Europe, and probably also in Siberia, resembled one another in their general characteristics. In both cases the climate was warmer and moister than at present. The differences in comparison with present-day climate were seemingly more pronounced in the present continental climatic provinces than in the oceanic. On the whole the climate was more oceanic. This process must result, in part at least, from the greater expanse of the oceans, for this must have increased the effective influence of the ocean over land areas. A further ameliorating influence was that the significant rise in temperature caused a reduction of the ice cover on the North Polar Sea. Flohn [238] considers that the sea was ice-free during the Eemian warm period. If this assumption is correct, the marked amelioration of the climate in central northern Eurasia is readily understandable.

In the northern Mediterranean region, observations also show that precipitation was higher and probably differently distributed through the year than at present. This leads to the suggestion that the atmospheric circulation was not only quantitatively but also, in some regions, qualitatively different from today.

With the climatic optima of the two warm periods so similar, the question of differences in the vegetation stands out all the more clearly. The only feasible explanation is that the vegetation differences were related not to climate but to differences in the migrational history. The correctness of the early hypothesis of Środoń [709] and Selle [672] can be fully confirmed by this.

In fact the immigration of the spruce (*Picea abies, P. obovata, P. omoticoides*), the fir (*Abies alba*), hornbeam (*Carpinus betulus*), linden (*Tilia platyphyllos, T. cordata, T. tomentosa*), elm (*Ulmus* sp.) and oak (*Quercus robur, Q. petraea*) north of the Alps in Europe at the beginning of the Holsteinian warm period followed a different course from that at the beginning of the Eemian warm period [260]. At the end of the Elsterian cold period, spruce, oak, and elm occurred simultaneously in Europe and were able to migrate relatively undisturbed. West of the Elbe this led to the rapid establishment of the oak and elm over the spruce, whose competitiveness appears to have been lower than further to the east. East of the Elbe, however, the spruce rapidly became dominant over the oak and elm. Only much later did the hornbeam and fir advance from the region of the Carpathians and spread rapidly at the expense of the spruce. This kind of migration permits two important conclusions: on the one hand, plainly at the end of the Elsterian cold period, spruce, oak, and elm slowly advanced northward. They were thereby soon in a favorable position to later reoccupy areas formerly covered by glaciers. On the other hand, it follows that the climate from the end of the Elsterian cold period became continuously warmer and moister; there appears to have been no distinctive dry period immediately at the beginning of the Holsteinian warm period.

The beginning of the Eemian warm period seems to have been quite different. In contrast to the late glacial of the Elsterian cold period, at the comparable phase of the Saalian cold period in central Europe, in addition to birch and pine only spruce occurred, in this case the Siberian spruce (*P. obovata*) rather than the European (*P. abies*). It advanced together with the Siberian larch (*Larix sibirica*) from the central Russian uplands into northeastern Poland. Oak and elm had not by then reached the European lowlands north of the Alps and Carpathians. Somewhat later the Siberian forest species retreated rapidly eastward. Once more pine became the dominant forest type, with birch present on a modest scale. This was the time of the first influx of oak and elm. Siberian and European spruce were absent. Finally, oak and elm took over, and the oak-elm phase of the last warm period began. As previously noted it is probably during the latter part of the pine-birch phase of the last warm period that the July temperatures rose to, or even slightly above, present values. The winters, however, were presumably still colder than at present. From this it can be seen that in contrast to the beginning of the Holsteinian warm period, the climate was quite continental. Possibly the dryness increased from the birch-pine-spruce phase to the pine-birch phase. Furthermore, it seems that because of the dryness the European spruce (*Picea abies*) was unable to establish itself as an important member of the woodland flora. Only during the later, moister oak-elm phase did it begin to spread rapidly over a wide area.

It may be concluded from this that the difference in the development of the vegetation in the two warm periods was a result of the differences in the locations of the refuges of the various species during the preceding cold period and the migration routes open to them, and the different climatic changes at the beginning of the two cold periods. At the beginning of the Holsteinian warm period it became gradually warmer and relatively moist, without ever becoming significantly drier, while at the beginning of the Eemian warm period there was at first a relatively moist but cold phase (birch-spruce-pine), then even a warmer but drier climate. Only during the oak-elm phase were there conditions for optimum temperatures and precipitation. This climatic fluctuation at the beginning of the last warm period has recently been confirmed by Emiliani (199a) with the help of ^{18}O data.

As has been noted previously, the results of many paleobotanic investigations provide convincing evidence that during the climatic optima of the warm periods the rainfall in northern and central Italy, and possibly in other regions of the northern Mediterranean, was higher than at present. This climatic improvement seems to have affected the present coast of northern Israel in the region of Tel Aviv, for during the second Tyrrhenian transgression of the Mediterranean, during the Eemian warm period, the proportion of halophytes was appreciably reduced, while the significance of pistachio and oak were enhanced with respect to the preceding, regressive, phase, which probably corresponds to the Saalian cold period [636]. If, consequently, it can be accepted that the climate during the warm periods was more favorable in certain regions of the Mediterranean—that is, more moist—then it is natural to assume that arid areas of the southern part of eastern Europe and central Asia also received more water than at the present day. The Caspian Sea actually spread far beyond its present bounds several times during the Ice Age, but these transgressions were related not to the warm but to the cold periods (p. 187). Another criterion for the determination of former hydrological conditions of the current arid zone is the southern forest-steppe limit. This important phytogeographical boundary in both warm periods lay in about the same position as it would now occupy under natural conditions (Fig. 79). Nonetheless, during the Holsteinian warm period it appears to have lain somewhat farther to the south than during either the Eemian warm period or the postglacial [details: 260]. The difference, however, was not very great. Evidently the ameliorating effect of an increase in total precipitation was to a large extent compensated for by the simultaneous rise in temperature, without the present arid region profiting to any great degree. This suggests that the climate in the extratropical arid zones did not alter in comparison with present conditions in contrast to the marked changes in the present forest climate zone. The seeming contradiction is certainly a consequence of the fact that on methodological grounds the climate of the dry zones cannot be reconstructed with the exactitude possible for the climate of the cool, temperate forest belt.

At the present time there is useful information from many other regions of the earth about interglacial sediments, soil formation, flora, and fauna. The amount is still too small, however, to permit a comparable paleoclimatic analysis. The best area to study is still North America, although there remains much that is puzzling.

As in Europe and northern Asia, the climate in Alaska and northern Canada during the younger Pleistocene warm periods appears to have been appreciably warmer than it is now. In this connection observations recently made in western Alaska and Banks Island in the Canadian North West Territories are of greatest interest.

On the Seward Peninsula in western Alaska there lies between the sediments of the older Iron Creek glaciation and the younger Nome River glaciation organic sediments that contain a fossil woodland flora. The Nome River glaciation may correspond to the Saalian cold period, and probably the Iron Creek glaciation is

FIGURE 79. The position of the boundaries of the forest belt of northern Eurasia during the Holsteinian warm period (A), the Eemian warm period (B), and the postglacial warm period (C). 1. coastline; 2. northern forest limit; 3. southern forest limit. Details [260].

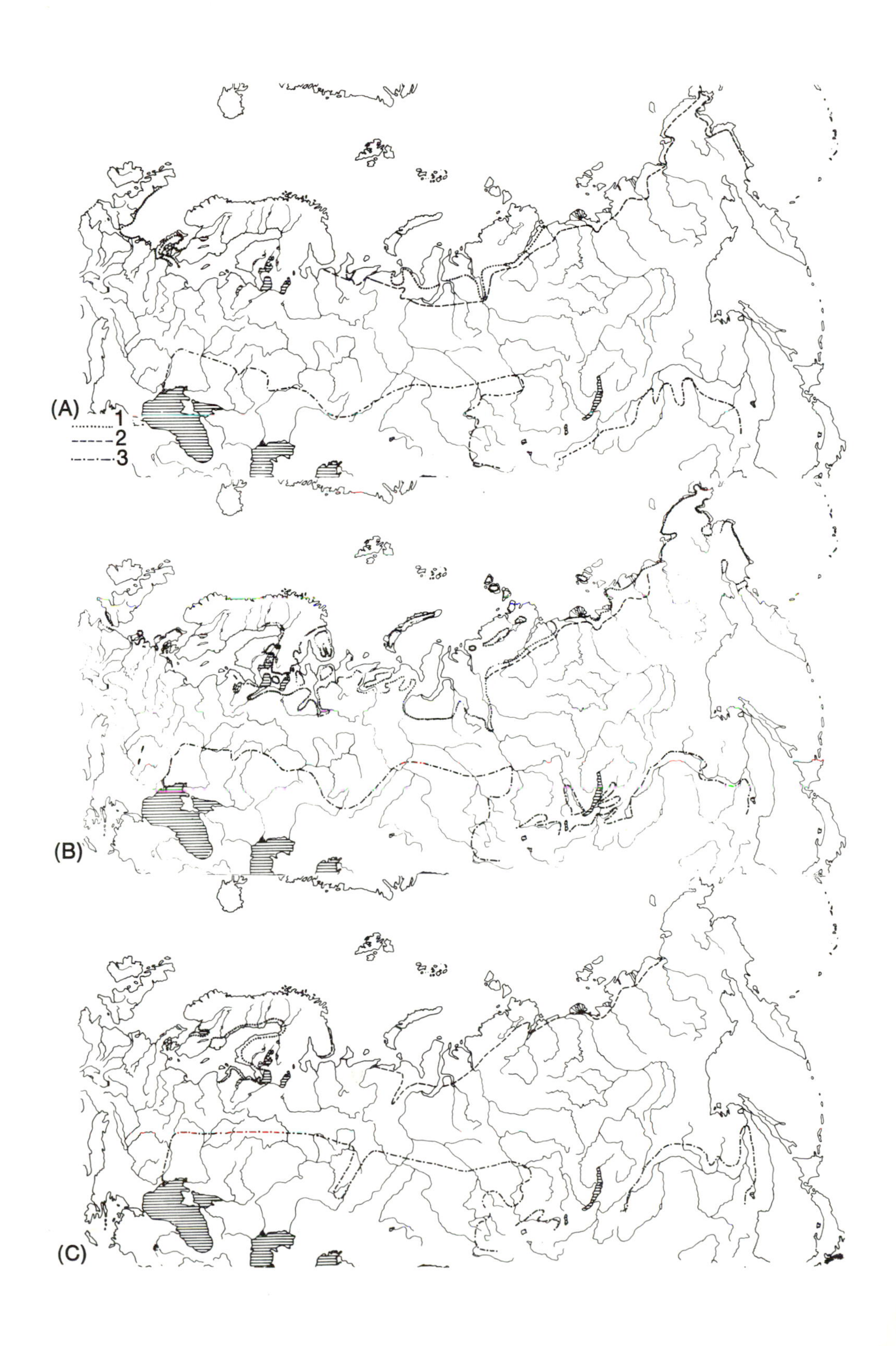
(A)
1
2
3
(B)
(C)

of the same age as the Elsterian cold period. The intervening warm-period woodland flora, which should fall into the Holsteinian warm period if the correlation is valid, parallels to a remarkable extent that now found in southeastern Alaska [586]. It would be difficult to conclude from this that the January temperature was milder; however, the July temperature was probably 4° to 5°C higher. A similar climatic improvement can also be deduced from the data of the same authors on the fossil flora and fauna also found in the sediments on Seward Peninsula but lying between the older Nome River glaciation (Saalian cold period?) and the younger Salmon Lake glaciation (the earlier part of the Weichselian cold period?). They must correspond to the Eemian warm period. Independent of whether the dating is correct, the observations show that during the younger Pleistocene warm periods woodlands advanced appreciably further north and northwest than at the present day. Something comparable at some unknown time in the younger Pleistocene occurred on Banks Island in northwest Canada. The careful pollen and macrofossil analysis of Terasmäe shows that there flourished at 72°N 120°W a former woodland flora consisting of spruce, pine, birch, and alder which may best be described as a woodland tundra. This tree limit lies some 200 miles north of its present position. The conditions of deposition show that it represents not the postglacial warm period but a much earlier one. The July temperature at that time appeared to be 4° to 5° warmer than at present. It is not possible to date exactly the age of this advance of the forest over the tundra region; yet on the basis of the observations on the Seward Peninsula and Banks Island, it may be suggested that this phase of favorable climate corresponds to one or even both of the last warm periods.

Unfortunately, the dating of organic sediments found in the western part of Washington State on Puget Sound—investigated by Leopold and Crandell [461] —is likewise uncertain. Here four important moraine horizons occur, separated from one another by peat and clay. The formation belonging to the second warm period, the Puyallup time, on the basis of its pollen flora permits the recognition of a change in the prevailing climate from cold, through cold but moister, passing into a cool, moist phase. Finally, a climate comparable to the present set in, which changed once more to cool and moist. It is different from the climatic changes at the beginning of the postglacial in that the temperature appears to have slowly and steadily increased in the warm period described, associated always with a higher precipitation. The stratigraphical relationships of the flora investigated led to the view that this represents the last or next to last warm period. The trend of climatic development has much in common with that of the European Holsteinian warm period but not with the Eemian (Sangamon) warm period of Europe and America, so that conceivably this formation may represent the last warm period but one. In contrast to the results obtained in Alaska and northern Canada, it in no way points to a climate in western Washington warmer than the present day, but the precipitation was apparently higher. Comparable observations for the previously considered tundra area do not exist. This contrast, however, does not signify that basically different climatic events are being considered, for in one case (Alaska and Canada) the presently available material permits only data concerning thermal conditions, while in the other, the data is primarily hydrological.

A better insight into the course of climatic fluctuation during the last warm period than the examples already described is furnished by certain observations made in the eastern and central United States and southeastern Canada.

As noted, the last warm period in North America is referred to as the Sangamon. This corresponds to the Eemian (Mikulino-Riss-Würm) warm period of Europe. At that time a marine transgression occurred along all coasts, and in eastern North America the eight-meter Pamlico Terrace was produced. Upon it rests the so-called Gardiner's Clay of Long Island. Its pollen flora indicates that during the climatic optimum of the last warm period a woodland developed chiefly composed of oak, beech, and hickory (*Carya*). This woodland very closely resembles the natural forest of the same region [166]. Perhaps the higher percentage of oak (50 percent of the arboreal pollens) leaves open the suggestion of a slightly more favorable climate. This would also be consistent with the occurrence of swamp cypress (*Taxodium distichum*) in a fossil peat near Washington. The latter thermophilous species reaches its northern limit there today. During the Sangamon warm period, however, it occurred particularly abundantly, so that Knox [419] was probably correct when he considered that the climate was warmer. Pollen analyses carried out in southeast Indiana in the vicinity of Richmond give many indications of the course of former climatic fluctuations. At first, at the beginning of the last warm period, pine and spruce woods predominated there, in which the proportion of spruce declined continuously. The climate appears to have been cold and moist (Phase I) [see 392]. Later, under a cold but dry climate, the pine became exclusively the dominant tree (Phase II). Finally, with rising temperature it had to retreat, and the hard woods and bushes migrated into the area (Phase III). It is noticeable that this development and the climatic changes deduced are strikingly similar to the corresponding processes in Europe but bear little resemblance to Puyallup time on Puget Sound. It seems particularly significant that after an initial moist phase a dry climate phase followed, which passed into a general warming trend. Finally, true deciduous trees reached their maximum distribution. The percentage of oak and hickory (*Carya*) pollen rose to 40 percent each, and in addition maple (*Acer*), beech (*Fagus*), ash (*Fraxinus*), walnut (*Juglans*), sweet-gum (*Liquidambar*), tulip tree (*Liriodendron*), hornbeam or hop hornbeam (*Carpinus* or *Ostrya*), and elm occurred; hemlock fir (*Tsuga*) and larch (*Larix*) appeared immediately afterward. The climate is said to have been warm and dry. Today, however, the limits of *Liquidambar* and *Tsuga* pass through the region, and the distribution limits of *Fagus, Liriodendron,* and *Carya tomentosa* are close. It can be inferred from this that the former climate not only was comparable with today's but was even warmer and moister. It seems improbable from the observations made that only the former maximum distribution limits of the tree species cited were determined. The remains of the following forest phase, the climate of which was again colder, were generally destroyed by erosion in the profiles under investigation. Finally, spruce and fir again became established as the climate became cold and moist (Phase IV). This is probably the transition to the last cold period.

The highly favorable climate of this warm period can be still more clearly proved by paleobotanical research near Toronto [737]. Here, at the optimum development of vegetation during the Sangamon warm period, a woodland flora was widespread which today is only found in any significant amount much farther south and east. A comparison of the macro- and microflora in the interglacial Don beds with flora of the present environment yields the results in Table 14.

TABLE 14. DIFFERENCES IN CERTAIN CLIMATIC FACTORS AT THE TIME OF FORMATION OF THE DON BEDS, FROM THE LAST WARM PERIOD (SANGAMON WARM PERIOD) NEAR TORONTO COMPARED WITH THEIR PRESENT-DAY VALUES IN THE REGION ON THE BASIS OF THE PRESENT AND FORMER EXTENT OF CERTAIN PLANT SPECIES (FIGURES IN THE TABLE INDICATE THE MINIMUM DEVIATION FROM PRESENT VALUES)

PLANT SPECIES	MEAN TEMPERATURE			ANNUAL PRECIPITATION (MM.)
	JANUARY °C	JULY °C	ANNUAL °C	
Liquidambar sp. (sweet-gum)	+3 to 4	+2	+3 to 4	+200 ?
Chamaecyparis thyoides (arborvitae)	=	=	+1?	+200
Gleditsia donensis (honey locust)	?	?	?	+400 ?
Fraxinus quadrangulata (ash)	+1	+2	+2	+250
Ilex sp. (holly)	+3 to 4	+2	+3 to 4	+250

A series of other plant species could be used as representative of warmer climates, in particular *Maclura pomifera* and *Asimina triloba.* Their present distribution is patchy, however, and restricted to the south. At the present time these species have not been able to occupy all the area climatically open to them, for because of the severity of the last cold period they were expelled far to the south. Subsequently, either migration routes were closed or because of the severe competition of other species, the species have not been able to advance. Consequently, they must be excluded from any determination of the former climate around Toronto. Nevertheless, the data presented are sufficient to show that during the last warm period, in those regions of North America examined, as in Europe and northern Asia, not only was the temperature appreciably higher than at present, but the precipitation increased. Thus in general the climate over the Northern Hemisphere during the last warm period appears to have been warmer and moister than today. The opposite view, presented by Rosendahl [634]—that at least during certain parts of the Pleistocene warm periods of North America an overall boreal climate prevailed—must have had its origin in Rosendahl's examination of floras predominantly from the close of the individual warm periods.

The scale of the relative differences in comparison with present climate appears to increase toward the centers of the continental masses. This is directly related to new ecological niches and migration routes opened to the more exacting deciduous vegetation even in southern Siberia, far outside their present distribution limits. These possibilities were rapidly utilized and individual thermophilous elements, thickets of oak, elm, and hazelnut from eastern Asia, and possibly from the southern Urals, advanced westward or eastward respectively.

This process must have been aided by the retreat northward of the southern limit of perpetually frozen ground (Fig. 21). A comparable climatic improvement was also brought about during the preceding Holsteinian warm period. The essential difference between the two warm periods lies in the fact that the climatic optimum of the last warm period occurred only after a period of probably very warm and dry climate, whereas during the Holsteinian warm period this initial dry period appears to be generally absent.

We are at present so ill-informed about climatic changes during the warm periods in other parts of the earth that no purpose is served by a detailed investigation of them. Certain data are to be found in discussions of problems of the possible course of climatic alterations during the Ice Age, as also in the discussion of the relation between cold periods and pluvials.

Saalian and Weichselian Cold Periods

If the Pleistocene warm periods represent the crests of a great fluctuating climatic development, the intervening cold periods form the troughs. The climatic conditions during the first of these cold periods have already been discussed. Since presently available material is not sufficient to investigate the climatic history of all cold periods of the Pleistocene without leaving gaps, in what follows our aim will be to concentrate on the last two, so that consideration may proceed from the better known last (Weichselian, Wurmian, Valdai, Zyrjanka, Wisconsin) cold period to the less examined last but one (Saalian, Riss, Dneprovsk, Samarov, Illinoian). This is justified because interesting results may be hoped for from an analysis of their climatic history: these two periods serve as examples of phases of *restricted* (Weichselian) and *intense* (Saalian) continental glaciation.

1. General. The glacial advances during the Saalian cold period were the maximum advances in many places on earth. Still greater advances are known from only a few localities, and they occurred during the preceding Elsterian cold period. From this it would seem that climatic conditions during the Saalian cold period were particularly conducive to strong continental glaciation. In comparison the spread of the ice sheets during the Weichselian cold period was relatively small. At that time and for the last time however, the continents in the present cool, temperate latitudes were under the influence of an extreme cold, arid climate. Thick loess deposits blanketed the old topography, and because of the unfavorable climate the plant and animal realms in the northern and southern temperate latitudes suffered critical losses.

The paleoclimatic problems associated with these significant climatic fluctuations are of differing nature. Consequently, the climatic conditions at the peak of the two cold periods will be considered next and this will provide a vivid contrast to the climates of the warm periods that have been investigated to this point. In Chapter V the course of the development of the climate during the last and next to last cold periods will be studied.

As was noticed in the Introduction, the magnitude of the former glaciations is ill-suited for paleoclimatic analysis. Much more can be estimated from the distribution of former permafrost as well as loess sedimentation. It is readily appreciated that the ground frost can work intensively only where the soil is amply supplied with water. In the dry regions of the earth traces of ground frost are rare, even where the winter climates are extremely cold. This makes it appear likely that the climatic variations of the cold period affect dry regions less than those regions which during warm periods had a comparably moist climate. With increasing soil dryness the apparent fall in temperature appears to be progressively less. From a study of the literature it would seem that this false conclusion has been drawn repeatedly. In addition, in the arid regions biological aids fail, so that paleoclimatic reconstruction of these areas is very difficult or even impossible.

Difficulties of another kind concern the regions of present woodland climate in temperate latitudes at those points where mountains are widely distributed rather than lowlands. In lowland regions during the cold periods, the unfavorable climate took full effect. Rarely were local situations to be found, such as steep-sided river valleys, south-facing slopes of small hills and similar situations in which a more agreeable climate may have prevailed. In contrast, the strong orographic relief of mountainous areas favored the development of various and varied local climates. As the vegetation in general was very dispersed, local differences in exposure of slopes, the form of valleys, thermal conductivity of soil, their color and many other factors must have played a decisive role in the creation and modification of very different microclimates comparable to those in high mountains above the tree limit or in clearings at lower elevations. This varied division of mountainous regions into qualitatively very different, narrowly defined microclimates must have favored tree growth in those latitudes bordering on extensive lowland tundra or cold steppe. The degree and scale of forest growth in a mountainous region can be determined, paleontologically or paleobotanically, only imperfectly. For here individual stands of timber, but not general trends in the distribution of the most important types of vegetation, can be investigated. It is understandable that preference in research is given to sediments in which a high fossil content can be expected. This is most usually the case where, in former times, particularly favorable local climatic conditions existed, so that in later reconstructions the mountain regions appear more densely forested than was actually the case. If now on the basis of these determinations the tree limit is reconstructed in the mountain regions and in the lowlands and then the attempt is made to estimate mean July temperatures, errors creep in whose magnitude is unknown. For in the one case (lowlands) the calculation is based upon the unfavorable gross climate, and in the other (mountains) on a favorable microclimate. The same difficulty, moreover, also affects reconstructions of the climatic conditions of warm periods, particularly when the position of the southern limit of permafrost is taken as a basis for the climatic reconstruction, for here local and particularly unfavorable conditions may permit permafrost to remain in a mountainous region long after the soil of the surrounding lowlands has fully thawed out (Fig. 21).

In regard to an estimate of past climate it must be recorded that the most continental regions of northern Eurasia, namely central and eastern Siberia, are almost exclusively occupied by uplands, for the limitations of each of the paleoclimatic methods are soon reached, and more detailed analysis of this climatically interesting region is therefore extraordinarily difficult.

2. ***The mean annual temperature.*** With regard to the Weichselian cold period, the maximum, as far as is known, fell between about 22,000 to 18,000 years B.P. In the following paragraphs the attempt will be made to outline the most important features of the climate at that time.

As has already been noted above (pp. 31 ff.), the former position of the southern limit of permafrost provides a particularly good starting point for the reconstruction of mean annual temperature, provided that too high an accuracy is not demanded. Several times in the last few years attempts have been made to determine the position of this important boundary at the peak of the last cold period in Europe, and to a lesser extent in northern Asia. In this connection mention should be made of the maps of Poser [600, 601] and Kaiser [391] for central and western Europe, and of Tumel' [756] for northern Asia. Subsequently, such investigations spread to eastern Europe [771], to the western Siberian lowlands [820], as well as to northern Eurasia [256] and North America [98] in general. Pingos (hydrolaccoliths, bulgunnjachi, bugry vspuchenija, Fig. 17) are a particularly characteristic form of permafrost regions. At the present time only in the rarest instances do they occur outside the permafrost zone. The present distribution of pingos in northern Eurasia, the present southern limit of the permafrost zone (southern limit of islands of permafrost within a predominantly thawed soil), and the distribution of fossil pingos probably belonging to the last cold period are shown in Figure 17. If it can be assumed that the present southern limit of permafrost approximately coincides with the −2°C annual isotherm, then the following figures at the peak of the last cold period can be deduced (Table 15).

TABLE 15. MINIMUM VALUE FOR THE FALL IN ANNUAL TEMPERATURE AT THE PEAK OF THE WEICHSELIAN COLD PERIOD, CALCULATED FROM THE FORMER DISTRIBUTION OF PINGOS

Wales	−12°C	Central Hungary	−13°C
East Anglia	−12°C	North Ukraine	−9°C
Paris basin	−12°C	Middle Volga	−8°C

Earlier, Kaiser [391] estimated a fall in the mean annual temperature in Europe during the last cold period of 15° to 16°C, making use of the distribution of frozen earth forms which do not require such extreme conditions as pingos but which are generally related to the permafrost zone. The assumption of Kessler [399] that the mean annual temperature during the last cold period was probably appreciably below −2°C is thus fully confirmed. The latest determined values agree very well, particularly so when the different starting points are considered. They can scarcely be reconciled, however, with the values for France calculated by Poser [600]. According to this, the mean annual temperature was between +6° and +7°C—that is, very little below present values and about 9°C above the values estimated by Kaiser and the author. The difference can be traced to differences in the premises upon which the calculations rest. Poser made the physically untenable assumption that then as now in France the mean January temperature is about 1/5 to 1/10 of the mean July temperature measured in degrees centigrade. Based upon material then available, Poser assumed that the mean July temperature in the Seine basin during the last cold period was about +10°C; on the grounds of this assumption he necessarily obtained the positive January temperature and the above-mentioned mean annual temperature. As the figures are based upon incorrect premises, their divergence from the much lower values obtained above need not be discussed further.

Insofar as is known at present, the southern limit of permafrost during the last cold period ran approximately through the Kazak Mountains to the south of the western Siberian lowlands. If in this case, too, the same assumption can be made—that is, the southern limit of the permafrost zone coincides with the −2°C annual isotherm—it follows that there was a decline from present-day values of at least 3°C. Probably this figure is misleading, for because of the great aridity no strong frost phenomena developed. Furthermore, it appears that frost fissures and desiccation cracks are easily confused [437]; hence many of what are taken for desiccation cracks may actually have been caused by frost action.

A somewhat greater depression of the mean annual temperature at the peak of the last cold period in the neighborhood of Irkutsk can be deduced from the data of Golubeva and Ravskij [285]. According to this during summer on the lower Belaya only the upper 150 centimeters of soil thawed. The same is seen today along the west-to-east stretch of the lower Tunguska, so that, with some reservations, an approximately similar temperature can be assumed—that is, a mean annual temperature of −8°C. It must therefore have been at least 5° to 6°C below present values. For a somewhat younger segment of the last cold period Golubeva and Ravskij [285] gave a depression of the mean annual temperature of 6° to 8°C on the lower reaches of the Lower Tunguska, and a 9° to 10°C depression on the middle reaches of the Lena near Yakutsk and from 10° to 12°C on the Angara, in comparison with present temperatures.

There is no reliable source of data from the Far East on the former distribution of permafrost. In place of this the approximate northern limit of thermophilous deciduous trees in the coastal provinces around the Japanese Sea for the last cold period is known [565, 257]. Today these trees have advanced just north of the mouth of the Amur River. During the last and next to last cold periods the most northerly outposts lay south of Lake Chanka—that is, west or northwest of Vladivostok [260]. This position suggests an 8° to 9°C lower temperature for January, so that the mean annual temperature must have been at least 4° to 5°C below present-day levels, and probably lower still.

In North America the position of the southern limit of permafrost during the last cold period has long been debated. Recently, Brunnschweiler has assembled observations and data [98] which shows, despite the views of certain authors, an appreciable southward displacement of the zone of perpetual freezing (Fig. 80). If the map represents actual conditions to some extent, then the mean annual temperatures in northern Tennessee and the bounding regions to the west must have been about 13° to 15°C lower than today. The correctness of this figure, which agrees approximately with those estimated for Europe, can be confirmed paleobotanically. In contrast to conditions in Europe, the advancing ice of the last cold period in North America overrode forests at many places, the remains of which, as so-called "Forest Beds," are often found in excavations. These re-

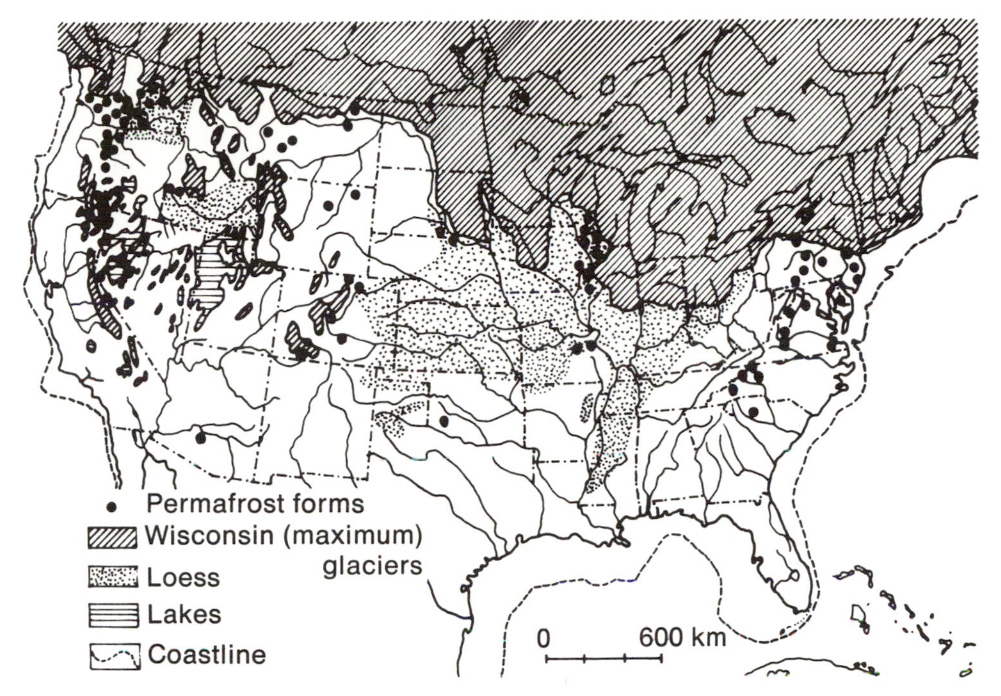

FIGURE 80. The distribution of periglacial phenomena in North America south of the Wisconsin Ice Front. Adapted from Brunnschweiler [98].

markable beds in western Ohio belong to three age groups, namely greater than 32,000 years, from 27,500 to 16,600 years, and from 14,500 to 8,500 years [282]. Within these age groups, samples belonging to practically every thousand-year period are known. These former forests were still living at the time of the approach and overriding of the ice. It is today possible to establish that the annual rings became thinner year by year until the ice reached the locality and flattened the still living trunks [104, 517]. As such buried forests are quite common in North America, it has been supposed that the ice front was girdled by a coniferous forest belt which rapidly passed over into the deciduous zone [see, *i.a.*, 104]. This seemed to be confirmed by the late glacial migration history of forest vegetation, for it appears that the forests immediately followed the retreating ice. It must be admitted that this impression is certainly in part due to the fact that until a few years ago no attention was given to the pollen of shrubs and plants in pollen analysis [570], so that the degree of former afforestation is largely unknown.

The idea of the ice front advancing into a coniferous forest zone is, however, scarcely compatible with distribution of loess of the same age. The loess thickness and distribution suggest a region of wide steppes with tree growth along the river courses [459]. In fact, the above-noted forest beds occur in isolation, to the extent that Martin [517] supposed them to represent a wooded tundra-like vegetation such as is found at present at Knob Lake, Quebec, at 54°N. "To me it indicates that the advancing Wisconsin sheet swept through a zone of scattered trees, relatively dense along drainage ways, but perhaps enclosing patches of tundra on hill tops" (Fig. 81). Goldthwait [283] calculated on the basis of these data that

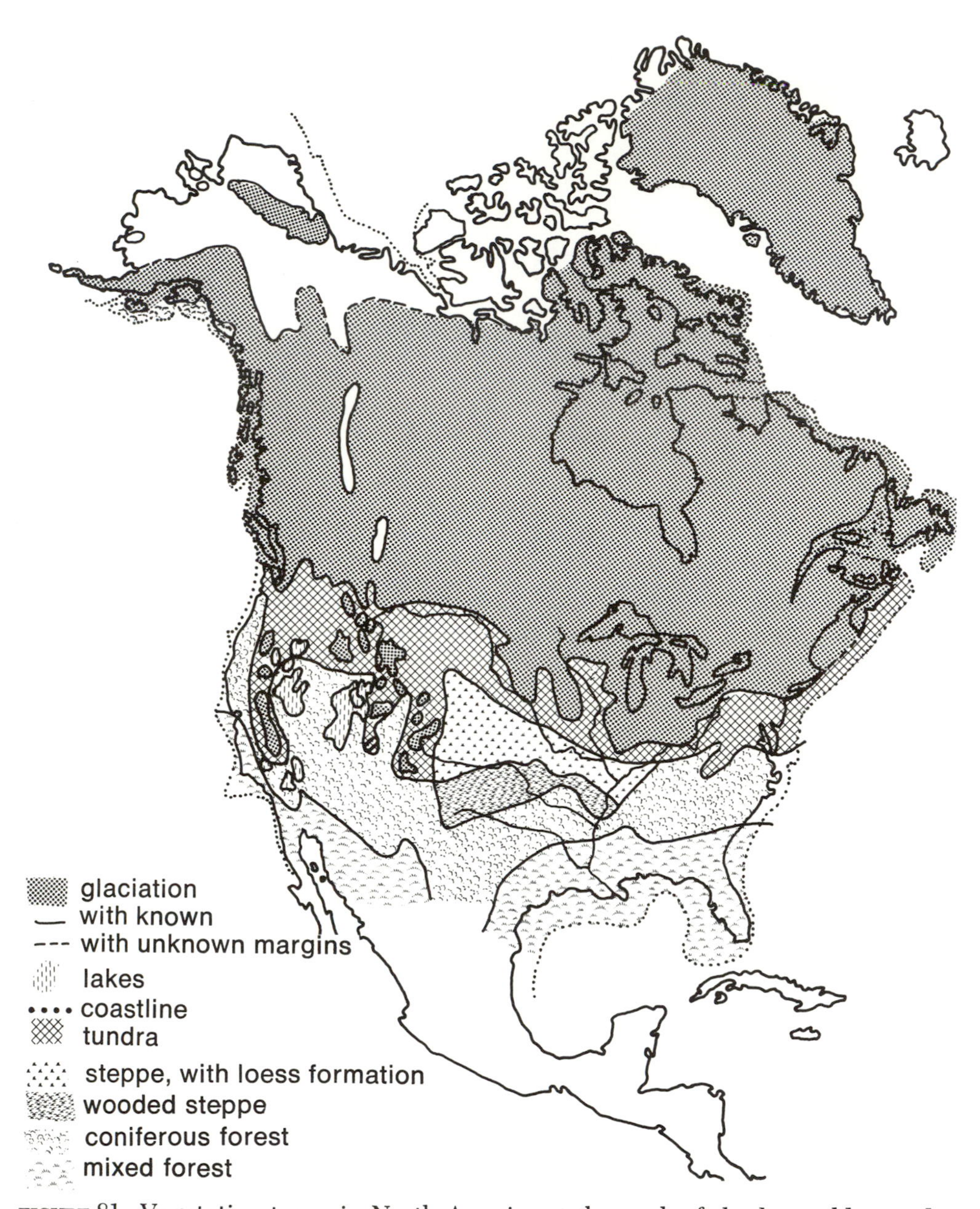

FIGURE 81. Vegetation types in North America at the peak of the last cold period. After Brunnschweiler [98] and others; cf. text.

the mean annual temperature in western Ohio was some 15°C and the mean January temperature about 20°C below present levels. These figures correspond very well with those calculated from the former spread of permafrost and are identical with the European. Comparable values result from Ogden's pollen analysis research on Marthas Vineyard, Massachusetts [570]. Here, at the time of the late glacial Buzzard's Bay advance, there was tundra in a region now occupied by thermophilous plants of a deciduous woodland. From the occurrence of distinctive species of the present tundra—as, for example, thrift (*Armeria siberica*)

—we may postulate a reduction in temperature at that time of at least 10°C for January, July, and the annual mean, although the area lies directly on the ocean.

The thermal conditions of the tropics during the peak of the last cold period are of especial interest. Flohn [236] assumed that because the temperature there was only 4°C lower, its effects on the biosphere could be nearly ignored.

In East Africa the tree line on Ruwenzori lay at least 1,000 to 1,100 meters lower than today; on the western rift slopes on the Cherangani hills (1°N, 35° 28′E) even during the cold period it was 500 to 600 meters below its present position, and at the peak of the last postglacial it must have been depressed much further [140, 824]. The reafforestation of Ruwenzori at 2,200 meters began at approximately 10,600 B.P.—that is, about the beginning of the postglacial in Europe. At the present time the tree limit is at 3,350 meters. Coetzee [140] calculated a fall in mean annual temperature of 8° from the depression of the tree line. Possibly an even greater value must be assigned, for the winds rising up the high mountain must have transported arboreal pollen in much greater amounts than would have been the case had the tree limit at the peak of the last cold period, or even at the beginning of the glacial retreat, been only 100 or 200 meters below the 2,440 meters elevation of the sampling site. Instead of this, the pollen flora of the oldest lacustrine sediments (15,862 ± 185 B.P. [140]) reflects only the pollen rain of the ericaceous zone, which currently lies above the tree line. Similar figures for the depression of the mean annual temperature in the Colombian Andes are obtained from the observations of van der Hammen and Gonzalez [332, 333]. Here the Páramo vegetation with the characteristic *Acaena* community was found at locations where during the last warm period and at present a forest predominated. The temperatures determined by van der Hammen and Gonzalez [332] for the Sabana de Bogotá and for the *Acaena* zone of the Páramo show that the mean annual temperature was at least 6° to 7°C below the present level. Temperature determinations of the surface waters of the Caribbean by the oxygen isotope method give the same amounts [197, 199]. At the peak of the last cold period the water temperature dropped 7°C below the present figure (21°C as against 28°C today); during earlier cold phases of the last cold period the differences in temperature of surface water amounted to about 5° to 6°C with respect to recent conditions. The far-reaching effects of these climatic changes on the vegetation even in present wet tropical lowlands is illustrated by the expansion of a savanna, poor in species, to the coast of British Guiana, where a tropical rain forest now prevails at the height of the last cold period. Van der Hammen [330] compared the savanna with florally related types which today are found only at high altitudes in Central America. If the comparison is valid, it clearly illustrates that in South America at the edge of the wet tropics, not only did the duration of the dry period increase, but simultaneously the temperature decreased appreciably. This cooling also left its traces in Africa, in the mountains of East Africa already mentioned, and also in Angola and Rhodesia. Formerly the mountain rain forests that today are patchily distributed were linked to one another; they were thus at considerably lower elevations, and mountain forests now only found in mountain climates with regular mists migrated during the interval from about 30,500 to 10,000 B.P.—that is, during the maximum of the last cold period [822, 823, 827].

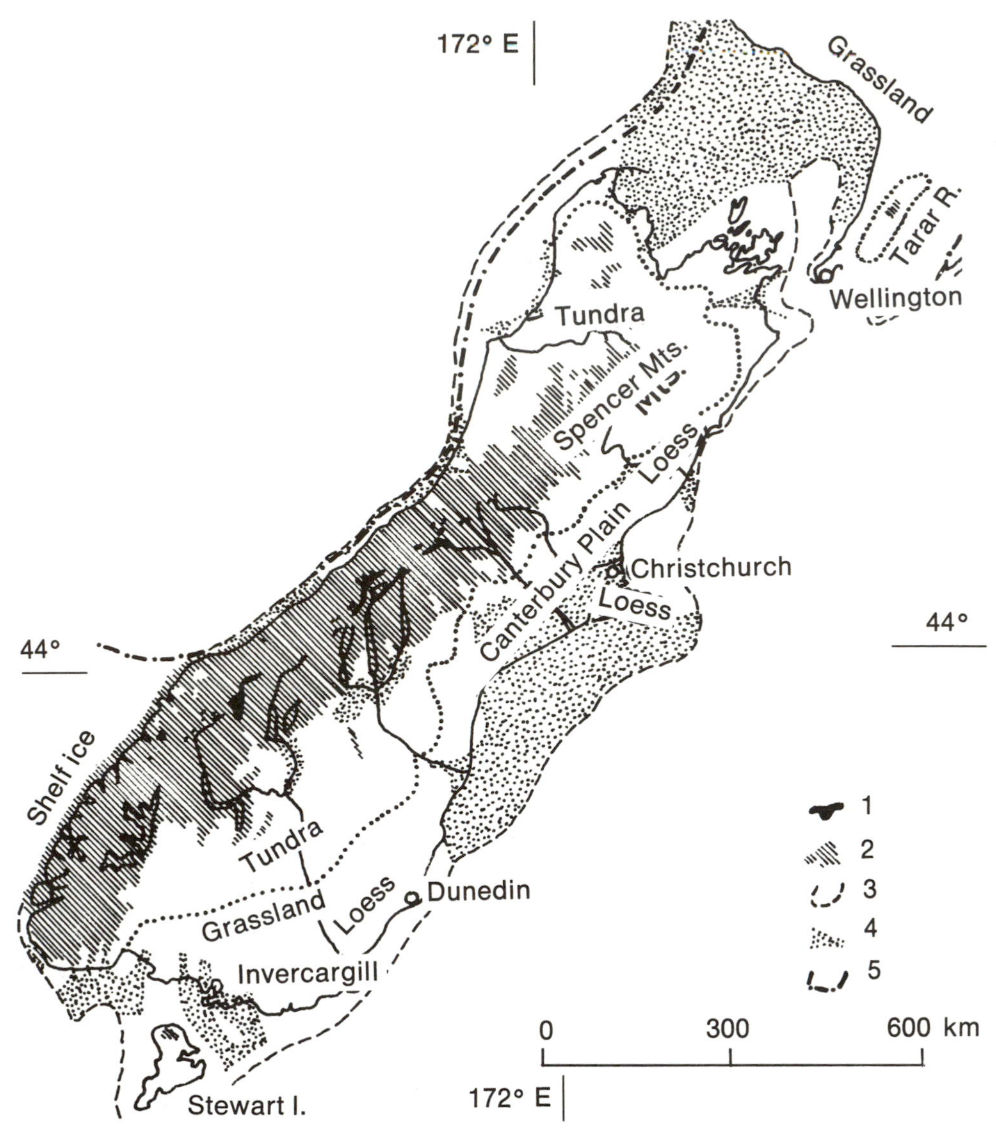

FIGURE 82. The physiographic regions of South Island, New Zealand, during the last cold period: 1. present glaciation; 2. glaciation at the peak of the last cold period; 3. coastline; 4. zone of deposition; 5. southern tree limit. From Woldstedt [806].

In New Zealand, Cranwell and von Post [145] have observed that a steppelike vegetation occurred in place of the present *Nothofagus* woodland in the southern half of South Island at the end of the last cold period. The most important plants were grasses (*Gramineae*) and sedges (*Cyperaceae*). Immediately after the beginning of the glacial retreat, the climate there was so dry that salt effloresced, and loess dust accumulated. Later investigations have confirmed these conclusions, so that South Island, New Zealand, at the peak of the last cold period should be represented as ice-covered in the west with a tundra or steppelike floral cover in the east. Only in the southwest part of North Island was woodland able to persist ([806] and Fig. 82). The climatic conditions at the present polar tree limit in the Southern Hemisphere have received considerably less attention than in the Northern. However, at the present time the position of the polar tree limit in the Southern Hemisphere is principally determined by strong winds and the generally marked oceanic climates. This raises difficulties when the displacements of the woodland zone determined for the last cold period are used to evaluate temperature changes. The former wide distribution of steppe on South Island shows that the influence of the oceanic climate upon the former position of the tree limit should be disregarded. If it can now be assumed that the polar tree limit during the last cold period approximately coincides with the 10° isotherm of the warmest month as in oceanic regions of Eurasia, then the mean January temperature of the time must have been at least 8°C colder than today. It is, of course, highly uncertain whether the assumption made it permissible. The temperature today at the upper tree limit in the mountains of southern New Zealand in forests formed of cold resistant *Nothofagus* species should not be much higher, so that the figure of 8°C can be taken, in the absence of anything better, as a preliminary estimate of the temperature fall of the warmest month. It is at least in good agreement with estimates made in tropical highlands. The development of loess and the occurrence of salt efflorescences clearly demonstrate the appreciable increase in continentality of South Island, New Zealand. As a result, the seasonal temperature fluctuations ought to have been greater than at present. With the mean temperature of the warmest month already about 8°C below the present level, the mean value of the annual temperature must have fallen at least 8°C and probably as much as 10° to 12°C.

The present exposition must remain incomplete in many aspects, for despite the countless observations of traces of the Ice Age in various corners of the globe, the number of reliable results from which the magnitude of the mean annual temperature can be estimated for the various cold periods is still very small. Yet in the author's opinion, it is clear that during the last cold period, the mean annual temperatures were appreciably lower over the whole earth. This applies equally to the former glaciated regions in the north, the northern edge of the Mediterranean, the present wet tropics, and the upland regions of the tropical mountains, and obviously is found in the Southern Hemisphere as well. It would appear that the greatest declines in temperature were in areas adjacent to the oceans, so that the luxuriant vegetation and faunas of the warm periods were catastrophically affected by the cooling. How strong the cooling was in the continental regions is not actually known, nor can it be assumed with any assurance from the presently available material.

Even less is known about the value of the mean annual temperature during the maximum ice advance of the Saalian cold period than about the Weichselian. Figure 21 shows the probable position of the southern limit of permafrost in northern Eurasia. Although on the map the total evidence available today was used, it remains uncertain whether the actual course of this important boundary on the map is correctly sited. If it is correct, the mean annual temperature during the Saalian cold period cannot have been much below that during the Weichselian cold period. At first, this does not seem likely, for the much stronger glaciation and the sometimes much lower elevation of the climatic snowline

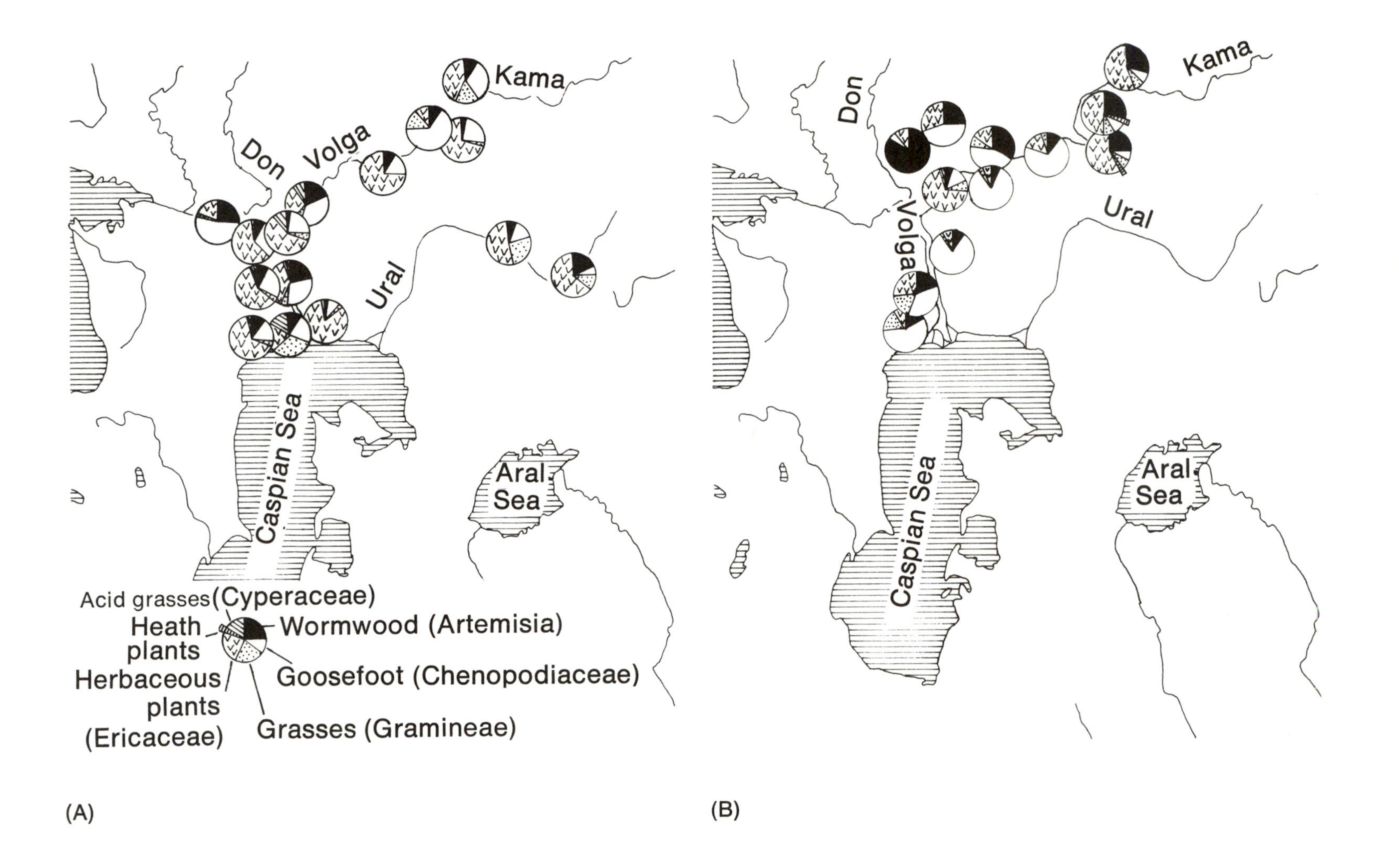
Kama
Don
Volga
Ural
Caspian Sea
Aral Sea
Acid grasses (Cyperaceae)
Heath plants
Wormwood (Artemisia)
Herbaceous plants
Goosefoot (Chenopodiaceae)
(Ericaceae)
Grasses (Gramineae)
(A)
Kama
Don
Ural
Volga
Caspian Sea
Aral Sea
(B)

FIGURE 83. Types of steppe and semidesert vegetation on the lower Volga during the Saale-Dneprovsk cold period (A), the Weichsel-Valdai cold period (B), and the present (C). According to the sector diagrams, the nonarboreal patterns of the three phases distinguish three distinctive kinds of open vegetation. This is particularly true with respect to the proportion of *Chenopodaceae* (white) and the various herbaceous plants (v symbol). Where samples were obtained from within the area of the cold-period transgressions of the Caspian, the pollen flora are a little older or younger, or are derived from the transgressive sediments [260].

observed at many localities leads one to suppose that the mean annual temperature had sunk much below the levels of the Weichselian cold period. Yet there is no indication of this in the available surface water temperatures from tropical seas [197, 199, 206, 207, 192, 193, 659]. Rather the results of oxygen isotope analyses show for the last cold period but one overwhelmingly the same values as for the last one (Fig. 30). In contrast again the pollen analysis research in the Andes near Bogotá appears to refer to an appreciably lower temperature, for the upper limit of forest occurred at markedly lower altitudes than during the last cold period. The elevation of the tree line in mountains, however, depends not only upon temperature but also upon the amount of precipitation and the available radiation, which is related to the degree of cloud cover. More important is the fact that the composition of the tree species of the Colombian forests changed appreciably since the next to last cold period. Only then was there a slow immigration of oak (Fig. 31). Yet in the Eemian warm period and during the last cold period it completely replaced the older woodland. This was certainly influenced by the elevation of the tree line, for the physiological requirements of the individual tree species are different. In just such a case, the varied elevations of the tree line during the two cold periods have no great significance as a possible indicator to measure differences in climate.

It is possible that observations made on the lower Volga may help to answer the question raised. There the pollen flora of the present steppe is composed primarily of a high proportion of *Chenopodiaceae* in which many grasses (*Gramineae*) and various herbaceous plants as well as wormwood (*Artemisia*) occur (Fig. 83). During the last cold period the significance of *Chenopodiaceae* in the same region increased to an extraordinary extent, so that their relative proportion in the total flora was about as great as it is today in the most southerly and driest part of the Caspian depression. Yet the open vegetation of the next to last cold period differed considerably from this. At that time other herbaceous plants predominated; *Chenopodiaceae* and *Gramineae* were also weakly represented, although they produce large amounts of pollen. The *Chenopodiaceae* flourish today in dry areas and prefer saline locations. It may be supposed from this that the climate of the lower Volga during the last cold period was particularly dry, but in spite of this, the approximately equal transgressions of the Caspian Sea during the next to last (U. Chazar transgression) and the last (Lr. Chvalyn transgression, Fig. 94) cold periods show that the hydrological conditions during the next to last cold period could not have been better than during the last, so that the different distributions of *Chenopodiaceae* during the two cold periods cannot have been caused in that way. It is noticeable, however, that *Chenopodiaceae* are almost completely lacking in the present tundra ([595, 393, 355]; flora of the Murmansk district III, 1956). Even in the vicinity of the new settlements in the tundra where *Chenopodiaceae* occur and many other weeds are unintentionally sown, *Chenopodiaceae* are unable to establish themselves [651, 173]. It thus seems that the distribution of these plants in the far north is limited by thermal conditions. As the water budget of the various localities on the lower Volga during the last two cold periods was probably equally bad, the difference in the distribution of the *Chenopodiaceae* can scarcely have been caused by hydrological conditions. Thus it is reasonable to assume lower temperatures during the next to last cold period than during the last cold period. In this connection it must be noted that in the northern Caspian depression significantly more plants of the present tundra or alpine plant communities are found in the sediments of the next to last cold period than in the last. The magnitude of the difference cannot be evaluated in this way. Nevertheless, the difference cannot have been great. This was already indicated by the temperature determinations using the oxygen isotope method, and the same is shown by pollen and diatom analyses of deep-sea borings from the Sea of Okhotsk [835, 837, 838]. On numerous occasions near the top of the sea-floor sediments of two appreciably warm periods and two distinct cold periods have been found. From their stratigraphical position these can only represent formations of the last two cold periods and the Holsteinian and Eemian warm periods. The pollen flora in a marine sediment remote from the coast gives a good representation of the vegetation that borders the sea over a wide region. The air and ocean current pollen contents are mixed, so that local peculiarities of the flora from individual regions are to a large extent eliminated, as well as errors induced by wind or current transportation. The pollen flora from the above-mentioned borings in the center of the Sea of Okhotsk, however, indicate no notable difference between the sediments of the last cold period and the one preceding it. The same result is obtained from the diatom flora of the sediments. In consequence, the coastal vegetation in the area of the Sea of Okhotsk during the last two cold periods must have been very similar, and it may be assumed that the climates there during the two cold periods must have been quite similar.

It follows from this that the mean January temperatures at the peak of the Saalian cold period were at best only insignificantly lower than those obtained during the Weichselian cold period.

3. The mean temperature during the warmest months. Until now the attempt has been made to determine the fall in the mean annual temperature during the last two cold periods. The results so obtained cannot be used directly as a starting point for the determination of the mean temperature of the warmest months. Rather this must be found in other ways. Brusch [cited in 99] has shown that the mean July isotherms of Europe, reduced to sea level, at present parallel the isolines of the climatic snow line, to the extent that the 10° July isotherm corresponds approximately to the 1,000-meter snow line; the 14° July isotherm to a great extent follows the 1,500-meter snow line, the 17° to 18° July isotherm the 2,000-meter line, and the 21° to 22° July isotherm the 2,500-meter line. These can be traced to the fact that the elevation of the snow line is primarily determined by the mean summer temperatures [462]. This provides a welcome aid in estimating the July temperatures during the cold periods, insofar as the position of the climatic snow line is known. However, to avoid serious errors it is necessary to remember the assumptions upon which the method is based. In the ideal case, the possible link of a certain reduced July temperature in the Northern Hemisphere with a given projected isoline of the elevation of the climatic snow line on the surface of the earth depends upon the amount by which temperature decreases with increasing altitude being constant in the area considered. If now the reverse process is attempted, and from the former position of the climatic snow line the former July temperature at the surface or sea level is deduced, then it has to be assumed that the magnitude of the noted vertical temperature gradient has remained the same over a long time. This will only be true when the climatic character of the area investigated has remained unchanged. Yet during the cold periods a more continental climate prevailed over wide areas, replacing that of the preceding warm period. It must be concluded from this that the connection between the July temperature figures and the height of the climatic snow line noted above was not in its present form. Mortensen [539] has considered in detail the difficult problem of the vertical temperature gradient during the cold periods for different areas. At present the correct estimate of this relationship is still very distant. Therefore, every attempt to read the mean July temperature from the former height of the snow line is extremely hazardous. Even in those areas of Europe where Quaternary geology is already very well known, it is not possible for now to estimate the magnitude of the vertical temperature gradient within individual regions. Yet, the above-noted relation between the reduced July temperature and the height of the climatic snow line, in the absence of any better figures, does provide a basis, however poor and uncertain, for the estimate of July temperatures. The position of the climatic snow line during the peak of the Weichselian cold period can be seen in Figure 84. The July temperatures determined in this way are given in Table 16 in comparison with those values deduced from the former position of the polar tree limit. It is suggested that on thermal grounds tree growth in lower Austria and Hungary was no longer possible, for at the peak of the last cold period in the lowlands of lower Austria no wood or even a bush was to be found where ground was wet enough, as in dips or along the river March, where, in other words, former dryness of the location does not enter the picture as a reason for the absence of tree growth [285, 289]. The same assumption can be made with respect to the lower Volga. For at that time, steppe had replaced woodland, although the Caspian Sea advanced far to the north during this, the Lower Chvalÿn transgression of the last cold period—that is, various locations along the shores of the enlarged Caspian must have been wet enough to support tree growth. When, despite this, the northern coniferous forest which existed only a very little earlier had disappeared, it would seem probable that lethal temperatures were responsible [301, 747]. The same also holds for the southern part of the western Siberian lowland.

In view of the large sources of error, the values determined agree very well. They show that the mean July temperatures in the areas of northern Eurasia investigated were about 8° to 10°C lower than today, and that the difference in the present-day oceanic areas seems to have been somewhat greater than in the high continental. Weichseł [786] earlier remarked on this. Yet the figures he calculated for Siberia are probably too small, because the great difference between the climate of lowland western Siberia and eastern Europe, on the one hand, and the mountains of central and eastern Siberia, on the other, is not adequately considered in the calculation.

The data in Table 16 can give only a very poor picture of what the anticipated conditions are likely to have been. Regional differences remain entirely unconsidered, although they were certainly as important then as now. Klute [408] several years ago attempted to pursue these questions in detail. His efforts, which resulted in an isotherm map for the months of July and January during the last cold period in Europe, will not be repeated here, for in the writer's opinion the time is not yet ripe for them. It seems far more important to compare the values determined for northern Eurasia with other areas.

As has been noted, the North American woods that were overridden by ice can be used for the determination of July temperatures as well as for mean annual temperature (above, p. 144). They show that the July mean temperature during the last cold period in western Ohio was about 11°C below the present figure [283]. The same figure is obtained from the climatological analysis of the late glacial spread of tundra along the Atlantic coast of what is now Massachusetts [570], while Leopold [462] calculated a reduction of 9°C in the mean July temperature of New Mexico. The basis for this was the depression of the climatic snow line by about

FIGURE 84. Elevation of the climatic snow line at the height of the last cold period in northern Eurasia. After Frenzel [256], modified by the addition of information subsequently available.

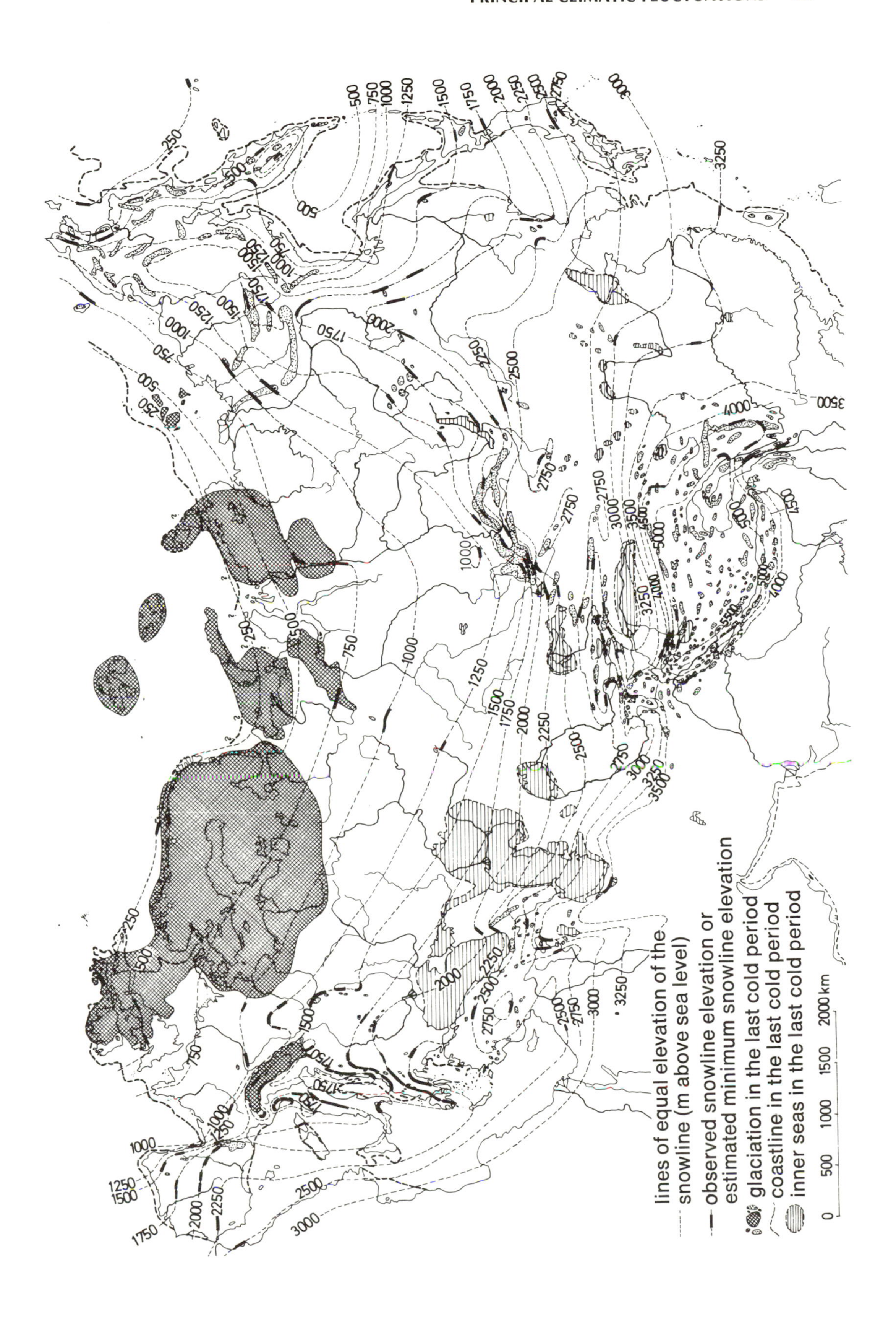
lines of equal elevation of the snowline (m above sea level)
observed snowline elevation or estimated minimum snowline elevation
glaciation in the last cold period
coastline in the last cold period
inner seas in the last cold period
0 500 1000 1500 2000 km

TABLE 16. DIFFERENCES IN THE MEAN JULY TEMPERATURES COMPARED TO PRESENT-DAY CONDITIONS, DETERMINED ON THE BASIS OF THE FORMER ELEVATION OF THE CLIMATIC SNOW LINE AND THE POLAR LIMIT OF TREE GROWTH ON THE PLAINS

REGION	DIFFERENCE IN JULY TEMPERATURES COMPARED TO THE PRESENT DETERMINED FROM:	
	ELEVATION OF THE CLIMATIC SNOW LINE	POLAR LIMIT OF TREE GROWTH IN PLAINS
Southern France	−10°C	−10 to −11°C
Northeastern and central Europe	−7 to −8°C	?
Vienna pforte	−9°C	−10°C
Southern Carpathians	−8 to −9°C	8− to −10°C
Southern Ukraine	−8 to −9°C	−10°C
Southern Urals	−5°C	?
Middle Yenessi and west Siberian lowlands	−6°C	−7°C
Baikalia	−7 to −8°C	−6 to −10°C
Northern Yakutia	−5°C	?
Northern Sakhalin	−8°C	?

1,500 meters [462] and the probably additional 800-meter depression of the upper tree line in the same region [42]. In a critical review of the older literature by Flint [232] it was concluded that in southwestern North America the July cooling amounted to about 6° or 7°C. The earlier, much lower, value of 3.5°C [232] deduced from the former distribution of certain mollusks has its origin in the fact that from these animals only the climate of the stratum of air close to the surface of the soil can be deduced, and this ill-suited to general climatic conclusions.

As was noted on p. 147 above, the former zones of vegetation of the South Isle of New Zealand suggest a diminution of the temperature of the warmest months by about 8°C. From the observations just indicated, the general result deduced is that the mean temperature of the warmest months at the peak of the last cold period over the whole earth was, on an average, 8° to 10°C lower than it is today. There were certainly regional variations, but up to the present time insight into them has been denied.

Even less is known of the warmest months of the Saalian cold period than of the Weichselian cold period. The position of the polar forest limit—that is, of the furthest forest outposts—seems to have been comparable to that during the Weichselian in northern Eurasia [map in 260], if the central Russian highland, which was ice-covered during the Saalian glaciation but which during the Weichselian cold period provided a refuge for many somewhat demanding plants, is excluded from this comparison. Probably about the same July temperatures were reached at that time as during the Weichselian cold period. The correctness of the assumption cannot be tested on the basis of the height of the climatic snow line, for this is not well enough known to permit such important conclusions. If it can be assumed that the thermal conditions in the two cold periods compared were not seriously different, then the significant contrast in the scale of glaciation between the two glaciations becomes more puzzling, and an explanation may be sought in differences in the water budgets.

4. The water budget. One of the best established paleoclimatological facts is that during the cold periods, when the glaciers had their greatest extent in Europe, the climate was not only significantly cooler but also extremely dry. This was already shown at the end of the last century by Nehring's [564] research on the vertebrate faunas, which showed that in central Europe at the same time that the wind was depositing a thick loess blanket over a distant treeless landscape, a fauna occurred whose center of distribution lies at present in east European and the central Asian steppes. Generally, during the cold periods, steppes seem to have been one of the most important types of vegetation, not only in Europe but also in Siberia, North America, and certain regions in the Southern Hemisphere. The climate, judging by that evidence, must have been characterized by low rainfall over wide areas. To what extent this also applied to near equatorial zones will be examined below (pp. 184–204), for although not of moment here, it is important to recognize the possible differences in the water budget, and through this the marked differences in the extent of glaciation during the last two cold periods.

Present-day steppe animals roamed the loess regions of Europe, Asia, and North America during the last cold period just as in the cold period preceding that. In both cases the climate must have been cold and dry. Yet it is conceivable that gradual differences arose between the aridity of the climates of the times considered, which could have been the cause for the difference in the spread of the continental ice sheets. This question will be examined in more detail in the following section beginning with the conditions of the last cold period.

It may appear surprising that consideration is given the water budget and not the amount of precipitation, yet in this case, too, the peculiarities of the methods applicable impel the greatest caution. In general only the budget can be determined, not the individual factors. This is particularly true of faunas and floras as indicators of past climates. There is, furthermore, another difficulty whose effect is appreciably greater in the present arid regions: the water budget of a region can be improved equally by thermal or by hydrological processes, if a fluvial regime or a change in the accumulation of sediment are excepted. Under the heading of hydrological changes are increases in total rainfall or a more favorable distribution of an unaltered total rainfall over the course of the year. The effect in the two cases is the same, but the causes differ significantly from one another, since the atmospheric circulation follows different patterns for the two examples given. Whether a climatic improvement can even be attributed to one process or the other is only possible in the rarest cases. The same is unfortunately true of the influence of temperature changes on the water budget. If this decreases, transpiration (evaporation from plants) and evaporation (from water or soil surfaces) follow in parallel. Consequently, the water budget is improved without the need for higher precipitation. It is therefore possible in such a case to have the water budget improve in a region where the climate is becoming colder, even though the available precipitation may actually decrease (below, p. 190, and [200]). The same is true here that was true for a possible change in the distribution of annual rainfall. Because it is generally not possible to distinguish clearly between the two influences, much must remain in the dark.

On the other hand so many paleoclimatic observations are available that it is always tempting to estimate the precipitation of past times. Klein [406] recently did so for Europe, with particular reference to the peak of the last cold period. At first the relationship between the summer temperature and the amount of snowfall of the recent climatic snow line was determined. Then Klein estimated the temperature at various places during the last cold period in central Europe and calculated the former precipitation on the basis of these data. This provided the basis for a rainfall map, which showed that the overall precipitation was lower than today, with the reduction much less in near oceanic regions than in eastern Europe.

All observations up to the present time point to a climate appreciably drier and more continental at the peak of the last cold period in Europe. The reconstruction just noted, therefore, strikes to the heart of the matter. On the other hand, it must also be considered that as a consequence of the greater dryness and continentality of the climate, the then vertical temperature gradient had some value other than the present and should have been higher. This was not sufficiently considered by Klein, although it could decisively change the results, as shown clearly in the following example: Klein based her calculations of the cold period precipitation on the Pic du Midi upon a vertical temperature gradient of 0.7°C/100 meters. If instead about 0.9°C/100 meters were used in the calculation and the calculations were carried out in the same way as by Klein [406], the impression is created that the former precipitation was the same as today or even exceeded it. That impression is certainly incorrect. The explanation of the discrepancy is that the estimated temperatures used by Klein for various locations are generally too high by a few degrees. The problem will not be discussed further here, for it is more important to establish that the reliability of the material available is clearly not yet sufficient for such calculations down to the minutest details, and an attempt must be made to look for alternate ways to obtain such information. For this purpose the character of the former vegetation is the most inviting.

It was already noted that in the place of woodland in central and western Europe, loess steppes occurred over wide areas. They comprised very different types [258–60]. In the immediate vicinity of the Alps and in northern Eurasia various wormwood (*Artemisia*) species predominated. The drier regions of the southern Ukraine and the southern part of western Siberia, however, were characterized by goosefeet (*Chenopodiaceae*). In central Europe beyond the Alpine foreland, a low-growing plant community dominated the scene, in which numerous plants—such as cinquefoil (*Potentilla*), ribgrass, the ribwort (*Plantago*) species, many species of the mustard family (*Cruciferae*), several plants of the daisy family (*Compositae*) (liguliflorae and tubuliflorae), and of the pea family (*Papilionaceae*), and grasses (*Gramineae*)—were particularly important. Knowledge of this plant community is derived from pollen analysis, which in the last few years has provided considerable insight into the vegetation of a high glacial, particularly in Russia. Moreover, the spread of the steppe was long known from sedimentological work and from vertebrate paleontology. It is generally supposed that the northern Mediterranean region at that time received a higher rainfall than is now the case. This hypothesis is based upon the assumption that an increase in the cold air mass in the north had as a consequence the southward displacement of the polar front, so that the Mediterranean region and particularly its northern part was much more influenced than it is now by wandering depressions and the precipitation was thereby increased appreciably. How far it is true that the polar front was displaced in the way indicated will not be examined here. Considerably more important is the momentous result that during the peak, and at the end of the last, cold period, contrary to the hypothesis noted, the northern Mediterranean received not higher but considerably lower precipitation, so that here, too, steppes and not woodlands were the characteristic element. This can be recognized with great clarity from the pollen flora of the late glacial sediments in the Pyrenees and their foothills [244, 245, 38, 780, 573], along the southern Alpine margin [40, 57, 828], and in the Zagros Mountains of southwestern Iran [358]. Similar *Artemisia* steppes dominated the scene in the northern Mediterranean during the maximum of the last cold period. It was reported from northern Spain by Florschütz and Menéndez Amor [245] and by Donner and Kurtén [167], and in southern Spain by Menéndez Amor and Florschütz [528, 529]. Bonatti [70–72] described nearly treeless *Artemisia* steppes which existed about 50 kilometers north of the present position of Rome between 24,460 ± 1300 and 17,040 ± 350 B.P. at an elevation of only 237 meters. The proportion of arboreal pollen in this pollen flora is so small that clearly no trees can have existed within a wide radius. A similar type of vegetation was described by Firbas and Zangheri [226, 227] from the Po basin near Ravenna. At the same time along the southern edge of the Alps and near Zetinje in Montenegro loess accumulated [253, 507, 514], and near Ljubijana an impoverished northern coniferous woodland developed which was broken up by large islands of open vegetation in which *Chenopodiaceae*, *Selaginella selaginoides*, and joint pine (*Ephedra*)—which is today found in dry, rocky locations in southern and eastern Europe [680]—are found in profusion. Even in southern Macedonia between more than 40,000 and 12,600 ± 200 years B.C. an *Artemisia* steppe predominated in the Drama plain, one in which large numbers of grasses and *Chenopodiaceae* occurred. At the present time the natural vegetation which flourishes at this locality is a Mediterranean oak woodland, which on approaching the coast is rich in pine. Oak and pine produce large quantities of pollen, which is released into the wind. Their pollen is thus easily transported and is found at many places far beyond their region of growth. As oak and pine pollen types are almost never found in the sediments of the last cold period, these tree species must have been almost entirely displaced from the region [334]. During the peak of the Weichselian cold period there was thus not only in northern Eurasia north of the Alps, but also in the northern Mediterranean region herbaceous and *Artemisia* steppes in locations where, during the preceding warm period, forests were the characteristic vegetation. Woodland could persist simultaneously only in the most protected and better-watered locations.

The vegetation just described flourished north of the Alps and in eastern Europe on ground that was either perennially frozen or at least frozen for periods of years. It thus resembled—not only florally but also in situation and ecology—the modern plant communities described by Richter, Haase, and Barthel [625] as well as by Haase [322] in Outer Mongolia. Other steppe regions, of southern Siberia and the central Asiatic mountainous regions, could also be used in this connection for comparison, as has recently been demonstrated by Matveeva and

Moskvitin [524]. In the regions of Outer Mongolia, the mean annual temperature at present fluctuates between 0° and −4°C. The mean July temperature runs from about +14.5° to about +19°C. These estimates are thus somewhat higher than for the last cold period in central Europe. However, as in general the climatic and phytogeographical conditions of the regions compared appear to resemble each other very closely, and as here and there loess was formed and is forming, it seems justifiable to adapt present-day precipitation figures of the Mongolian region as a basis for the determination of the magnitude of the precipitation in Europe during the last cold period. This is about 200 millimeters per year. Of a somewhat different composition floristically, the steppes in the southern part of the western Siberian lowlands, with a mean annual temperature corresponding to that noted, enjoy appreciably warmer summer months and today receive between 300 and 350 millimeters of precipitation per year. This datum demonstrates that the mean annual precipitation at the peak of the last cold period in central Europe and in the present woodland zone of eastern Europe did not exceed 300 millimeters per year. Thus from northern France in the west to the middle Desna in the east, at least 300 millimeters less rain must have fallen per year than at the present time.

In the present steppes in the extreme southeast of Europe, in Turan, middle and central Asia the extreme aridity of the climate favored an upward movement of the ground water. Consequently, the uppermost soil horizon becomes progressively more saline. It forms saline steppes in which essentially only a very few *Artemisia, Cruciferae,* and *Chenopodiaceae* species can survive. A similar situation has been described in the extremely arid climates of the Arctic [67, 263, 527, 359, 749]. Often efflorescence crusts of gypsum, magnesium sulphate, or common salt form there (Fig. 85). Kessler [399] earlier supposed that the same conditions must have pertained in central Europe during the cold periods, so that the former loess steppes can be designated as saline steppes. Actually, even today the loess of many central Asiatic localities is rich in gypsum, common salt, and other salts at great depths [506, 115, 567]. In central and eastern Europe, however, save in the immediate vicinity of the Black Sea and the Sea of Azov [627], no high salt content of the loess has been discovered, if the well-known richness in calcareous material can be excepted. Yet Kulczynski [443] and subsequently Florschütz [243] described fruit and seeds of halophytes from sediments of both the Saalian and Weichselian cold periods in Poland and the Netherlands. These plants seem also to have occurred along the eastern margin of the Alps; at least their pollen is found in the sediments [258, 259]. They were, however, only the so-called "potential halophytes," and so do not indicate highly saline locations. On the other hand, Monoszon [536–38] has reported several extremely interesting pollen analyses, observing the occurrence of the true halophytes in the cold period steppes of eastern Europe—that is, of plants which appear to flourish only on saliferous ground. The palynological descriptions of Monoszon have been followed since then by many Russian Quaternary geologists. It now seems quite certain that the soils of eastern Europe and the southern part of western Siberia were saline at the height of the last cold period, so that at least regionally, salt steppes formed an important vegetation type. This assumption is in agreement with the widespread distribution of *Chenopodiaceae* in the steppes of the cold period. The

FIGURE 85. Tundra salt efflorescences. Upon a frost-heaved loamy base (Gärlehm) dolomite efflorescences occur (white specks between the rocks). Dolomite determined by Sandberg.

paleoclimatic consequences go beyond what has just been said. For it is probably associated with a significantly reduced precipitation or in general with a still worse water budget, in cases where the salt steppes actually involved regions of Europe and western Siberia now occupied by the forest belt. The assumption of widespread saline steppes, however, is based upon pollen morphology of certain species and genera of *Chenopodiaceae*. In this connection it would seem that too much is being demanded of pollen analysis, for the morphological variability of pollen grains in a single species of *Chenopodiaceae* is as great as in the whole family, so it is impossible, at least provisionally, to determine the individual species or genus of *Chenopodiaceae* from the the pollen morphology [260]. It is therefore highly uncertain whether salt steppes were actually so widespread during the cold periods, and the observations noted of the former occurrence of "potential halophytes" says very little in this context. The plants even today occur not only in weakly saline environments but also in virgin soils. In short, very little of paleoclimatic interest can be deduced from the marked occurrence of *Chenopodiaceae* during the cold period steppes of Europe and western Siberia.

In North America the land surface in Alaska, in northern and southeastern Canada, and in the central United States slopes away from the ice margin of the former ice-sheet region. Then as now the water could drain off unhindered. Conditions in Europe and the northern part of the western Siberian lowlands were basically quite different, for the ice advanced towards uplands, from which large streams were draining. Their courses were blocked by the advancing ice, and

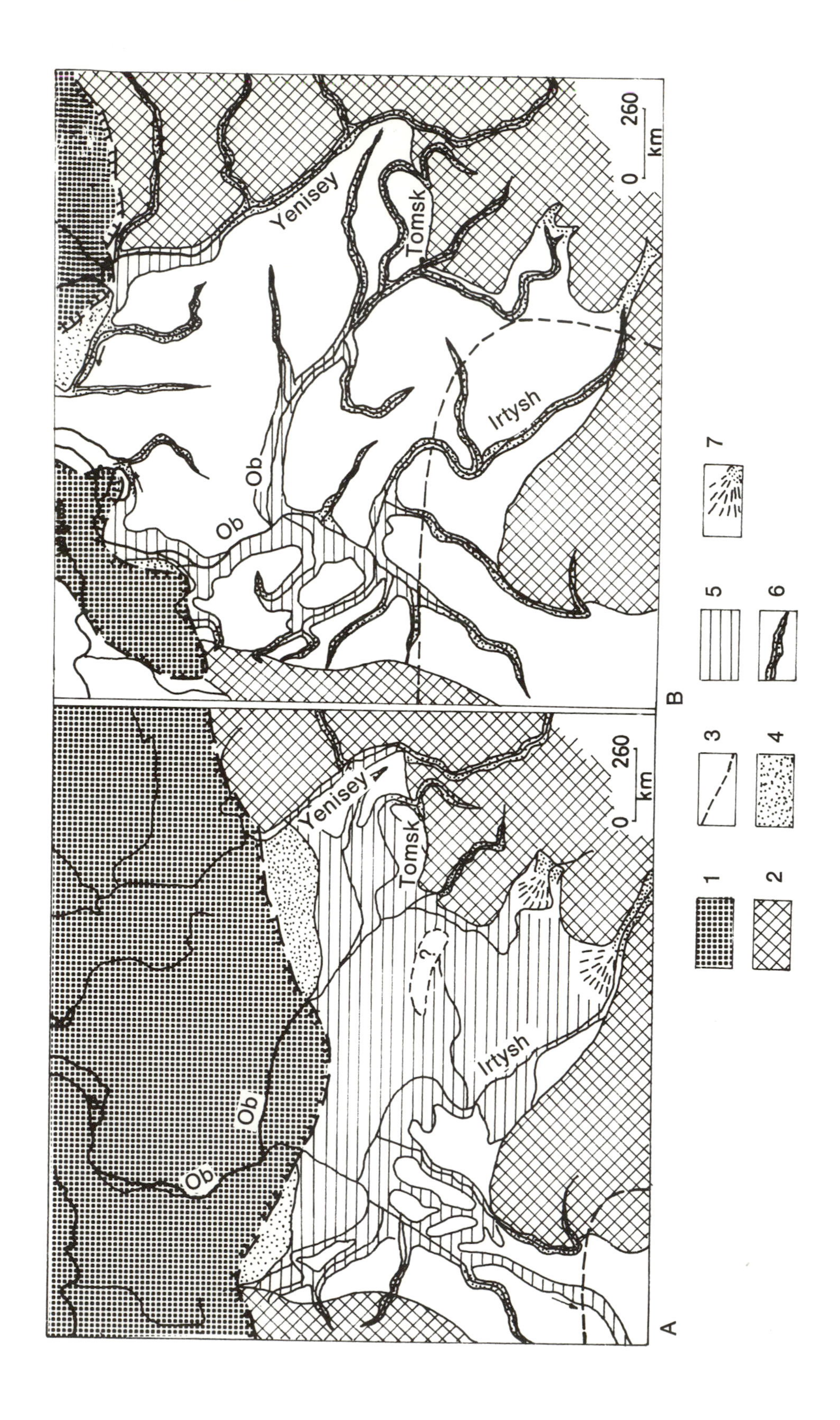
A
B
Yenisey
Tomsk
Ob
Ob
Irtysh
0 260 km
Yenisey
Tomsk
Ob
Ob
Irtysh
0 260 km
1
2
3
4
5
6
7

their channels filled or diverted by ice, moraine, and sandy material. Genieser [270] and Eissmann [183, 184] described these events in detail with reference to the examples of the rivers of Saxony and Thuringia. Besides this, in Europe the large ancient valleys parallel to the main ice front were used by melt water draining south from the ice front and by the streams draining northward in the remaining part of the ancient valley [802, 804, 805]. Instead of this it was very often suggested that in the northern and central parts of the western Siberian lowlands masses of ice dammed water to form lakes in front of the ice sheets which lay to the north. An idea of these lakes can be gained from Figure 86, based upon data from Zarina, Kaplanskaya, Krasnov, Michankov, and Tarnogradsky [820]. I have elsewhere [260] attempted to show that the assumption of such great glacial lakes in the western Siberian lowlands is probably incorrect, and consequently the problem will not be examined again here. Even in the event that the existence of these giant lakes is denied, the existence of smaller lakes both during the last cold period and particularly during the next to last cold period should not be overlooked in this part of northern Eurasia. They appear to provide an appropriate basis for an estimate of the water budget during the last two cold periods. For given that the size of the lake basin and the drainage area of the streams entering the lake are known, as well as a knowledge of the thermal conditions, information can be obtained relating to the water budget. These requirements, however, pose insurmountable difficulties. On the one hand, the sizes of the former lake basins in fact are insufficiently well known, and on the other, the amount of water draining to the north from the ice is unknown. If it can be assumed that the southward advance of the ice in central Germany or in the northern to eastern part of Europe and in the western Siberian lowlands halted all northerly drainage, then possibly a beginning could be made on the calculation. If, however, the discharge is not completely blocked, then a source of error of unknown magnitude is introduced. It is unfortunate that at the present time no distinction can be made as to what actually happened, so any attempt to estimate the water budget accurately during the period of the maximum ice advance in this way is thwarted. However, it need not necessarily follow that the same volume of water was transported northward under the ice as at the present time. When the former ice-dammed lakes in the Saxony-Thuringian basin—along the margins of the German Mittelgebirge and in the western Siberian lowlands at the peak of the next to last cold period and in part during the last cold period—were not very large, the conclusion seems to be that the amount of water flowing northward in the rivers in front of the ice margin must have been small. The observations of Golubeva and Ravskij [285] on certain rivers of western and central Siberia point to the same conclusion. The fluviatile sediments of the last cold period which form the present second terrace of the Yenisey, Angara, and Stony Tunguska rivers and certain tributaries is usually composed of a thick silt, which was probably an overflow silt. Other

FIGURE 86. A hypothetical sketch of the supposed ice-dammed lake which existed in front of the ice sheet in the western Siberian lowlands. A. Saale-Samarov glaciation; B. Weichsel-Zyrjanka glaciation. 1. continental ice sheet; 2. marginal mountains; 3. southern permafrost limit; 4. outwash sands deposits; 5. ice dammed lake; 6. river valleys; 7. alluvial fans. After Zarina, Kaplanskaya, Krasnov, Michankov, Tarnogradsky [820].

fluviatile sediments, such as gravel and coarse sand, are generally absent from the terrace. However, coarse, irregular blocks which could have been brought in only by floating ice commonly occur in them. Golubeva and Ravskij assume from this, with good reason, that the volume of these rivers at the peak of the last cold period during winter and in part also during the summer must have been extraordinarily low. During spring, however, with the melting of the snow, the rivers flooded the surrounding regions with the resultant deposition of the silt and large blocks. The flooding was probably devastating, because downward erosion was impeded by perennially frozen ground, and because the open vegetation could not stop run-off or hinder soil erosion. This observation is in very good accord with a large body of other data gathered about the above-mentioned tundra and steppe vegetation of these regions. Moreover, it shows clearly that the extreme aridity of the last cold period operated even in the most continental regions of Siberia. Also, it is incorrect on these grounds to assume a special position for Siberia in the climatic history of the last cold period. Only south of the Sajane and Altai mountains does the water budget of the last cold period appear to have been better than the present. This is suggested by the considerable length of the former glaciers in the Mongolian Altai [540], probably a coeval pollen flora of a pine-spruce wood (*Pinus sibirica, Picea* sp.) on the southern slopes of the western Tannu Ola in the valley of the Juzhnij Targalyk [523] as well as the high level of Lake Baikal and adjacent lakes [446]. To what extent these undoubted indications of a more favorable water budget indicate a higher precipitation or only a reduced temperature and consequently reduced evaporation remains at present an open question.

In North America, too, loess accumulated over wide areas (Fig. 81) during the last cold period, and the constantly improving paleobotanic observations of the late glacial vegetation in the northern part of the present United States give an insight into the former vegetation, which during the last cold period formed a zone more than 100 kilometers wide bordering the ice front. This region is currently populated by a varied deciduous and coniferous woodland [396, 669, 570, 571, 677, 658, 800, 376, 262, 261, 212, 146, 24, 471, 153, 150, 516]. Unfortunately, present knowledge does not extend far enough to estimate the amount of former precipitation. The reasons are twofold. First, it appears that in the molluscan fauna of the Peoria loess of Missouri [459] no adaptations to a dry climate can be recognized; it appears much more that an animal community lived in a sufficiently thick, moist, organically rich layer which may have dried out a little in a few weeks but which in general, even during the summer, was very moist. The temperatures, particularly in summer, are said to have been somewhat lower than at present in the Great Plains. If these observations could be generally applied to other North American loess deposits, then the water budget must in fact have been more favorable than in the contemporaneous cold climate phase in Europe. Supporting this contention is the fact that the continental ice sheet in North America during the first phase of the last cold period had already advanced far to the south, whereas a similar advance in Europe does not appear to have occurred. Such conclusions however, rest upon two assumptions whose validity has not been proven. First it must be doubted that the observations made in the vicinity of the Missouri can also be applied to a wide region at greater distances from the river. Leonard and Frye themselves remark that beyond the confines of the valley, dry prairies prevailed. Secondly, there is no certainty that the molluscan fauna dates from the height of the last cold period. In Europe, too, snail faunas during the last cold period existed in relatively moist, wooded steppes as well as in dry steppe locations. The first named was to be found particularly during the first part of the cold period, while the other populated the steppes during the height of the glaciation. The possibility of such a faunal change may also be considered in North America during the course of the last cold period, so that the observations of Leonard and Frye should not be applied too hastily.

The second difficulty, which particularly affects the estimate of the amount of precipitation in North America, has its roots in the fact that the North American steppes, the prairies, unlike those in Eurasia, were affected to a great extent by the grazing of vast herds of large animals. Of these, the enormous bison herds, which were annihilated only in very recent times by Europeans and American Indians, were the most important. The area of the North American prairie was thus limited not only by climatic but also by biological factors, and the steppe appears to have existed in part in a true woodland climate. So the postglacial area of prairies in North America cannot provide a sure basis for determining the climatic conditions under which forests are replaced by steppes.

These difficulties are so profound that it is impossible even today to make any statements about the water budget of these parts of North America during the cold periods. It cannot be doubted, however, that the precipitation during the last cold period was considerably less than at the present time. The same is also true of South Island, New Zealand (above, p. 146).

Insofar as is known at the present time, during the Saalian cold period loess accumulation did not cover a significantly greater area than during the Weichselian cold period. Yet it must be admitted that there is still insufficient information concerning phytogeographical conditions, and present conceptions of the former climate may change in many aspects. The available data suggest that the amount of precipitation must have corresponded approximately to that of the Weichselian cold period. It can scarcely be supposed to have been less. On the other hand, there are no signs that appreciably more rain fell in the present cool, temperate latitudes than during the Weichselian cold period. This circumstance is remarkable, for it carries with it the implication that the difference in the scale of glaciation at the peak of the glaciations cannot be explained by distinctly different regimes of precipitation. For the present, every basis for understanding contrasts in the history of glaciation during the two cold periods in Eurasia seems to fail. Possibly two other observations may be more successful in giving that insight.

The snail fauna of a small region can be taken as nearly fixed to that area, for these animals do not wander great distances. The consequence is that land snails must be particularly well adapted to the conditions of their environment (biotope). If the ecology of the individual snail species is known, then the character of the ancient biotope can be determined from the snail fauna. Next to moisture, thermal conditions play a great role in the life of land mollusks. From the work of Czech scholars, in particular Ložek, we have many analyses of snail faunas found in the Czech loess. It could be shown that there existed remarkable differences between the faunal assemblages of true warm periods, interstadials, and cold periods. Within these groups the loess snail faunas prove to be independent, to a certain degree, of the age of the community. There are clear and appreciable differences, however, between the faunal communities of the different climatic regimes. In the cold climate in Czechoslovakia [488 and the literature reviewed there], in Bavaria [91, 97], in central and southern France [76, 77, 526], and in western Russia [445, 767, 768] animal communities characterized by *Columella edentula columella* are particularly important. This community always occurred at the time of the coldest and driest climates. It lived simultaneously in southern Siberia. In the loess belonging to the Saalian cold period it is notable that the snail community does not occur so profusely [488]. This difference may be related to relatively little research on Saalian loess or alternatively to a climate not as severe as that during the height of the last cold period. I refer only to the water budget, for, as was shown above, the temperature during the Saalian cold period can scarcely have been higher than in the last cold period.

Possibly the climate of northern Eurasia during the Saalian cold period was less dry than during the Weichselian cold period. If this difference existed, it must have been extraordinarily small, for the spread of steppe and the zone of loess formation in the two cold periods were so similar that no great difference can be assumed. A small increase in the total rainfall or, as it is generally expressed, an improvement in the water budget may have favored increased growth of the glacier. It is not possible to overestimate the significance of these factors.

The other significant observation concerns fluctuation of the climate during the two cold periods. After the Holsteinian warm period ended, the climate in Europe became cold and dry. By that time loess had developed in many places in Europe. This segment of the Saalian cold period, of unknown duration, was finally succeeded by a phase of warmer climate, for not only had the loess ceased to accumulate, but thick soils were formed [642–44]: the upper soil of the PK IV of Czech Quaternary geologists, the Muglinov soil in the Moravian Gate [686], the fossil soil above the Lichvin interglacial sediments and below the moraines of the maximum glaciation [542]; and the forest returned to central Europe [818, 88, 14, 303, 371, 388, 418, 641, 500, 716, 753]. The latter was composed principally of conifers, but nearly everywhere minor quantities of oak (*Quercus*), elm (*Ulmus*), hornbeam (*Carpinus*), hazelnut (*Corylus*), and other demanding tree species occurred. That the warming trend was significant is indicated also by the molluscan fauna of the thick upper fossil soil of PK IV in Czechoslovakia [488]. The last cold period began in a comparable manner. In both cases in central Europe no glacial advances during the first cold dry climatic phase are known for certain, even if they have been variously claimed for the Saalian cold period. After the significant early glacial interstadial in the Saalian cold period, the maximum glacial advance occurred. It took the form of several waves whose history has been convincingly reconstructed in the Netherlands by Ter Wee [738]. During a later significant interstadial, woodland migrated back into eastern Europe and probably into central Europe. This was primarily a pine-spruce woodland in which, here and there, fir also flourished. Thermophilous deciduous trees were nearly lacking. The interstadial probably corresponds to that which occurred between the maximum Saalian advance and that of the much weaker Warthe Stadial, in the terminology of northern Germany [260].

Fundamentally different from the glaciation history of the Saalian cold period indicated above was that of the last cold period. In it, after the first significant interstadial (Brørup interstadial and the minor warm fluctuations that occurred close to this in time), which can be compared with that at the beginning of the Saalian cold period, as far as is known there was no glacial advance in Europe. The maximum advance of the ice sheet in Europe set in much later, after the Stillfried B interstadial.

Thus the glacial history of the two cold periods differed significantly one from the other, and that may be one of the reasons for the difference in the scale of glaciation. Why there should be a difference is presently unknown. We have no indications of the distribution of rainfall during the year, or climatic conditions in the regions above the snow line of the great ice sheets.

The Holsteinian and the Eemian warm periods as well as the postglacial are thus separated from one another by two cold periods, during which not only the temperature but also the amount of precipitation dropped appreciably. The same holds true for the northern Mediterranean, which, during the next to last warm period, and possibly also during the last warm period (as in the more northerly regions of northern Eurasia and as in North America), received more precipitation than today. At the same time, it appears that in the northern Mediterranean instead of the dominant winter rains of the present time, summer rains occurred. On the other hand, it must be supposed that during the cold periods in this region, winter rains were particularly important, so in the change between warm and cold periods not only the temperature and amount of rainfall but also the atmospheric circulation altered appreciably.

THE PROBLEM OF CLIMATIC CHANGES DURING THE ICE AGE

General

Thus far, attention has been concentrated on the scale of change in the climate of the Ice Age. Climatic conditions at the peak of the warm and cold periods have been compared with one another. Despite the many existing uncertainties which must be attributed to the difficulties of establishing reliable dates of the individual events, the scale of the fluctuations is today clearly recognized. The knowledge obtained in this way permits another question to be broached—namely, whether the climate during the Ice Age fluctuated about a constant mean value or whether the climatic fluctuations had superimposed on them an additional climatic change.

It was early found in paleobotanic studies that the flora of any region was progressively impoverished during the course of the Ice Age until it reached the present condition. Reid [623] was the first to attempt to comprehend this quantitatively and attempted to use it as a means of dating. Subsequently, knowledge became more precise through other investigations, so that at present the development of the European flora during the Pleistocene warm periods is well known. Figure 7 (above, p. 9) gives an impression of the geographic and historic fluctuations of the floral elements of three regions of Europe during the Quaternary. The thermophilous, more demanding elements obviously diminish from warm period to warm period. From this it might be presumed that the climate since the end of the Pliocene in sequence of warm periods became progressively less favorable. Yet only rarely has this conclusion been drawn, since the floral impoverishment shows less a decrease in the favorable nature of the warm periods than a selection during the preceding cold periods. In this connection a further question is justified: whether the climate of the cold periods, perhaps during the

TABLE 17. FAUNAL CHANGES DURING THE ICE AGE IN CENTRAL EUROPE

SPECIES	1	2	3	4	5	6	7	8	9	10	
Equus stenonis	+	−	−	−	−	−	−	−	−	−	
Archidiskodon planifrons	+	−	−	−	−	−	−	−	−	−	
Cervus falconeri	+	−	−	−	−	−	−	−	−	−	
Anancus arvernensis	+	+	−	−	+?	−	−	−	−	−	
Dicerorhinus etruscus	+?	+	+?	−	+	−	−	−	−	−	
Equus robustus	+	+	−	−	−	−	−	−	−	−	
Leptobos elatus	+?	+	−	−	−	−	−	−	−	−	
Ursus etruscus	+?	+	−	−	−	−	−	−	−	−	
Crocuta perrieri	+?	+	−	−	−	−	−	−	−	−	
Sus strocci	−	+	−	−	−	−	−	−	−	−	
Cervus dicranius	−	+	−	−	−	−	−	−	−	−	
C. rhenanus	−	+	−	−	−	−	−	−	−	−	
Eucladoceros tegulensis	−	+	−	−	−	−	−	−	−	−	
Pannonictis pliocaenica	−	+	−	−	−	−	−	−	−	−	
Dicerorhinus kirchbergensis	−	+	+?	−	+	−	+	−	+	−	
Trogontherium boisvilletti	−	+	+?	−	+	−	−	−	−	−	
Macaca sp.	−	+	+?	−	+	−	+	−	−	−	
Hippopotamus cf. *amphibius*	−	+	−	−	+	−	+	−	+	−	
Archidiskodon meridionalis	−	+	+?	−	+	−	−	−	−	−	
Equus Caballus	−	−	+?	−	−	+	+	−	−	+	
E. süssenbornensis	−	−	−	−	+	+	−	−	−	−	
Cervus elaphus	−	−	−	−	+	+?	+	+	+	+	
Palaeoloxodon antiquus	−	−	−	−	+	−	+	−	+	−	
Mammontheus trogontherii	−	−	−	−	+?	+	+	+	+?	+?	
Sus scrofa	−	−	−	−	+?	−	+	−	+	+	
Bos primigenius	−	−	−	−	−	−	+	+	−	+	
Ursus spelaeus	−	−	−	−	−	−	−	+	+	+	
Alces alces	−	−	−	−	−	−	−	−	+	+	
Cervus(Megaceros)giganteus	−	−	−	−	−	−	+[1]	+?	+	+	[1] *antecedens*
Rangifer tarandus	−	−	−	−	−	+	−	+	−	+	cold
Lemmus sp.	−	−	−	+	−	+?	−	+	−	+	cold
Coelodonta antiquitatis	−	−	−	−	−	+	−	+	−	+	cold steppe
Mammontheus primigenius	−	−	−	−	−	−	−	+	(+)	+	cold
Bison priscus	−	−	−	−	−	−	−	+	+	+	cold steppe
Saiga tatarica	−	−	−	−	−	−	−	+	−	+	cold steppe
Dicrostonyx sp.	−	−	−	+	−	+?	−	+	−	+	cold
Alopex lagopus	−	−	−	−	−	−	−	+?	−	+	cold
Lepus timidus	−	−	−	−	−	−	−	+?	−	+	cold
Microtus gregalis	−	−	−	−	−	−	−	−	−	+	cold
M. nivalis	−	−	−	−	−	−	−	−	−	+	cold
Equus hydruntinus	−	−	−	−	−	−	−	+	−	+	steppe

+ signifies an occurrence in the time indicated; − indicates no occurrence is known.
1 = Pretiglian cold period; 2 = Tiglian warm period; 3 = Eburonian cold period and Waalian warm period; 4 = Menapian cold period (?); 5 = Cromerian warm period; 6 = Elsterian cold period; 7 = Holsteinian warm period; 8 = Saalian cold period; 9 = Eemian warm period; 10 = Weichselian cold period. Cold = fauna of a severe winter climate.

course of the Pleistocene, became progressively less favorable, so that the selection pressures became even more efficient. This question is particularly apposite when changes in the plant and animal kingdoms are both involved.

Table 17 shows that during the Ice Age the fauna underwent a change comparable to that of the flora. It is particularly noticeable that animal species which have lived in extremely cold winter climates have been found only in central Europe since the Saalian cold period. Their forerunners appeared first during the Elsterian cold period. The fossil floras show the same pattern: the plant species found today, particularly in tundra and alpine regions, first established themselves in central Europe after the Elsterian cold period [260]. Certainly the tundra element occurred earlier, but its significance was never so great. It appears that the intense cold began with the Elsterian cold period. The preceding cold periods seem to have been milder.

As has been noted, in the course of time changes in the flora and faunas of the warm periods occurred which may be the result either of a progressive decrease in temperature from one warm period to the next or of an increasingly arid climate. Since these suppositions are not very likely, it seems necessary to fall back on differences in soil formation. In the impressive clay pit of Červeny kopec near Brno, many thick beds of loess and fossil soil horizons are intercalated (Fig. 87). As so often occurs, most of the fossil soils closely follow one another, separated by thin bands of clay, loess, or hillside wash material. Each group of successive soil horizons is referred to as a Pedocomplex (abbreviated to PK) by the Czech Quaternary geologists. If only the most highly developed soil in each pedocomplex is compared with the others, then from oldest to youngest (PK VIII to a postglacial soil) the following section is obtained [441] (INQUA-Subcommission for loess stratigraphy excursion, August 1963):

Postglacial soil	(claypit in Modřice)	:	chernozem
PK III	(claypit in Modřice)	:	brown earth
PK IV	(Červeny kopec)	:	dark brown earth
PK V	(Červeny kopec)	:	dark brown earth
PK VI	(Červeny kopec)	:	brown loam
PK VII	(Červeny kopec)	:	dark red brown rubified brown earth or a brown loam type parabrown earth
PK VIII	(Červeny kopec)	:	red loam

The red loams (*terra rossa*) and brown loams (*terra fusca*) show not only a greater degree of weathering than the brown earths or even the chernozems, but also the results of higher temperatures and greater and more irregular seasonal rainfall. It appears from this that the Červeny kopec profile indicates a deterioration of the warm-period climates during the course of the Ice Age. Butzer [111] described similar differences in the fossil soils of the Mediterranean and came to comparable paleoclimatic conclusions.

When the problems of possible climatic changes during the Pleistocene are to be discussed, cognizance must be taken of the fact that along the sea coasts in many places several fossil marine terraces occur one above the other in such a way that the oldest is the highest. Insofar as these terraces date from the Pleistocene, they were very early linked to the rise in sea level during the warm

FIGURE 87. Views of part of the Červeny kopec profile near Brno. A thick blanket of loess covers the western slopes of the Svratka Valley. In it several fossil soil horizons can be recognized, each group of closely succeeding soils forming a pedocomplex. The excursionists are standing on one pedocomplex, to the left of them a relatively poorly developed pedocomplex dips into the valley, and at the bottom right of the photo a very old pedocomplex can be recognized.

periods, for with the melting of the ice sheets the level of the oceans had to rise eustatically. It might be deduced from these marine elevations that during the first part of the Ice Age larger volumes of water were returned to the oceans than during later times, so that in this case, too, the warm periods must have become thermally more unfavorable. Since this hypothesis does not accord with the facts, tectonic processes as a cause of the progressive fall in sea level were discussed. It was imagined that the ocean floor slowly sank so that the sea level during climatically comparable periods was gradually lowered. Recently, another hypothesis has come to the fore. It posits a fall in sea level caused by the gradual growth of the Antarctic ice sheet as more water was drawn from the oceans and locked up as ice. This presupposes that in Antarctica throughout the entire Ice Age the climatic process progressed unidirectionally and, to an increasing extent, affected the weather conditions in low latitudes in the Southern Hemisphere as well as the Northern.

The study of many loess outcrops provides an additional basis for the determination of possible climatic changes during the Ice Age. These rocks cover wide areas of the Northern Hemisphere, and part of the Southern Hemisphere too, but they date essentially from the last two cold periods as well as from late and postglacial times. Older loess is only rarely found. This is true for the Elsterian cold period and to an even greater degree for the still earlier cold periods. Thus the climate of the cold periods seems to have become progressively more favorable for cold, steppe vegetation during the course of time. Supporting this are the observations made by Brunnacker [95] near Ratisbon. The oldest loess there, which Brunnacker refers to as the "Gunz ice age," was decalcified under a relatively moist climate during the course of sedimentation. The next youngest loess ("Mindel ice age") still shows signs of only reduced moistness, but the loesses of the Riss and Würm ice ages are calcareous and indicate the existence of a dry climate at the time of their formation. From one cold period to the next according to this, the amount of available moisture decreased. This may also be used as an indication of a Pleistocene climatic change superimposed on the climatic fluctuations.

The attempt will be made to discuss the questions outlined in the following pages.

The Warm Periods

The attempt was made in Chapter 4 to determine the minimum climatic change in Europe during the various warm periods with respect to present-day conditions. The results obtained can be summarized in Table 18.

TABLE 18. MINIMUM CHANGE IN CERTAIN CLIMATIC FACTORS WITH RESPECT TO THE CORRESPONDING PRESENT-DAY VALUES IN EUROPE

	MEAN TEMPERATURE °C			ANNUAL PRECIPITATION
WARM PERIOD	JANUARY	JULY	ANNUAL	(MM.)
Northern Germany				
Postglacial warm period	=	+2	+2	=
Eemian warm period	+1 to 2	+3	+2 to 3	=
Holsteinian warm period	=	+2	+1	=
Reuverian B	=	+3	=	higher
Poland				
Postglacial warm period	+1	+2	+1 to 2	somewhat higher?
Eemian warm period	+3 to 4	+3	+3	+50
Holsteinian warm period	+3 (to 4)	=	+1 to 2	+50
Reuverian B	+5 to 6	+4 to 5	+3 to 4	+350 to 400
Central Russia				
Postglacial warm period	+2	+2	+2	+50?
Eemian warm period	+9 to 10	+2	+4 to 6	+100
Holsteinian warm period	+8 to 10	+2	+5 to 6	+100
Reuverian B	?	?	?	?
Middle Volga				
Postglacial warm period	+2	+2	+2	+50
Eemian warm period	?	?	?	?
Holsteinian warm period	+12?	=	+6°	+300
Reuverian B	+6 or +10 to 15	+2 to 3	+5 to 6	+350 to 400

During the Reuverian B the climate was clearly warmer and more moist than during the two Pleistocene warm periods considered. The climates of the latter compared to one another show no significant differences. Only the postglacial warm period appears to have been climatically less genial than the preceding warm periods. It has already been made clear that it is not possible to determine the climate of the older warm periods at the present time. Yet it appears that the climate, at least of the Tiglian warm period, was more favorable than that of the Holsteinian or Eemian warm period. This supposition is based upon former plant distribution. The difference, however, cannot have been great. For example, Zagwijn [819] gives a temperature in the Netherlands for July of about +2°C higher during the warmest phase of the Tiglian warm period than at present. This estimate is in complete agreement with that calculated for the Holsteinian warm period. In the Cromerian warm period approximately the same phytogeographical conditions were established which later characterized the Eemian. Today no appreciably greater advance of woodland over the steppe region of the present Ukraine or more exacting forest types extending further north than during the Eemian can be established for the Cromerian. It may be assumed as a working hypothesis that at the thermal optimum of the Cromerian warm period the climatic conditions were approximately as favorable as those during the Holsteinian and Eemian warm periods, though we have no sure foundation for the assertion. The climatic history of the Waalian warm period is still the least clear at present. In the Netherlands, Zagwijn [819] estimated a mean July temperature of about +2°C for the optimum of this period, a figure that agrees with those for the Tiglian and Holsteinian warm periods. From the standpoint of floral history, this period occupies a medial position between the Tiglian and Cromerian warm periods. Nothing is known about the contemporary fauna, for it has not yet been possible to correlate definitely the faunal divisions of the older Pleistocene with the floral. The results of further research must be awaited.

In general it may be assumed that the climatic optima of the Pleistocene warm periods were somewhat less favorable than of the Reuverian B but warmer and moister than those of the postglacial warm periods. Furthermore, the Tiglian warm period probably lay climatically in a position intermediate between conditions in the Reuverian B and those determined for the Cromerian to Eemian warm periods. Yet the differences between the individual warm periods cannot have been very great.

The fossil soils in the Červeny kopec clay pit seem to show a progressive climatic deterioration from one warm period to the next. In contrast, in central Franken terra rossa was formed only up to the end of the upper Miocene. In the later parts of the Tertiary only soils formed which could also be formed during the Pleistocene warm periods [93]. Paleopedological research in Czechoslovakia established the same result [487, 693, 694]. The youngest terra rossa in that limestone region usually is found upon Pliocene travertines, and only exceptionally is it found on perhaps the oldest Pleistocene limestones. During the later warm periods a brown earth (terra fusca) developed in their stead, and their development in the older warm periods is greater than during the Eemian warm period. Finally, in the postglacial, rendzinas—relatively immature soils—commonly formed. To judge from this, the climatic conditions of the individual warm periods operating on comparable substrates and at corresponding localities in central Europe have influenced soil formation in a like manner. Significant climatic changes apparently cannot be deduced from them. Only in the last warm period do somewhat anomalous soil types occur. Brunnacker [96] linked this with a lower precipitation at the end of the Eemian warm period than during the Holsteinian warm period. This small difference was also observed in an earlier case (above, p. 128). Only in the postglacial soils does there seem to be evidence for a strongly divergent climate (cf. Table 18).

The fossil soils of the last two or three warm periods on the middle Main and on the Danube near Ratisbon correspond approximately to those being formed there at the present time [94, 95]. They thus show when compared with one another no change of climate during the latter half of the Pleistocene. Instead of this, the soils of still older warm periods both here and in the Alpine foothills are often extraordinarily thick. It was because of this that Brunnacker [293] called them "Riesenböden." Earlier it was often thought that they had formed under climatic conditions quite different from the later ones. Brunnacker pointed out, however, that the Riesenböden were formed out of a material different from that out of which the younger, less thick soils developed. Hence, the difference in thickness of the soils reflects most of all the influence of the substrate and not of a changed climate. Typologically, the Riesenböden considerably resemble the fossil soils of the younger warm periods. In them, however, the tendency to pseudogleying is more strongly developed than in the younger soils. This may have been caused by a somewhat higher precipitation. The comparison of the climatic optima of the Pleistocene warm periods with one another suggests that their climate became somewhat drier [96]. Typologically different from the fossil soils of the warm periods are most of those dating from the postglacial period in central Europe. As noted above, there appeared to be an intensive development of terra fusca during the last warm period for the last time upon limestone in Czechoslovakia. In the loess of Modřice near Brno the last brown earth was formed at the same time (above, p. 134); the present soil is a typical black earth. The same observations can be made in many parts of central Europe, particularly in the present dry regions. The character of the snail fauna points in the same direction: that with respect to the faunas requiring warmth and moisture, there are fewer such communities in the postglacial period than in the Pleistocene warm periods [488]. However, they also show that in the present dry regions, even during the postglacial, instead of the present heath-steppe there was woodland [485, 486, 491, 431], which was replaced with an open steppelike vegetation only by human influence. The present soils of these central European dry regions, on the other hand, suggest that steppe existed there throughout the postglacial. In this respect there is a clear and important difference between results from biology and pedology. Both the pollen flora and the snail fauna show that in the modern dry regions of central Europe, from which Gradmann [292] described the subsequently famous steppe-heath, shortly before human interference a thick cover of warmth and moisture-loving deciduous, mixed forest was widely spread, under which a brown or parabrown earth might be expected. There are several possible explanations for these contrasts:

(1) It might be that a steppe black earth can form under a forest cover.
(2) The black earths of the present dry regions of central Europe might have been incorrectly interpreted.
(3) Possibly these steppe black earths may date from the time of the destruction of the forest—that is, they developed within the last 5,000 years, replacing the former forest soils.
(4) The soil sequence of development from rendzinas to brown earths may be diverted to other sequences as a result of human interference.
(5) The biological indications of woodland may have been greatly exaggerated.

At present it is not possible to distinguish which of these explanations is correct. In my opinion the biological data are at the root of the matter, but human interference has affected the course of soil development, diverting it in other directions, for as a result of human activity microclimates have developed there which in their essential aspects resemble a steppe climate.

Independent of these difficulties, there are other observations which show that during the postglacial warm period the climate was no longer as favorable as during the preceding warm periods. It can be assumed from this that with respect to the climatic optima of the warm periods the first clear change was at the transition from the Tertiary to the Quaternary and that the second occurred between the last warm period and the postglacial. Comparing the Pleistocene warm periods with one another, however, we see that no marked changes occurred. With respect to temperature, nothing definite can be stated. Only the amount of precipitation appears to have diminished a little in many places.

These climatic changes can scarcely have caused the significant transformation in the flora and fauna of central Europe during the individual warm periods, and something else must be responsible for the spontaneous disappearance of the hippopotamus during the postglacial, even though it occurred in central and western Europe during the earlier warm periods (Fig. 88), for the indicated climatic differences between warm periods are very small. The decisive factor was much more the damaging effects of the cold periods. Before the question of climatic changes during the cold periods can be taken up, it is necessary, however, to refer once more to changes in the water budget during the younger Quaternary warm periods. In contrast to what has been said up to this point, Selle [674] emphasized that the climate during the last two warm periods (Eemian, postglacial) apparently became more moist, for the growth of peat bogs increased from the Eemian to the postglacial. The attempt has already been made (pp. 104, 112) to reconstruct the climatic conditions of the last two warm periods of the Ice Age (Holsteinian and Eemian warm periods) with particular reference to the available precipitation. The result of this was to show that the water budgets of the two warm periods were not very different from one another. Furthermore, it could be established that the apparent increase in wetness at the end of the last warm period was probably only an effect of falling temperature. Yet the notable fact remains that seemingly the number of bogs in the postglacial was appreciably greater than during the last warm period, which, again, had a greater number of bogs than did the Holsteinian warm period. It must be questioned, however, whether this contradiction is more apparent than real.

Bogs develop not only as a result of the silting up of lakes and ox bows but also on mineral soils between water courses. Their number is particularly great in regions where the climate is humid throughout the year and where there are numerous closed depressions in the topography. Thus bogs are especially numerous in regions of central Europe which are dominated by an oceanic climate and in the region of the last glaciation, where the retreating ice left numerous depressions, which offer a good starting point for bog development.

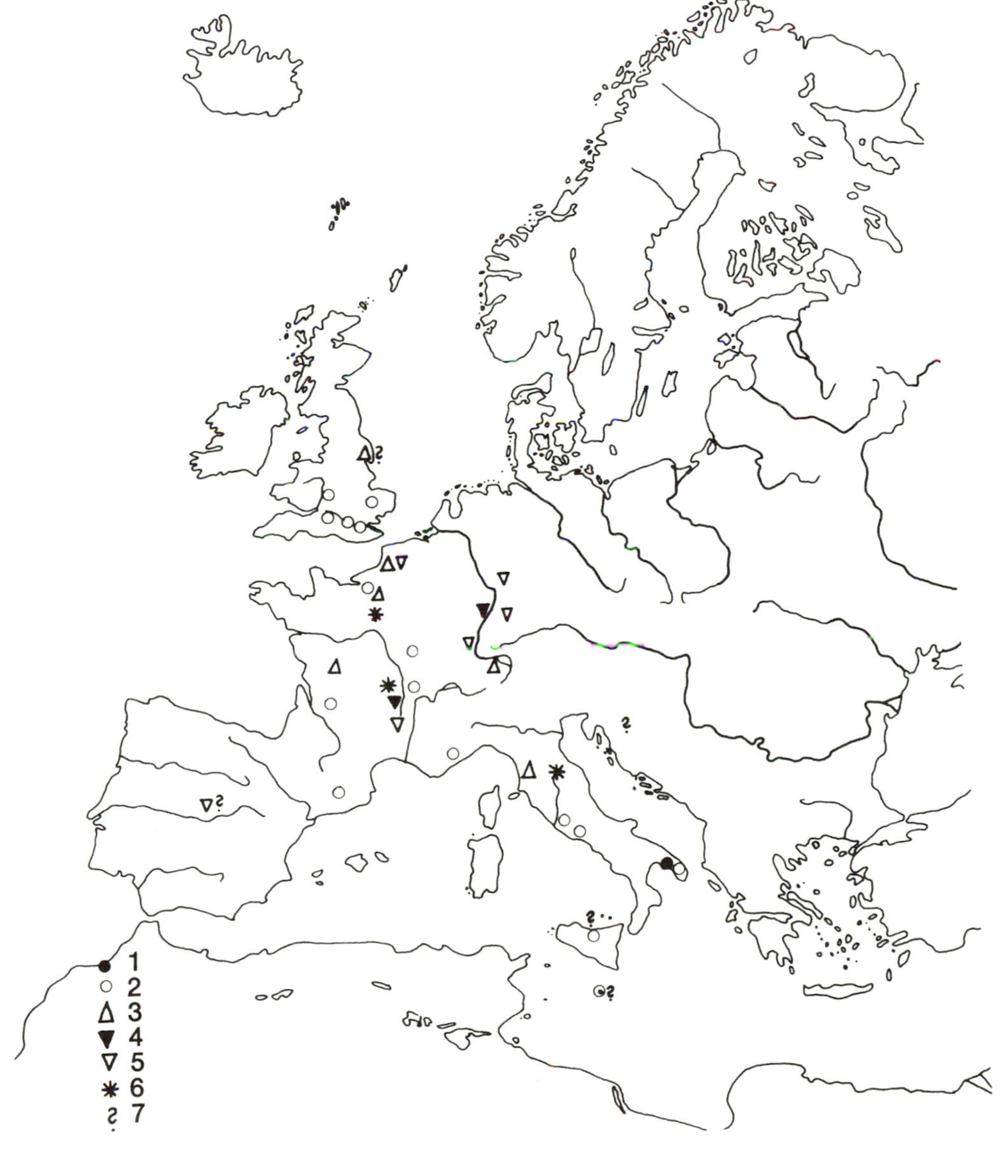

FIGURE 88. Hippopotamus (*Hippopotamus* sp.) fossil localities in the Pleistocene of central, western, and southern Europe: 1. last cold period; 2. Eemian warm period; 3. Holsteinian warm period; 4. Elsterian cold period; 5. Cromerian warm period; 6. lower Pleistocene; 7. age unknown [260].

At the beginning of a cold period, frost-induced soil-creep is particularly noticeable on the slopes of hills and mountains. It is thus easy to imagine that under the influence of frost, the bogs on the watersheds are destroyed first. The warm-period bogs of the so-called "Herning" type are a persuasive witness. They developed in depressions and after the beginning of the last cold period were covered by peat material from the surrounding slopes as a result of frost action. Consequently, evidence of the former existence of bogs of earlier warm periods will be found only where former basins occurred. On this evidence the apparent area of warm period bogs must appear smaller than the postglacial.

After the melting of the Saalian ice sheet there remained in northern, central, and eastern Europe a relief type which had all the characteristics of a typical young moraine topography. In the countless small lakes numerous bogs formed in the course of time. In more humid climates bogs also developed on the watersheds. At the beginning of the last cold period these were largely removed by very active slope movement, so that their remains occur only here and there under the most favorable local conditions. Yet the distribution of Herning-type bogs occurring from northwest Germany and Denmark in the west to central Russia in the east gives the impression of an appreciable development of watershed bogs during the last warm period. These slope and watershed bogs in the periglacial region and especially in the regions later occupied by the ice sheet itself were destroyed during the last cold period. Most of the bogs in old lake basins or depressions fell victim to the abrasive action of the ice sheet, so that the remains of warm-period bogs are usually only found beyond the margins of the last ice sheet (Fig. 76). The small number of bogs of the last warm period in comparison to the postglacial is thus a consequence of periglacial and glacial denudation during the last cold period, and it is unnecessary to attribute it to a contrast in the climate of the last warm period with respect to the postglacial. The same is true for the Holsteinian warm period, yet the conditions are still worse. The Saalian glaciation was on an appreciably larger scale than the Weichselian, so that beyond the margins of the last glaciation there remained a zone of glacial relief produced by the Saalian glaciation. Kettle holes in this zone were partly filled with peat during the course of the last warm period and subsequently were not destroyed, although covered by loess and solifluction layers during the last cold period. It is in this zone that most of the bogs that provide a valuable indication of the former climate and vegetation history were later found. The scale of the Elsterian glaciation was, however, generally not greater but usually less extensive than the Saalian glaciation. Thus the relief of the young moraine left upon the retreat of the glaciers of the Elsterian cold period was, later and almost without exception, ground off by the Saalian glaciers. On this basis alone the number of fossil bogs of the Holsteinian warm period must be appreciably smaller than during the Eemian warm periods and very much smaller than during the postglacial.

Thus from the number of bogs found in the various warm periods, nothing can be deduced of possible climate changes. The above-mentioned researches have done much more to show that the precipitation during the Holsteinian and Eemian warm periods was about the same. In a few rare cases it is possible to infer a wet closing phase of the Holsteinian warm period. This set in, as in the case of the Eemian warm period, immediately at the end of the warm period, since from other evidence it can be seen that the onset of cooling was already marked. The wet phase continued into the beginning of the following cold period to a greater extent than happened at the end of the Eemian, so that the Saalian cold period seems to have begun less suddenly than the Weichselian cold period.

THE COLD PERIODS

It has been established that the climatic conditions of the Pleistocene cold periods must have been the decisive factors controlling selection within the animal and plant kingdoms during the Quaternary. On the other hand, during the earlier Pleistocene cold periods, many plant and animal species, though from an apparently more favorable climate, were nevertheless able to persist. This gives rise to the impression that the severity of the climate increased from the older to the younger cold periods, so that we must consider climatic changes not only with respect to warm periods but also with respect to cold periods. That obviously raises a very important paleoclimatic problem. For if the impression were to correspond with the facts, it would have to be assumed that the climate, at least of the Northern Hemisphere as a whole, during all phases of its Pleistocene history, underwent a large-scale, unidirectional change which affected all factors controlling the form of the climate. In other words, this might provide a starting point for a deeper understanding of Pleistocene climatic history.

To probe further into this question, it is necessary to go back to some earlier achievements (Table 19). Regrettably, only the Pretiglian and Weichselian cold periods can be compared, for knowledge of the other cold periods, in a quantitative sense, is still too imprecise.

TABLE 19. MINIMUM VALUES OF CHANGE OF CERTAIN CLIMATIC FACTORS IN COMPARISON WITH THEIR PRESENT-DAY VALUES DURING THE PRETIGLIAN, AND WEICHSELIAN COLD PERIODS IN CENTRAL EUROPE AND WESTERN RUSSIA

COLD PERIOD	MEAN ANNUAL TEMPERATURE °C	MEAN ANNUAL PRECIPITATION (MM.)
Weichselian	−13	−300 to 350
Pretiglian	−8 to 10	−150 to 250

It must once more be repeated that these values serve only as guidelines. The absolute value itself is not reliable. However, the data show (cf. Fig. 89) that the climate at the height of the Weichselian (and the climatically very similar Saalian) cold period was appreciably colder and drier than during the Pretiglian cold period. This holds even if formations and fossils of different ages should be united or confused under the general heading of Pretiglian cold period. Always material from the oldest Pleistocene, however, is being considered, and if the observations are not valid for the Pretiglian cold period, then they refer to a somewhat younger cold period still within the oldest Pleistocene. Even if the actual age is different, the marked climatic change expressed in the figure remains unchanged.

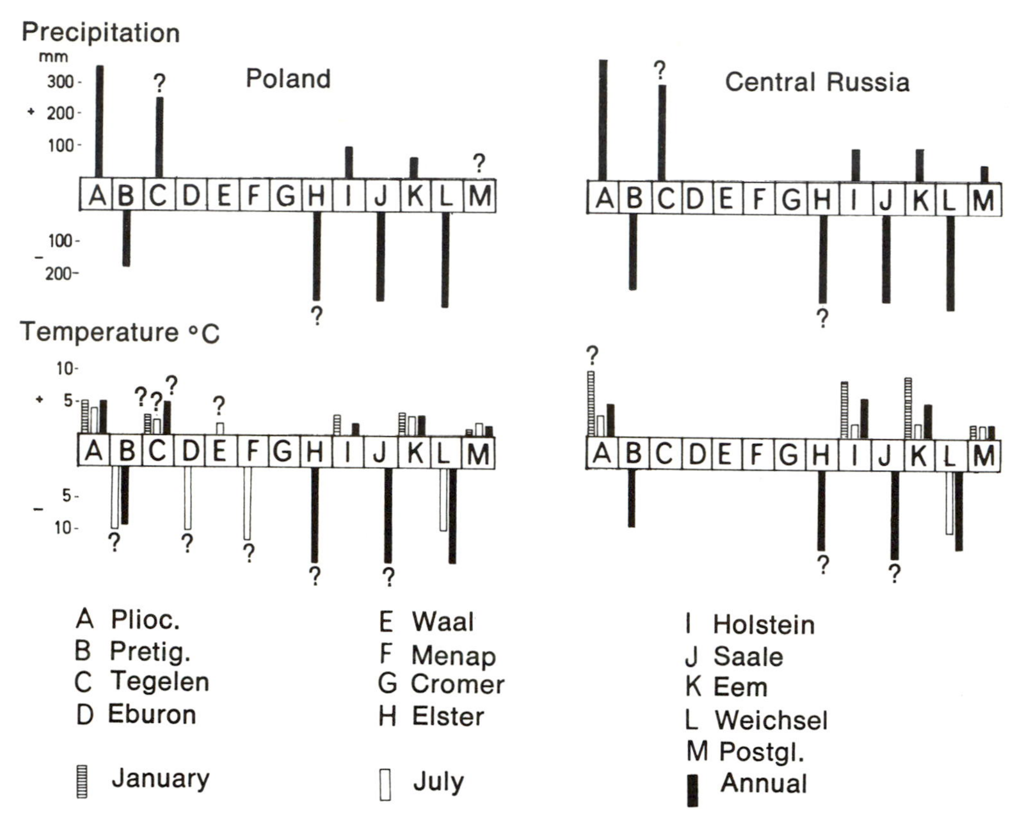

FIGURE 89. Review of the scale of climate fluctuations and climatic changes in Poland and central Russia during the Quaternary. Upper row: precipitation; lower row: temperature. Only the minimum difference from present-day figures is shown. Positive deviations above the line, negative deviations below the line.

Thus the climate of the cold periods, at least in central Europe, became progressively colder and drier. This fact does not signify that it was a continuous process. It was determined earlier that the climate at the height of the Saalian glaciation with respect to temperature and dryness could be compared to that of the Weichselian cold period, although perhaps a little colder and a little wetter. It can be seen from Table 17 that during the preceding Elsterian cold period representatives of animal communities adapted to extreme winter cold began to appear. Thus the climate of the Elsterian cold period was probably very cold and dry. This is supported by the former distribution of loess and tundra flora (Fig. 90) in Eurasia. There is still a long way to go before the distribution and character of the various types of vegetation of this cold period can be reconstructed; nevertheless, Figure 90 shows very clearly that the climatically induced spread of open-steppe vegetation resembles in general the characteristics of the later cold periods. The maximum of the Elsterian cold period, according to this, cannot have been appreciably different from that of the Saalian or Weichselian cold periods, even if many deviations occurred which may be individually significant but which in sum alter the general picture very little. The distribution of still older loess and tundra flora is less convincing, the more so when it is considered that, in Figure 91, localities from the whole of the lower Pleistocene—that is, from the Pretiglian to the Menapian cold period—are represented. Woldstedt [806] drew attention to the fact that probably a greater part of the older Pleistocene sediments have been eroded away in the course of time, so that only a very incomplete picture of the former biogeographic and climatic conditions can be obtained. Yet it appears questionable that the difference represented, between the Pretiglian cold period on the one hand and the Elsterian, Saalian, and Weichselian cold periods on the other, can be attributed only to this. The already numerous observations on the spread and magnitude of the oldest ground frost evidence as well as the evidence of the character of the vegetation during the Pretiglian cold period [260 and Fig. 49], show that the climate during the least favorable segments of the oldest cold periods was somewhat warmer and moister than during the last three cold periods.

A very welcome confirmation of this statement is provided by researches into the surface water temperatures in tropical seas [199]. It can be shown that the lowest mean extremes of temperature from the last three or four cold periods are similar, and that they are clearly distinguishable from the higher mean minimal temperature of the preceding Pleistocene cold periods.

It must be assumed from the above observations that the somewhat different climatic character of the Pleistocene cold periods concerns climatic changes which essentially affect the first part of the Pleistocene, up to about the Menapian cold period of Dutch terminology. Since then, no further temperature changes at the peak of the cold periods can be determined with certainty. It is possible, however, that changes occurred in the water budgets of the continents. It is understandable that these climatic differences must have increased the chances for survival of the less resistant plants and animals during the earlier cold periods, for the number of available ecological niches must have been much greater than was later the case. It should be added, furthermore, that the less severe cold periods permitted a more rapid and extensive remigration of organisms into newly accessible regions with the improvement of climate than the much more extreme younger cold periods.

The Problem of "Stable Climate" During the Ice Age

The climate of the oldest cold period at the time of its maximum was warmer and moister than during the later events. A strong continental glaciation is first recognized in the Northern Hemisphere at the time of the Menapian-Nebraskan cold period. Occasionally, still earlier glaciation is reported. Figure 49 gives the possible conditions for the Pretiglian cold period. The scale of these glaciations in no case appears to have exceeded those of the second half of the Ice Age. On the other hand, the thermal difference between the climatic optima of the warm periods is not so great that it must be assumed that during the older warm periods more ice must have melted from the colder regions of the earth than during the young warm periods or during the postglacial. If we accept this condition, the various levels of the marine terraces formed during the warm periods cannot be related to progressively less melting of the continental ice sheets, as alleged. On the other hand, the assumption of a slow sinking of the ocean floor during the

FIGURE 90. Indication of the localities where open vegetation of the Elsterian cold period has been found in northern Eurasia: 1. floral localities; 2. loess; 3. faunal localities [260].

FIGURE 91. Indication of the localities where open vegetation of the Pretiglian, Eburonian, and Menapian cold periods has been found in northern Eurasia. The localities must all be combined on the one map, for the dating is not sufficiently certain. 1. floral localities; 2. loess; 3. faunal localities [260].

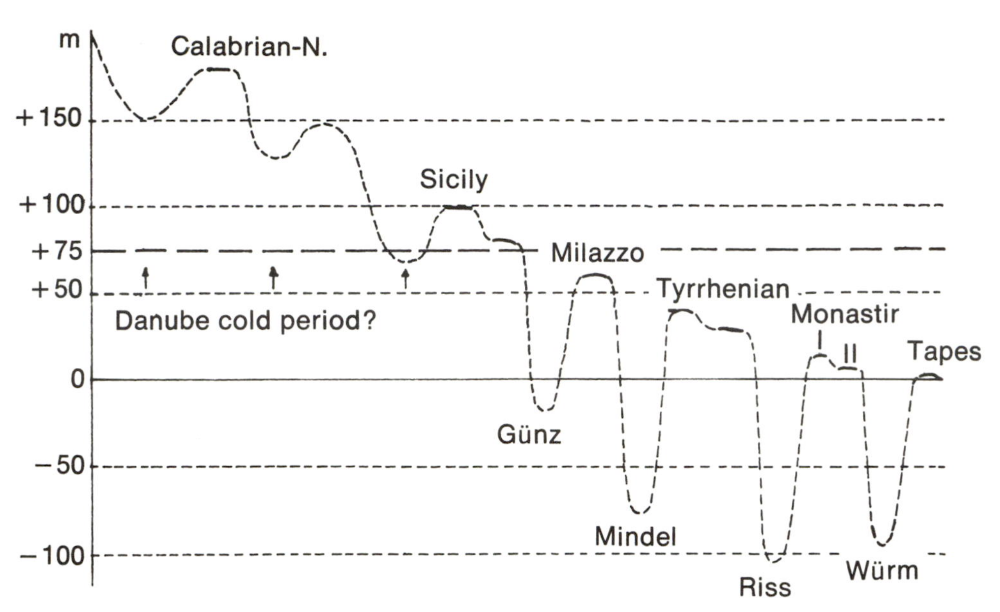

FIGURE 92. Elevation of the raised beaches of the Ice Age and the probable elevation of sea level if the ice in Greenland and Antarctica were melted (heavy broken line about + 75 meters). Modified from Woldstedt [804].

Ice Age [102, 238, 806, 807] must also be seriously doubted, and it seems improbable that the general fall in sea level during the Ice Age (insofar as it appears from the decreasing elevations of the warm-period marine terraces) can be assigned to a tectonic process. This contradiction was the reason for giving increasing attention to the history of the Antarctic continental ice sheet. It has been calculated that the melting of this ice mass and that of Greenland would lead to a rise in oceanic level to about the height it attained during the earliest Pleistocene transgression of Sicily (Fig. 92). If we apply the idea to the explanation of the elevation of the younger sea levels, we may infer that either the Antarctic ice, which is by far the most important store of the present ice, during the whole of the Pleistocene grew more or less uniformly independent of the climatic fluctuations that occurred further north—in which case Antarctica was a region of stable climate [786]—or the Antarctic ice sheet was subject to a marked change between ice accumulation during the cold periods and a diminution during the warm periods, but with the intensity of the warm periods continually decreasing. It always follows from this that the changes and climatic fluctuations in Antarctica must have run a different course than in the regions considered up to now.

It was earlier demonstrated that the climatic fluctuations of the last cold period occurred simultaneously, not only in the Northern Hemisphere but also at the equator and in the Southern Hemisphere. This is true for New Zealand, Patagonia, and Tierra del Fuego [27]. Even each relatively insignificant climatic fluctuation which caused the retreat of glaciers in the distant parts of the globe during the last 100 to 150 years was a simultaneous phenomenon. The process at the present time operates also in Antarctica, though it is notable only in the most northerly marginal regions [585]. In the region of McMurdo Sound in Victoria Land the position of the glacier front has remained constant for the last fifty years, as a comparison with older photographs shows [584, 585]. Radiocarbon dating of seal skeletons found there from 100 meters to about a mile from the glacier front gave ages from 200 to 2,600 years old. Dating of dried-out algae from a small pond lying on a moraine which still had an ice core provided a figure of 2,500 to 6,000 years B.P. The radiocarbon dating in this region is really uncertain, since the present atmospheric radiocarbon concentration is unknown in Antarctica. Nevertheless, the reported values, even when regarded as only minimal, show that the position of the glacier front in McMurdo Sound for several centuries if not for a few thousand years has remained constant. The same is shown by the relatively old frost fissures which cut up the soil immediately in front of the glacier [584]. Those glaciers mentioned, independent of the climatic fluctuations of which traces can still be found around the periphery of Antarctica with the same rhythm found in regions further north, show no change in the position of the glacier front. The same conclusion holds even where only short mountain glaciers which are not connected with the main ice mass are being considered. Thus the present climatic fluctuations do not affect the ice budget of the continental ice sheet to any notable extent. On the other hand, traces of an extensive glaciation have been described there. The former upper ice surface lay in places 300 meters higher than today [347]. Büdel [102] and Flohn [238] have assumed that this glacial high stand occurred at the beginning of the last cold period or even in the Eemian warm period. It was thus supposed that the warm periods of temperate latitudes were phases of strong glaciation in Antarctica, and during the cold periods the glaciation must have receded, or at least did not increase. It does not follow from this that the warm periods of temperate latitudes corresponded to cold periods in Antarctica. Rather, the explanation is that even though the warm and cold periods occurred everywhere simultaneously, an intensified cold climate in Antarctica did not lead to an increase in the ice mass there, for the supply of snow was even more strongly restricted.

It seems that during the Pleistocene in Antarctica climatic fluctuations did not play the same dramatic role as in northern temperate latitudes [836]. If this is true, then the main mass of the Antarctic ice sheet must actually be very old if its volume even in the change between warm and cold periods increased or decreased to what is still today a largely unknown extent. The gradual accumulation of the present ice mass certainly must have led to a fall in sea level over the course of time. Yet it is questionable whether the different elevations of the strand line can be explained in this way, for the terrace of the Sicilian transgression is not the oldest of the Ice Age. Still higher lie terraces of the Calabrian stage formed at the beginning of the Pleistocene, which cannot be explained in this way, for the ice mass of Antarctica and other glaciated areas is too small. Furthermore, Emiliani's research [198] on the decrease in temperature of the deep waters of the equatorial Pacific during the immediate geological past has shown that the beginning of the decisive cooling lies considerably farther back than the older Pleistocene. Ericson, Ewing, and Wollin [205] have also found ice-transported material in the Pliocene sediments of the Indian Ocean, which indicates at least a partial glaciation of Antarctica at that time. It must therefore be concluded that the climate even prior to the beginning of the Quaternary was favorable for a continental glaciation of Antarctica. Here is a region in which the climate has remained comparable and stagnant over a long time. Whether such an impression is really justified must be determined by further research. That this region with stable climate was able to influence the climate of the low latitudes in the southern and northern latitudes [102] seems to me very unlikely, in view of the more recent changes in the glaciers around the periphery of Antarctica which were synchronous in the same sense as those in the Northern Hemisphere. It is also clearly unlikely that the continually decreasing elevation of the marine terraces of the warm periods is intimately connected with the increase of the Antarctica ice mass. For if the origin of this ice lies further in the past than Büdel or Flohn supposed, then the strongest field argument in favor of that hypothesis fails. Possibly Woldstedt [807] was correct in referring to an opinion of Arambourg's according to which the increasing elevation of the marine terraces of the warm periods with increasing age is related to the rise of the continents caused by the unloading of the land masses as a result of denudation during the youngest geological past.

Weischet [786] has drawn attention to the fact that outside Antarctica another region of cold climate occurs in which a similar stable climate may be recognized. This is central and eastern Siberia. The history of Pleistocene vegetation seems to show that no distinctive changes occurred as in Europe; instead, during the cold periods far to the north, woods or thickets of spruce, pine, and larch are recognizable which to a considerable extent resemble modern vegetation. Vasil'ev [763, 764] earlier expressed the same ideas based upon his investigations of the Pleistocene flora of central and eastern Siberia. Essentially, however, he appears to have compared only the flora of the sediments formed during the warm periods and unfortunately, did not consider the fossil fauna, which even in central Siberia very clearly shows several advances of the tundra to the south. It must nevertheless be questioned whether Weischet's assumptions were correct even excluding Vasil'ev's attempts at reconstruction. The concept of the stability of the climate of central and eastern Siberia was based upon an estimate of July temperatures from the position of the polar forest limit. As previously noted (above, p. 140), the comparison of the position of the polar tree limit in a mountainous region (central and eastern Siberia) with an adjacent lowland (western Siberia, eastern Europe) carries with it errors of undetermined magnitude when an attempt is made to extrapolate July temperatures from this reconstruction. The 10° July isotherm so determined combines regions in the mountains with equally favorable microclimates, but it does not indicate the overall climatic conditions. It is thus only a supposition that during the cold period the temperature decreased by only a small amount. The true scale of this climatic deterioration is unknown. However, the preceding studies have repeatedly brought out that the decrease in temperature and precipitation in the continental regions was appreciably less than in the oceanic region. Yet it does not follow that the climate in the center of the continent was always particularly stable. The basis for a necessary determination of the climate of the cold periods is far too uncertain and insufficient. There are, on the contrary, many indications that the climate of the warm periods in comparison with the present day was considerably different. Just as in the oceanic climatic regions of northern Eurasia, so too in the continental there were decisive climatic fluctuations during the Ice Age. In the continental regions, however, these fluctuations were particularly apparent during the warm periods, while in the oceanic regions they were most obvious during the cold periods. The presently available material is insufficient to establish how strong the fluctuations actually were in the center of the continent, or to indicate whether climatic changes occurred there also. The changes in composition of the plant communities caused by climatic fluctuations were relatively insignificant compared with the changes in Europe and in the oceanic regions of the Far East. Yet this small alteration in the vegetation is a result of the fact that in Siberia since the end of the Tertiary a woodland type adapted to extreme winter cold was widely distributed. Thus the high resistance to cold of these plant (and animal) species means that the climatic fluctuations do not appear clearly. The picture is considerably different when the history of the thermophilous deciduous tree species in southern Siberia, or of the permafrost region, is considered (Fig. 21). They show that climatic fluctuations in Siberia were of greater acuteness—that is, Siberia was not a region of stable climate—during the Pleistocene.

Summary

With Europe as an example, it was shown that the climate at the peak of the cold periods became progressively colder and drier. Comparison of the optima of the warm periods showed that at best temperatures fell only insignificantly, but precipitation diminished appreciably. This is particularly true when the postglacial is included. Thus in general the continentality of the climate in northern Eurasia increased. The explanation of this process lies in part in the developmental history of northern Europe. Vosktesenskij and Grichuk [777] as well as Büdel [102] have pointed out that as a result of the uplift of mountains the continentality of the climate of northern Eurasia during the Ice Age, particularly in the interior regions, must have increased. The uplift of the central highlands during the Pleistocene did not in general exceed 300 meters [260]; the high mountains of alpine character, however, were elevated appreciably more. It is readily understandable that the continentality of the interior regions would be increased by this means and that on the weather side of the mountains the oceanity of the climate increased. The tectonic movements of various regions of northern Europe thus led to a sharpening of climatic contrasts, and in general the climatic development during the Ice Age may have been pushed in this direction, which has already been recognized. It appears that the time of the more significant climatic changes—that is, the greatest increase in aridity and cold during the individual cold periods and the clearest decrease in temperature and precipitation when the warm periods are compared with one another—fell within the first part of the Ice Age, at a time shortly prior to and simultaneous with the time of greatest tectonic movement during the Ice Age [260]. With the end of the strong tectonic movement the magnitude of the climatic changes (but not the fluctuations) diminished. Is this an accident, or did the tectonic movements during the first part of the Ice Age cause the sharpening of the continentality of the climate? In this connection it must be stressed that it is quite unlikely that the Pleistocene climatic fluctuations and changes were caused by tectonic movements in individual regions of the earth. Schwarzbach [667] has very convincingly presented the arguments against a tectonic origin of the Ice Age, but for other causal factors the tectonic movements clearly operated as a modifying influence.

COLD PERIODS AND PLUVIALS

In the course of the investigation up to the present, consideration has been restricted to regions outside the present arid zones, as if the Pleistocene climatic fluctuations and changes did not affect the region of present-day desert and steppe regions. This impression is deceptive, however, for the paleoclimatic problems there are so difficult and the obstacles opposing a reliable but limited estimate of the past climate in the deserts and steppes so numerous and so great that it is necessary to devote the following section to the questions associated with them.

Basic Problems

In the arid regions of the earth, particularly in the Sahara and the deserts of Asia Minor, traces of an apparently moister climate in the past are repeatedly found. Ruins of settlements pushed far out in the present most arid zones, dried-out valleys, gravel beds of former rivers, rock paintings of animals in regions where they may no longer live (Fig. 93), and many other signs seem to indicate that earlier the climate there was much better. Particularly at the limit of the inhabitable zone in regions which can no longer be used by men, a climatic deterioration in the water budget must have played a decisive role in maintaining possible human living conditions. It has often been suggested that the greatest political events of the past, such as the Mongolian invasion, might be attributed to climatic fluctuations. It is clear, however, that the expansion of mankind from the still usable regions to the limits of the dry zone were not only caused by climate but also associated with changes in the course of river beds [Tarim Basin, 840; lower Amur darja, 750, 751], and to a very great extent by the extent of political struggles and ambitions of specific persons or times. The extent of former agricultural plains and the position of settlements are thus no safe indication at the limits of the arid zone of past climatic conditions.

On the other hand, pedology, Quaternary geology, and paleontology (in the widest sense) provide so many indications of an improved water budget in earlier times in the present deserts and steppes that the reality of the existence in former times of an improved water budget in the dry regions of the earth cannot be doubted. Yet there are fundamental difficulties in attempting to understand the associated problems. These are, excluding the inaccessibility of many arid regions, that the dryness of an arid region may be the result of too small, too irregularly distributed rainfall or of too high an evaporation. A reduction in temperature just as much as an increase in precipitation or a difference in rainfall distribution through the year can improve the water budget of a region. Moreover, the water balance can be improved by a reduction in radiation; thus an increased degree of cloudiness can reduce the aridity of a dry region. According to this, the evidence of increased settlement of a dry region by man, animals, or plants may depend upon very different thermal as well as hydrological causes; or both factors may interact. The individual datum is, from the climatic standpoint, generally ambiguous.

The second difficulty lies in the fact that in the arid regions of the earth the rain showers that occur from time to time have a great geomorphological effect. Within the shortest of times giant floods roar down the dry valleys, transporting huge masses of sediment for shorter or longer distances. Thus in the arid regions, fluviatile sediments are not so rare. Yet their value for paieoclimatic purposes is very low, particularly when other means of estimating the time elapsed between two or more periods of precipitation are lacking. The accumulation of sedimentary layers of aquatic origin following immediately upon one another in some regions may thus only represent rare "rainy periods" which were interspersed with long periods of no sedimentation or even of aeolian denudation.

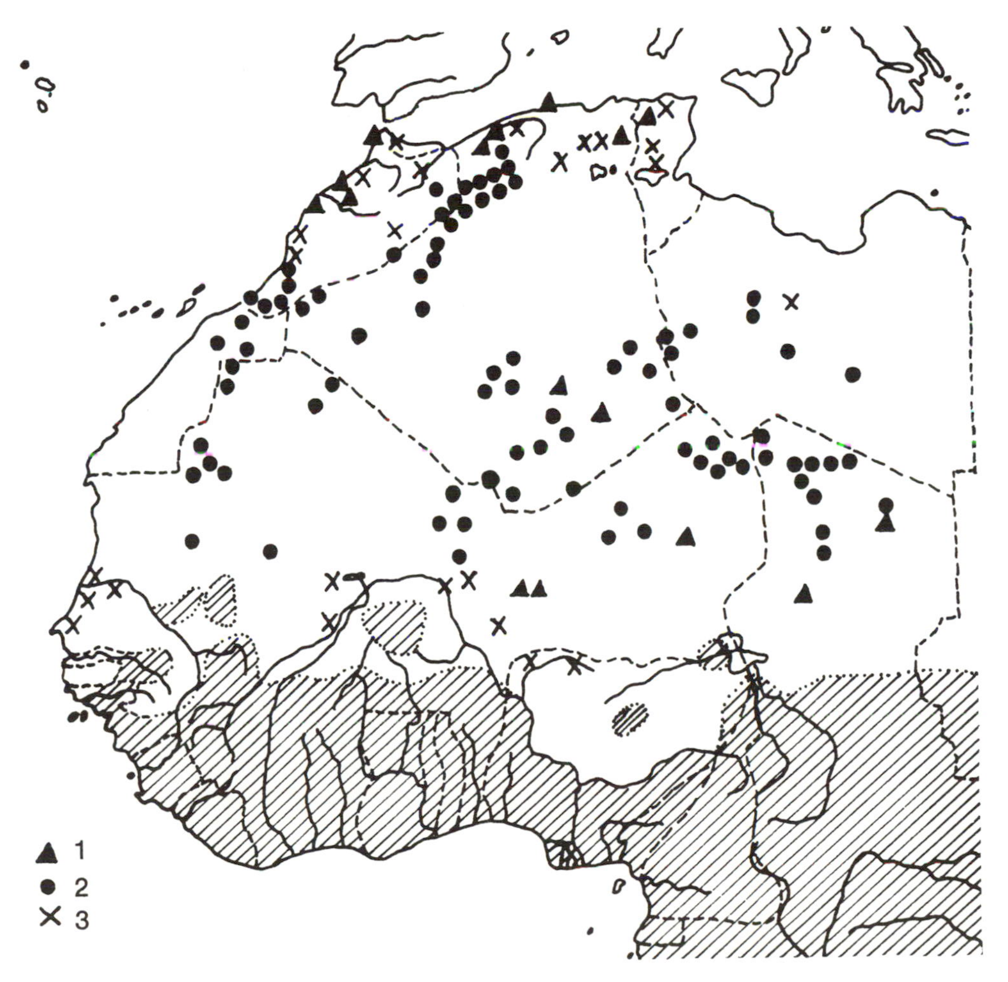

FIGURE 93. Indications of the former occurrence of elephants in northwest Africa. Ruled area: present occurrence. 1. skeleton remains; 2. rock paintings; 3. reports from the literature after Monod [535].

The third problem lies in the paleoclimatic interpretation of former plant and animal distribution whose center of gravity today lies in an appreciably more favorable climate (Fig. 93). The distribution limits of a certain taxon are due either to the fact that members of this taxon can no longer endure the increasingly unfavorable abiotic conditions or that they cannot withstand the strong competition of other living forms. In total, the vitality of the individuals of the species considered is significantly less there than at the center of distribution. If, now, man destroys the furthest outposts of a particular plant or animal species, a very long time must elapse before the species can again reach the original limits. And the time is increased the more man destroys not only individual species but whole biological communities. The subsequent regeneration of the region, furthermore, requires that man or his domestic animals decisively has interfered only once, and that since then the species attacked has been left in peace. In the arid regions this was undoubtedly not the case, for there are only a small number of hunting animals or useful plants available. From this it follows that these forms are intensively sought. So many of the rock drawings in the Sahara of elephants, giraffes, hippopotami, and other animals are not so much the impression of the abundant animal life there as a cultural act by a form of magic-powered wish that these animals should appear and be hunted. It is not my intention to enter into a discussion of the problems of rock drawings in the Sahara, but only to recall that the activity of man there in the last few thousand years must have decimated the plant and animal world, so that the impression is created that the climate became increasingly drier and hotter.

Finally, in the dry regions, most of the common aids for relative and absolute dating of deposits of the moister periods are lacking, for well-formed river terraces which can be used as a reliable criterion of age are generally absent, and organic matter that would permit radiocarbon dating is present only in very much smaller amounts than in the humid climates.

These critical remarks are not to deny the existence of wetter climatic phases in the present arid regions, but to warn against attributing too great a significance to preliminary conclusions. As Monod [535] has recently written, "The literature, however considerable, deals more with theories, opinions, and suppositions than with observed facts. These themselves are often subject to interpretations which are hazardous, ambivalent, or even impossible. . . ."

From the common occurrence in the arid regions of evidence of a wetter climate, the conclusion has frequently been drawn that the former precipitation was higher than it is today. This led to the term "pluvial" for the rainy periods that paralleled the glacials—that is, the ice ages in temperate latitudes [102]. It is clear that a general cooling of the atmosphere has resulted not only in a considerable spread of the glaciers in the north and south, but also in a cooling in the arid regions of the earth and a displacement of the weather-controlling polar front toward the south (and toward the north in the Southern Hemisphere). The present arid regions should thus have had rainy periods at the time when the cold temperate latitudes were enduring ice ages. How far this supposition can be substantiated must be tested in detail. Yet even when there is a positive answer to the question, we cannot say that the rainy periods on both margins of the arid region were simultaneous—that the southern boundary of the subtropical arid zone in the Northern Hemisphere (and the corresponding boundary in the Southern Hemisphere) advanced northward at the same time that the northern boundary moved southward. In other words, it is questionable whether the pluvial periods on the northern and southern sides of the arid zone were synchronous or were out of phase [102]. Furthermore, it is very reasonable to ask whether the pluvial climatic conditions prevailed during the entire northern ice ages or whether only individual segments of the Ice Age, such as the advance phases of the glaciers, brought the true pluvials in the south [105]. Thus the original simple picture of a relationship between the cold periods and pluvials loses itself upon closer consideration in a mass of varied, and at present very difficult, problems.

The Caspian Sea

The Caspian Sea lies approximately at the western boundary of the Asiatic desert region with the steppe and wooded steppes of southern Russia. It has at the present time no drainage to the ocean, so that changes in influx or evaporation cannot be compensated by water from the oceans. It thus would seem that the history of the Caspian Sea presents a very good insight into the problem of the pluvial periods. Figure 94 shows that during the Ice Age there were considerable changes in the area of the Caspian Sea. The number of figures could be increased, but in principle, simultaneous with the advance of the northern ice, massive transgressions always occurred on the lower Volga; the regressions coincided mostly with the warm periods. In the cold periods the water level rose repeatedly, so much so that the Caspian Sea overflowed into the eustatically lowered Black Sea through the Manytch Depression. It is suggested that we see in the cold-period transgressions of the Caspian Sea an expression of increased precipitation and reduced evaporation. There is an added difficulty, however, for the Caspian Sea in the most recent geological past was sometimes fed from the Amu darja. At some periods this great river emptied into the Caspian Sea, at others as now it flowed north into the Aral Basin. From the paleoclimatic point of view it is important that the changes of level of the Caspian between warm and cold periods were independent of course changes of the Amu darja; if so, they provide particularly good examples of the effects of pluvial climates.

During the last cold period the great lower Chvalyn transgression (Fig. 94) and the less important upper Chvalyn transgression affected the Caspian Basin. Between the two the water retreated somewhat, and a particularly significant regression occurred approximately between 2,000 to 4,000 B.C. [460] separating the upper Chvalyn transgression from the younger and relatively weak new Caspian transgression. For the last-named transgression, dates between about 1,000 B.C. and the beginning of modern times have been given [460, 214, 747, 367, 368, 397, 398], but the exact age is obviously not certainly determined. Of the transgressions noted, that of the lower Chvalyn was the greatest. Butzer [106, 113] assumed that this transgression occurred prior to any notable fall in temperature during the earliest part of the last cold period, so that this immediately would clearly confirm the assumption of a very wet phase at the beginning of the last cold period. The most recent Russian research, however, does not confirm these assumptions.

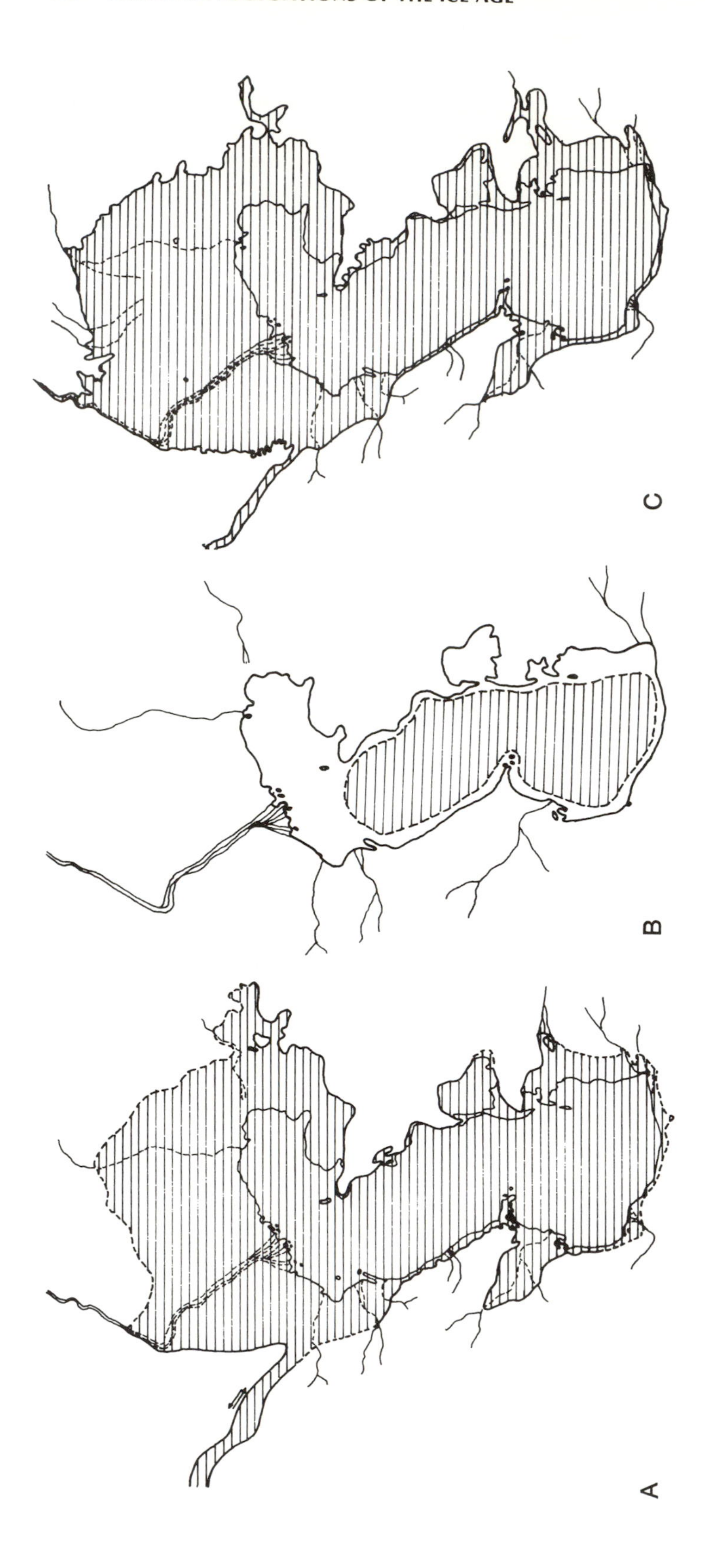
C
B
A

North of Volgograd (Stalingrad) in a tributary channel of the Volga a paleolithic station (Suchaja Mechetka) was discovered some years ago. It contained cultural remains of the late Mousterian [310, 311]. In the same horizon, remains of animals that were characteristic of steppe and fringing woodland were found resting on a fossil soil. The coeval pollen flora indicated an impoverished steppe flora. Man thus lived in a markedly dry steppe region. Subsequently, the fossil soil and the remains noted were covered by slope material that slipped from the neighboring hill slopes as a result of frost action [304, 550]. Probably of the same age are traces of fossil ice wedge nets found at many places in soils of the northern Caspian depression [545, 213]. Only when some 20 meters of loamy sediments had been deposited over the cultural horizon, a process twice interrupted by phases of weak soil formation, did the lower Chvalyn transgression reach the region of the gorge. During the transgression the Caspian Sea was bordered by a boreal spruce-pine forest with some fir which had replaced the steppe flora. Rather surprisingly, at the peak of the transgression the importance of the forest vegetation diminished and gave way to a forest steppe, in the groves of which the same tree species still thrived. This is also the case for those locations which are found immediately on the banks of the lower Chvalyn Sea [301, 747]. With the northward advance of the Caspian Sea the Volga and its tributaries were forced to deposit their sedimentary load. A thick river terrace lying some 12 to 14 meters above the present level of the Volga in the region of Kuibyshev is evidence of this process. Possibly of the same age is the 20- to 22-meter terrace which the Belaya cut in the western foothills of the Urals. This terrace has a lower 13 to 14 meters of fluviatile and lacustrine material and contains in its upper levels the pollen flora of a spruce wood. A very primitive flint scraper was also found there which Bader [32] assigned to the Aurignacian on the basis of the stratigraphical conditions, but which could equally well be much older. Above, these sediments follow old river formations in which spruce stumps were found; these stumps, according to the evidence of the coeval pollen flora, come from trees of a spruce-wooded steppe [137]. Radiocarbon dating of this material carried out in Heidelberg gave an age of 29,700 ± 1,250 B.P. [364], which made it seem to be material from a formation of the Stillfried B interstadial (Table 5, p. 67 above). Only above this bed does a thick layer of loess-clay occur, representing hillslope material that slid in as a result of frost action covering the deposits of the falling river.

If it can be assumed that the spruce woodland phase on the Belaya corresponded to the spruce woodland phase at the time of the advancing lower Chvalyn Sea and that the spruce-wooded steppe phase of the Belaya was contemporaneous with the equivalent phase of maximum expansion of the lower

FIGURE 94. Alterations in the area of the Caspian Sea during the changes between warm and cold periods: A. Upper Chazar transgression during the Saale-Dneprovsk cold period; B. probable area during the Eemian-Mikulino warm period by analogy with conditions during the postglacial warm period; C. Lower Chvalyn transgression of the Weichsel-Valdai cold period. At the time of maximum transgression the Caspian Sea was linked via the Manytsch Depression (northwestern area of the Caspian Sea) with the Sea of Azov and the Black Sea. After Fedorov [213: A, C] and Leont'ev and Fedorov [460: B].

Chvalyn transgression, then it follows that contrary to Butzer's view [106], this important transgression began only after a period of considerable frost action during the early stages of the Weichselian cold period. It appears to have reached its maximum extent shortly before or during the Stillfried B interstadial. Its retreat occurred at the time when extreme *Chenopodiaceae* steppes became widespread in the lower Volga, apparently during the peak of the last cold period.

There are three important consequences of these observations:

(1) The transgression of the lower Chvalyn Sea cannot be considered as an indication of a pluvial period prior to the setting in of strong frost action.

(2) A comparison of the development of the vegetation with the history of the transgression shows that an earlier time for the peak of the "pluvial period" must be deduced from the history of the vegetation rather than from the history of the transgression.

(3) In both cases, however, the pluvial period occurred essentially after the beginning of the last ice age and before the period of maximum glacial extent and greatest climatic contrast.

To judge from this, the climate of the last cold period in this arid region began with a cold dry phase; later the climate was still cold, but the water budget appears to have become relatively more favorable. After the Stillfried B interstadial, a renewed cold, dry phase set in, during which the sea began to retreat. The climatic character at the time of the upper Chvalyn transgression is unknown.

The fact that the woodland vegetation, before the maximum of the lower Chvalyn Sea, was replaced by forest steppe and subsequently by *Chenopodiaceae* steppes most probably can be traced to the increasing cold and possibly to the increasing salinity of the soil.

It is difficult to determine the grounds for the lower Chvalyn transgression. Was it formerly wetter or was the fall in temperature and with it the reduction in evaporation a sufficient explanation? According to Butzer [106] the Caspian Sea loses annually a layer about 1,000 millimeters in thickness; evaporation accounts for about 980 millimeters, and inflow to the Gulf of Kara Bogaz the remaining 20 millimeters. The water influx (likewise about 1,000 millimeters annually) is made up of about 180 millimeters of precipitation on the sea surface, the surface inflow (about 810 millimeters), and subsurface inflow (about 10 millimeters per year). If it can be assumed for the purposes of crude estimate of the minimum precipitation that the annual and July temperatures in the phases of interest during the last cold period were about 10°C below the present, it follows that the potential evaporation that is thermally possible from an open water surface was about 400 millimeters per year (Fiziko-Geograficheskij Atlas Mira, 1964). Probably the actual figure was considerably less than this, for the fall in mean temperature was probably greater than was assumed here. During the first part of the last cold period the catchment area of the east European rivers was not blocked by the continental ice. Their form deviated somewhat from the present, however. In general this difference cannot have been great, so that assuming the same amount of precipitation, approximately the same amount of water could have influxed below and above ground as today. As the ground in the region considered north of the Caspian Sea was by then permanently frozen, the influx must have been predominantly surface flow. The low temperatures reduced evaporation over the land surface as well as over the lower Chvalyn Sea, so that the necessary amount of precipitation to permit the massive lower Chvalyn transgression was in no case greater than at present and could well have been much less. How large it may actually have been cannot be estimated with certainty at the present time, as there are too many unknowns in the calculation. From Butzer's data [106] the yearly water loss from the Caspian Sea is 415 cubic kilometers. At the time of the maximum of the lower Chvalyn transgression the basin, with an area of 922,000 square kilometers, was about twice as large as at present. Despite this, the annual loss was only 368 cubic kilometers through evaporation if this can be taken as 400 millimeters. In other words, on the basis of the assumptions made here, a precipitation 15 to 20 percent less than that of today over the whole drainage area of the Caspian Sea would be sufficient to explain the lower Chvalyn transgression. In this estimate no consideration is given to the fact that the lower Chvalyn Sea drained west through the Manytch Depression. However, it must be noted that the temperatures upon which the calculations were based were probably much too high, so that the water loss through the Manytch Depression, which was not considered, was probably compensated for by the lower evaporation rate. On the whole, the impression given is that the lower Chvalyn transgression is traceable not to a higher precipitation but to reduced evaporation and lower rainfall.

The Mediterranean Region and Africa

The problem of pluvials in the Mediterranean and the northern Sahara raises particular difficulties, for the divisions of the Pleistocene established for the Near East, including the eastern Sahara, which appear to agree even in fine detail with those in central Europe [107], must be reconsidered. The grounds for this are that subsequently the divisions of the last cold period in Europe have changed considerably. As a result, a number of wet or dry periods that were formerly equated to certain segments of the young Pleistocene are no longer satisfactorily correlated. Another ground for the uncertainty of the reconstruction of the paleoclimatic history of this region is that knowledge of the phytogeographical conditions which prevailed in the Mediterranean and the Near East during the cold period has appreciably altered. It was assumed earlier [105, 257, 810] that the northern Mediterranean was then occupied by different types of woodland. It was thought that only at the southern feet of the Alps did loess steppes occur, which would permit one to recognize that the climate there must have been very dry. On the whole, it appeared that the zone of maximum precipitation in the Mediterranean and the Black Sea region lay at about 45°N [238]. Today it can be established with certainty that this assumption is incorrect. It has been shown (above, p. 157) that presently available evidence of paleobotanical and Quaternary geology from the northern margin of the Mediterranean for the last cold period, as well as from the Iberian Peninsula, southern France, and north central and southeastern Italy (in part also in northern Slovenia, western Montenegro, Macedonia, and southwestern Iran) indicates steppes in which woodland was almost entirely absent. In this connection it is noteworthy that Wright [811] was able to determine a pronounced humidity of the soils (fossil soil at a depth of 10.5 meters in Abri Ksar Akil) in western Lebanon only during the time prior to 28,500 ± 380 B.P. but not later—that is, not during the phase which corresponded to the period of maximum glaciation of the last cold period in Europe. At the peak of the last cold period, from which time most of the finds date, in the regions considered there was thus not the high rainfall of a pluvial period, but quite the reverse, a considerably diminished rainfall. The same seems to hold for Cyrenaica [343] if to a lesser degree. The winter climate there between 32,000 and 12,000 B.P. was wetter than today, but there was insufficient summer rain, so that the frost detritus was no longer cemented by calcareous matter. The frost action was in addition less active than during the preceding cold climate phase (50,000 to 43,000 B.P.). The reason for this may be found in greater available wetness during the first phase of cold climate than occurred later. The fall in the winter temperature to about 0°C (at present +14°C) from 32,000 to 12,000 B.P. shows, however, that the temperature was depressed to an extent similar to that found further north.

Thus the time of maximal cold and extreme climatic conditions in Europe was matched in Cyrenaica by a phase of very cold but not particularly moist climate. The preceding cold phase of the last cold period (50,000 to 43,000 B.P.) in contrast was accompanied by an obviously more humid climate. In this it resembled in many aspects the period of the transgression of the lower Chvalyn transgression in the lower Volga.

In northwest Africa several river terraces are known which can be related convincingly to glacial deposits in the Atlas [review of the data in 102]. The facts show, as do the observations reported by Hey and older data assembled by Butzer [105, 107], that the water budget during the last cold period in the northern Sahara was actually more favorable than at present. The same was certainly true of earlier cold periods. It is understandable that such conditions must have had a favorable effect on the biosphere, particularly because in the arid regions even the smallest improvements in the water budget, even today, can cause appreciable changes in the character of the plant and animal world. Just how great this influence is can be seen very clearly in the data of French workers assembled by Schwarzbach [668]. According to this at Tamanrasset in the Hoggar massif an annual precipitation of 150 millimeters supports very good pasture; with 100 millimeters the pasture is still abundant; and with 50 millimeters of rain they are still satisfactory under the conditions existing there. With 30 millimeters of annual precipitation the pasture is only moderate, and with only 20 millimeters formation of pasture is impossible.

Recently several attempts have been made to find traces of the pluvial periods by pollen analysis through the changes produced in the vegetation of the Sahara. These analyses refer to Tunisia and the northwestern Sahara [116, 120, 52] as well as to the Hoggar massif [596, 597, 612, 117, 118, 121, 122] and the southern Sahara between Adrar des Isforas in the west and Lake Chad and the Djurab Depression in the east [613, 535]. These results are in accord—the pollen flora of cold period sediments, said to be from Villafranchian [121, 122] to the last cold period, showing a marked colonizing of the mountains by trees. Thus the plants of the Mediterranean region during the cold periods seem to have advanced to the southern margin of the Sahara. Only in the postglacial is this flora said to have retreated and have been replaced by plants from the Sahel and Sudan. There unfolds before the reader an imposing picture of plant migration clear across the Sahara, which along with the persistence of older, more demanding floral elements in the mountains seems to be a persuasive argument for the favorable climate of this arid region during the cold periods of Europe. A further indication supporting this hypothesis is that in the Hoggar massif there is a widespread depositional terrace accompanying the present dry river beds. It contains here and there sediments of an old soil [101] and is at an elevation of more than 2,000 meters in these mountains and locally is clearly developed above 2,500 meters. Kubiena [438, 439] mentioned this as an earthy brown loam ("erdiger Braunlehm"), and Büdel on the basis of this result and the magnitude of the terrace tried to estimate the former annual rainfall.

	MEDIAN ELEVATION ZONE OF THE HOGGAR (1,000–1,500 m.)	ELEVATION OF THE HOGGAR (2,000–3,000 m.)
Present day	42 mm. per year at Tamanrasset	more than 100 mm. per year
Main Pleistocene wet period	about 500 mm.	about 1,000 mm.

According to Büdel the main Pleistocene wet period should correspond to the old Riss glaciation in the Alpine terminology (above, 59). Yet it seems to me that in estimating this interesting find the greatest caution is necessary. The pollen of wind-pollinated plants and in part of insect-pollinated plants can be carried by wind over very wide areas (above, p. 49), so that this distant transport can seriously falsify the pollen spectra of regions in which woods are rare or absent. The consequences of distant transport have been repeatedly investigated at the polar forest limit. Fortunately there are corresponding results available from the Turan Desert ([505]; Figs. 95, 96), which show that here, 600 to 700 kilometers from the nearest woodland, through the influence of wind, deceptively, there is a flora of the thickest of forest plants, requiring favorable conditions. It is furthermore very important that the most striking falsification occurs when the desert soil under consideration is moist at the very same time that the trees are in blossom in far remote forests. In this case the wind-transported pollen is retained immediately by adhering to the moist soil, and thus is preserved against subsequent redeposition or destruction. On a dried-out soil surface the inblown pollen does not lie long and is rapidly destroyed.

It cannot be doubted that these observations on the distant transport and retention of tree pollen in desert soils must be considered in any discussion of pollen analyses from the Sahara. In this case, probably none of the works named would stand criticism, although the proportion of arboreal pollen is always very small. Moreover, the fact that in the supposed cold-period sediments the overwhelming amount of the tree pollen is only of Mediterranean trees suggests that the pollen spectrum owes its origin to the prevailing northerly winds which blew more or less regularly in spring and early summer—that is, at the time when most of these plants were blooming. Only during the postglacial was this pollen

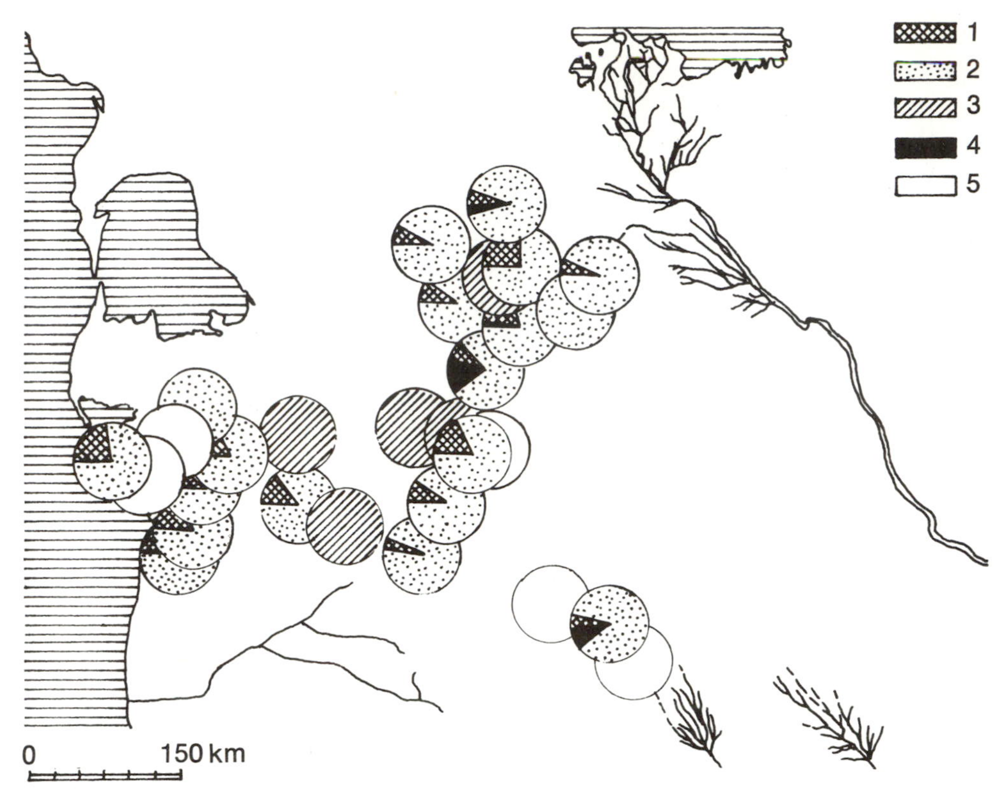

FIGURE 95. Recent pollen transport in the Turan Desert. Despite the prevailing desert vegetation, the proportion of arboreal pollen (crosshatched) in the total pollen spectrum of surface samples is very high. In the upper right of the diagram the lower Amur darja and its mouth in the Aral Sea are shown. On the left margin of the diagram is the eastern margin of the Caspian Sea and the Gulf of Kara Bugaz (horizontal ruling). 1. tree pollen; 2. shrub and nonarboreal pollen; 3. pollen in insignificant amounts; 4. spores; 5. sporomorphs absent. Simplified after Mal'gina [505].

FIGURE 96. Recent wind-blown pollen in the Turan Desert. A. The influence of wind direction and the position of the forest regions on the composition of the present pollen flora over the western Turan (pollen sampled from the air). The research area Nebit Dag (about 39.5°N 54.3°E) is about 300 to 400 kilometers from woodlands to the west and southwest. When the prevailing wind is from the west (up to 29 April) the proportion of arboreal pollen and spores in the total pollen flora is very high, but diminishes suddenly with the change to an easterly wind (30 April) to an extraordinarily low value, which then remains unchanged until 6 May, when circulating winds occur. 1. arboreal pollen; 2. spores; 3. nonarboreal pollen. After Mal'gina [505] B. Composition of arboreal pollens in surface samples of saline clay soils of western Turan, which are moistest at the time the trees are blooming in the Caucasus and Elburz, and thus the best time for pollen capture. Column 1: distance in kilometers from the Miankale Peninsula on the southeast banks of the Caspian Sea; Column 2: distance in kilometers from the Apseron Peninsula; Column 3: composition of the arboreal pollen flora as a percentage of the total tree pollen flora; Column 4: general composition of the sporomorph flora of surface samples. 1. pine (*Pinus*); 2. birch (*Betula*); 3. alder (*Erle*); 4. mixed oak forest (*Quercus*, *Tilia*, and *Ulmus*); 5. hornbeam (*Carpinus*); 6. beech (*Fagus*); 7. arboreal pollen; 8. spores; 9. nonarboreal pollen. Somewhat simplified from Mal'gina [505].

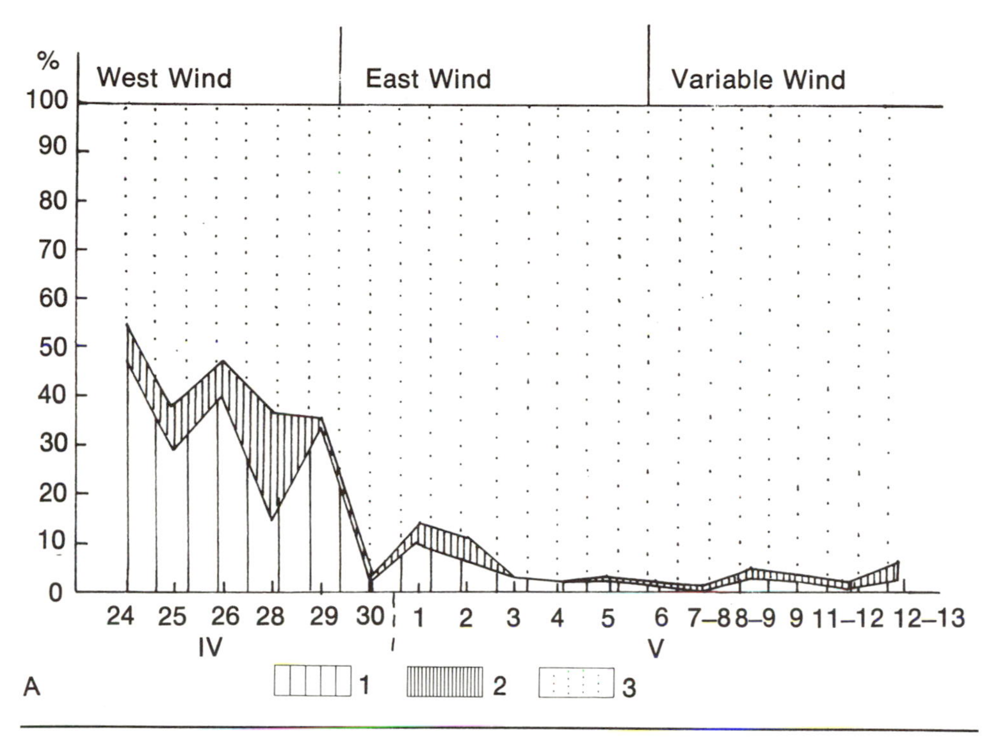
%
West Wind
East Wind
Variable Wind
100
90
80
70
60
50
40
30
20
10
0
24 25 26 28 29 30 1 2 3 4 5 6 7–8 8–9 9 11–12 12–13
IV
V
A
1
2
3

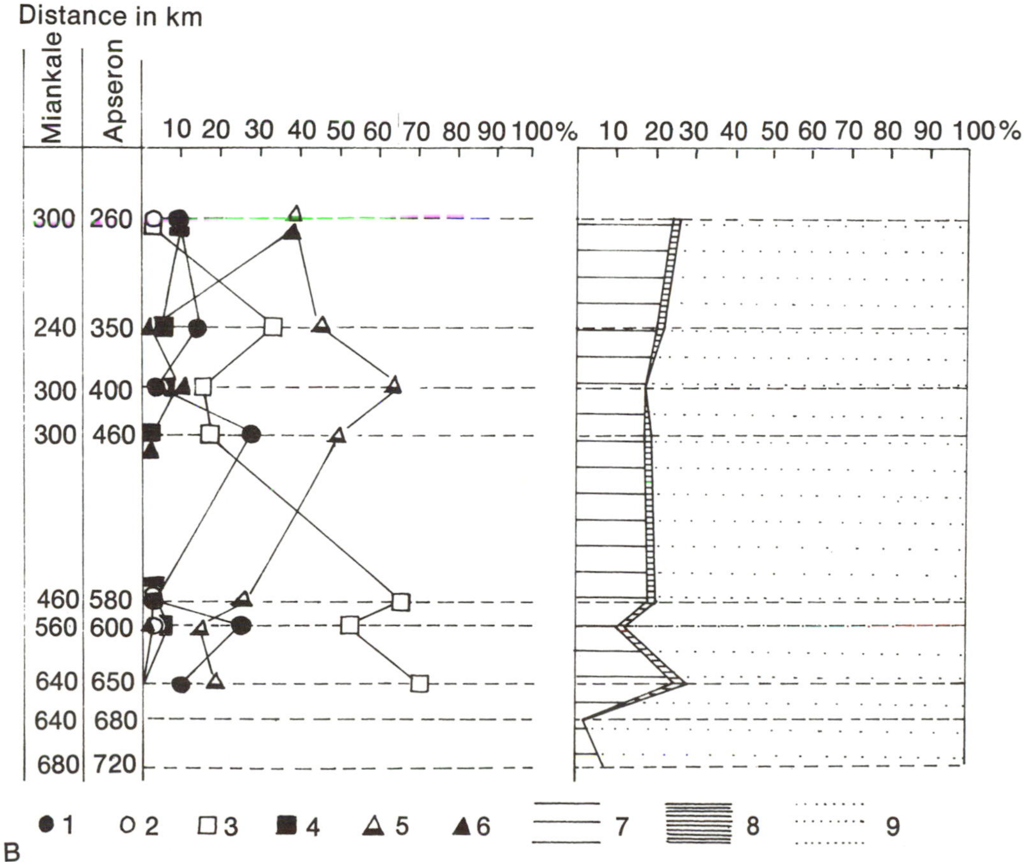
Distance in km
Miankale
Apseron
10 20 30 40 50 60 70 80 90 100%
10 20 30 40 50 60 70 80 90 100%
300 260
240 350
300 400
300 460
460 580
560 600
640 650
640 680
680 720
1
2
3
4
5
6
7
8
9
B

flora replaced by that from the Sudan and Sahel. Obviously, south winds occurred frequently, replacing the prevailing northerly winds of the cold periods.

If this interpretation, made with the aid of pollen analyses, proves to be correct, it must be assumed that atmospheric circulation during the cold periods deviated to a considerable extent from that of the present—an extent that cannot yet be correctly estimated. It correspond exactly, however, with the scheme presented by Balout [34] some years ago in an ingenious study.

There is a further question: whether the arid zone of the Sahara with its present width was displaced as a whole southward during the last and next to last cold periods, so that the wet tropical zone occupied a more restricted area than at present, or whether the wet tropical zones simultaneously expanded northward and southward over the arid zone, considerably narrowing the latter. If during the last cold period in the whole of the Sahara the prevailing wind in spring and summer was northerly, it is very difficult to see how the tropical rain belt could have expanded northward. In this connection the observations reported by van der Hammen [330, 331] from the coast of British Guiana and from the Llanos Orientales of eastern South America are of interest. According to this, in place of the present tropical rain forest an open grass savanna impoverished in species occurred in the vicinity of Georgetown at the maximum of the last glaciation. A similiar type of vegetation seems to have simultaneously covered the eastern Llanos. Only during the postglacial was it replaced by woodland, which itself was eventually destroyed by human activity. According to Lauer [451], at the present day the transition from the tropical rain forest to the wet savanna occurs when there are fewer than ten wet months in the year. To judge from this, the intensity and duration of the dry period on the coast of British Guiana and in the eastern Llanos must have been greater than at present. Thus the trade winds system at the maximum of the last glaciation must have been intensified. As the region lies at 6°N, it seems that in northern South America at least the tropical rain zone did not expand northward. This view contrasts with the information given by Conrad [142], according to whom in Africa the oldest lacustrine horizon of the Grand Chad (Fachi limestone) has a radiocarbon age of 22,000 years. The maximum expansion of Lake Chad thus corresponds exactly with the phase of wet climate in the northern Sahara, during which Wadi Saoura (Bou Hadid) was probably lightly wooded, from which wood a radiocarbon age of 20,300 ± 1000 B.P. was obtained. Yet this observation is not necessarily proof that the pluvial on the northern margin of the Sahara also corresponded to a pluvial on the south side of the Sahara. The former wider spread of the lake may have other hydrological and geological causes without having to make a climatic fluctuation responsible. Furthermore, a fall in temperature would have led to a reduction in evaporation and thereby an improvement in the water budget, whereas in northern South America the intensification of the trade wind system had the consequence of accentuating the dry season.

Yet as an indication of the simultaneity of the wet periods on both sides of the Sahara, we must consider the improvement of the water budget along the northern edge and at the southern limit of the Sahara simultaneously during the postglacial—namely between 7,000 to 4,000 B.P.—to better than present conditions [142, 613, 34]. Butzer [107] designated this moist phase in the eastern

Sahara and the Nile Valley "sub-pluvial II." Nevertheless, it appears only rarely possible to adduce certain proof that this was in fact a time of increasing moistness followed again by a slowly developing phase of increasing aridity, for the occurrence of the remains of settlements, rock drawings, and the like signifies very little (above, pp. 186 ff.). In specific cases the observations do indicate a former, wetter climate. These are related to the fossil soil in Wadi Saoura in the northwestern Sahara, which has a radiocarbon age of 6160 ± 320 B.P. [142] and a corresponding fossil soil with Neolithic pottery in Tenere north of Lake Chad [613], to a rich steppe flora in the Hoggar [597], or a mud-filled basin in which Bedouin "microlithicum" (5350 ± 350 B.C.) occurs [107]. If these are not special, local cases, the rainfall simultaneously in the north and south of the Sahara was higher than at present. Unfortunately, it is still not possible to give the rainy seasons for these remote times in the northern and southern Sahara. Perhaps the climatological record of Claudius Ptolemäus, which was prepared about A.D. 100 in Alexandria, provides this information for the northeastern margin of the Sahara. According to the record the significance of northerly winds was less than that of southerly and westerly winds. In contrast to present conditions (winter rain), the rainfall was uniformly distributed throughout the year, and there was a series of other notable deviations from the present. Hellmann [339] assumed from this that in the records, reports from different stations were mixed together. This need not necessarily be so, for the postglacial warm period resembled the last warm period in many respects. As may be recalled (above, p. 107) in the northern Mediterranean region considerably more rain fell than at present, and possibly there was more summer rain than at present. Furthermore, the composition of the postglacial pollen flora of the Sahara shows a prevailing southerly wind (p. 196). It can thus be assumed, with reservations, that in the postglacial warm period too (about 6,000 to 3,000 B.C.) a higher precipitation fell in the Mediterranean region, which may have been even higher than at present. A considerably higher proportion should have fallen during the summer months, and often southerly winds blew over the Sahara. This picture corresponds exactly with what Ptolemäus described. The climatic optimum of the postglacial warm period without doubt occurred long before the Christian era. However, the subsequent climatic changes were not great and did not occur suddenly. It is therefore conceivable that Ptolemäus described the dying stages of this climatic change, and so his observations agree very well with information obtained by other means.

The wet climate phase during the postglacial appears from this to have been simultaneous on the northern and southern sides of the Sahara. It must remain an open question whether the same holds true for the last cold period, for the relevant data available are insufficient for a final conclusion.

South of the tropical rain forest zone in Africa there is a second arid zone at the center of which lies the Kalahari. Here, too, during the Pleistocene an alteration between periods of wet and dry climate can be observed in which the wet phases coincided with periods of cooler climate [827, 154, 822]. The time of coolest and wettest climate has been called the Gamblian phase. It lasted from about 31,000 to about 10,000 B.P. and thereby corresponds approximately with the maximum of the last cold period in Europe. At that time woodlands of the upland stage and *Brachystegia* woods of the higher elevations were widespread under conditions of a cool moist climate in Angola. And between the now isolated forests of this type in the higher mountains there was a nearly unbroken connection. In the area of modern Leopoldville, however, a semi-arid vegetation replaced the present wet savanna (lecture of van Zinderen-Bakker, 1965, in Burg Wartenstein).

After the close of the Gamblian phase there was a "Makalian wet phase." Cultural remains in sediments of this age gave radiocarbon dates from 4,700 to 6,830 B.P. The Makalian phase must be taken, therefore, as the equivalent of the postglacial warm period in Europe and the coeval moist phase in the Sahara.

Prior to the Gamblian phase a dry climate prevailed for a few thousand years in Angola. Cultural remains from this time gave an age of about 38,000 B.P. (lower Lupemban). As the onset of the Gamblian wet phase began about 31,000 B.P., the earlier dry period may be viewed as the equivalent of that which occurred in Cyrenaica between 43,000 and 32,000 B.P., for the climate then was also much drier than before or after that time [343]. This dry period in Angola, still within the last cold period of Europe, was preceded by a further phase of moist climate. It has been referred to as "early Gamblian." Cultural remains of possibly the same age at the Calambo Falls at the southern end of Lake Tanganyika give a radiocarbon age from 41,000 to 43,000 B.P. A still earlier phase of warm, dry climate whose age is greater than 57,300 B.P. is very probably an expression of the last warm period of Europe [827]. The preceding observations show very clearly that the decisive climatic fluctuations of the last cold period in Africa north and south of the equator were simultaneous, a welcome confirmation of the correctness of the assumption that these fluctuations were synchronous in both hemispheres. The data available, however, permit no elaboration of the important problem of the water budget of the arid zones during the dry periods. It cannot be doubted that the water budget in the northern Sahara and in the above-mentioned dry region of Africa in the Southern Hemisphere was significantly better than at present. According to Knetsch [417] a greater part of the modern ground water of the Sahara dates from the last cold period. Can one deduce from this that the cold periods in the Sahara were true pluvials—that is, periods of increased precipitation? The widespread occurrence of steppe in the Mediterranean during the cold periods does not speak in favor of a true pluvial climate. The observations in the southern Ukraine and in the region of the Caspian Sea too are not consistent with a higher rainfall than at the present time during some phase of the last cold period. They show instead, with the exception of the immediate vicinity of the then enlarged Caspian Sea, an absence of phases of better water conditions which coincided with the cold periods. There, during the cold periods, loess accumulated, but periods of soil formation and of rich vegetation—that is, *ceteris paribus*, also periods of better water conditions—were only to be recognized during the interstadials, phases when plant and animal remains suggest higher temperatures. As will be shown in the following section, the coldest climate during the Weichselian cold period in Europe set in after the Stillfried B interstadial even if the temperatures during the preceding cold phase had already sunk to a low level. If it can be taken that the climate at the time of maximum cold prior to the Stillfried B was not as severe as subsequently, then it must follow that the evaporation was greater than later

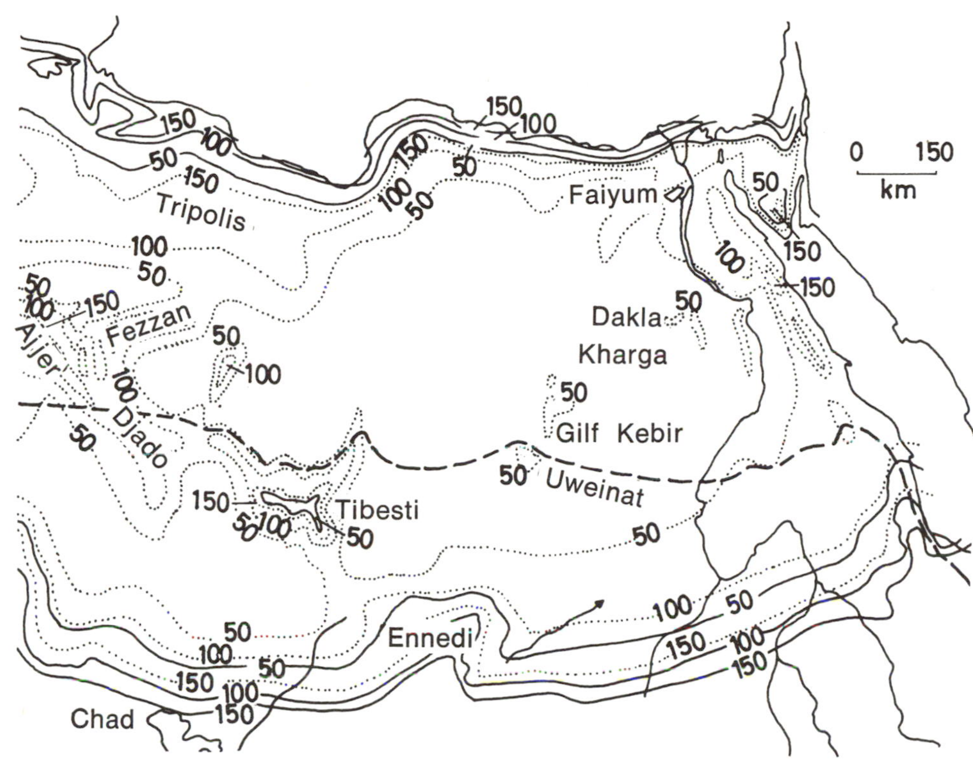

FIGURE 97. The amount of precipitation in northeastern Africa in millimeters. Continuous lines: isohyets of the last cold period; dotted line: present isohyets; heavy dashed line: limit of the northern predominantly winter rains against the southerly monsoon rains (present day). From Butzer [108].

on. In spite of this, immediately after the first really cold phase on the lower Volga, as well as in North Africa, traces of a better water budget are more distinctly marked than in the period following the Stillfried B. This suggests that during the first cold climate phase the available moisture was greater in amount than in the second, which was accompanied in Europe by a time of maximum aridity. Comparable observations could be made in North Africa. It has been seen that the first, apparently wetter part of the last cold period in the vicinity of the Caspian Sea cannot be regarded as a true pluvial, as the precipitation there was considerably below that available today. Perhaps conditions in North Africa were somewhat more favorable. Yet also here the eventual increase in precipitation cannot have been very great. Butzer [108, 110, and Fig. 97] calculated that there was an increase in the annual rainfall of 100 to 150 millimeters. It must be questioned, however, whether even these low figures can actually be demonstrated, for the evaporation, because of the already considerable fall in temperature, was greatly reduced. It appears impossible at the present time to make a reliable estimate. The potential evaporation in the region of North Africa of interest here according to the data in the Fiziko-geograficheskij Atlas Mira (1964), comprises today some 1,500 to 2,000 millimeters per year. The actual evaporation, directly on the coast of Cyrenaica, is 200 millimeters per year. A fall in the July and January temperatures of about 10°C would probably reduce the potential evaporation to about 600 millimeters per year. This is still appreciably more than the present actual evaporation.

Under these conditions, short-period positive rainfall—that is, perhaps several successive years of somewhat heavier rainfall—has a high geomorphological and phytogeographic efficiency. Still one cannot consider standing water or flowing rivers in the Sahara during the cold periods. Finally, the past distribution of rainfall over the year is unknown. Probably the major proportion of the rain fell during the winter months, so that the time of strongest frost action was simultaneously the period of greatest moisture, thereby intensifying frost weathering. In my opinion, it is not possible at the present time either to give the former amount of rainfall in the northern Sahara or to answer correctly the question whether, during the Pleistocene cold periods generally, pluvials occurred, even when at these times a higher rainfall may be assumed. The very interesting example of the southwestern mountain chains (Zagros Mountains) of Iran facing the lowlands of Iraq provides an excellent illustration of how difficult these problems are [810]. The climate snow line during the last cold period in the Zagros Mountains lay about 1,800 meters below its present level. If the vertical temperature gradient is estimated as 0.5°C/100 meters decrease with elevation, then the annual and July temperatures must have been about 9°C lower than today. At the present time, however, the mean annual and July gradient is about 0.7°C/100 meters. With the assumption that this figure also obtained during the last cold period, the calculated temperature fall is 12°C. Wright concluded from this: "It seems more likely that a modest depression in temperature was combined with an appreciable increase in snowfall in the mountains under consideration so that the glaciers could form at the relatively low levels indicated. The special location, trend, and height of these mountains with respect to more frequent storm tracks from the Mediterranean might explain the postulated great increase in snowfall" [810]. Thus Wright assumed a considerable increase in the amount of winter precipitation. This process fully corresponds with what is described under the term "pluvial." The improvement of the water budget, despite all the hazards of the summer half year, should have had expression somewhere in a rich vegetation. Yet on Lake Zeribar near Merivan at an altitude of 1300 meters both before and during 14,800 B.P.—that is, during the maximum of the last cold period—an *Artemisia-Chenopodiaceae* steppe of a cold dry climate occupied a region where oak forests now prevail. This steppe vegetation is plainly contrary to an increased rainfall, even if it is suggested that the oak on thermal grounds should have retreated from the area, for its place should have been taken by cold-resistant trees of this mountainous region if the available water budget were to favor tree growth. So it is obvious that no pluvial climate prevailed here.

It remains only to find an answer to the question whether the zone of increased precipitation lay still further to the south. The work of Büdel [101] and Kubiena [438] on the thick and locally well formed river terraces and on relicts of brown earths in the Hoggar might form a basis for a reply. Büdel placed this phase of wet climate in the central Sahara in the middle Pleistocene, since it seemed unlikely that the relatively modest glaciation of the Würmian-Weichselian cold period could have left behind such strong traces, particularly if there are only signs of a single cold period. If the estimates of Büdel and Kubiena are reasonable and in the Hoggar massif between 500 and 1,000 millimeters of rain fell per year (today there is significantly less than 100 millimeters), then there was truly a pluvial period. Yet from the information given by the authors, there is nothing to indicate what drainage or what process of soil formation must be considered if a mean fall in temperature of about 12°C can be assumed without a simultaneous increase in rainfall. There is a further difficulty in that the brown soils, so important paleoclimatically, are rarely found, and before this there was no possibility for a determination of the age of the paleosoils just mentioned. These interesting observations can thus scarcely be accepted at this stage as an indication that the zone of higher precipitation, at least during a middle Pleistocene cold period, lay in the central Sahara.

America

Van der Hammen [329] assumed, from the displacement in the Colombian Andes of the upper tree line during the last cold period to about 2,000 to 2,500 meters elevation, not only lower temperatures but also a higher precipitation than today. During the warm periods the opposite occurred. The area lies at about 5°N. The conditions there, consequently, cannot be used directly as an indication of the existence of pluvial periods. The height of the climatic snowline in the western part of North America during the last cold period seems to provide a more useful means. It increases from north to south, but between Pike's Peak (38°52′N 105°02′W) and Cerro Blanco (33°23′N 105°48′W) the snow line of the former climate seems to have remained constant at approximately 3,900 meters and not risen southward. According to Antevs [18] the effects of the increasing rainfall to the south are to be seen in the position of the snow line in this region, for a stronger cooling further south, which could relatively depress the snow line, is scarcely likely. This would signify that the precipitation increased above present levels in New Mexico during the last cold period. It also appears that it was essentially a winter rainfall, for plants characteristic of summer rains appear not to have migrated in until the beginning of the postglacial warm period [518]. Moreover, most of the pollen analyses carried out in the arid regions of the southwestern United States show that the rainfall during the cold periods was higher than today, for in place of the present wide steppes and semidesert there were formerly pine forests on many occasions in which those elements with higher water requirements such as spruce (*Picea*), fir (*Abies*), and Douglas fir (*Pseudotsuga*) were not lacking [518, 633, 325, 139, 519]. Particularly interesting in this connection are the observations of Stuiver [723], Roosma [633], and Hafsten [325] in California and in the Llanos Estacados. There is in southeastern California the basin of a dried-up lake, Searle's Lake, in which the sediments are 230 meters thick. Up to the present time only the uppermost 40 meters have been examined. They are composed of an upper salt layer 20 meters thick under which is a mud with organic matter. The results of several radiocarbon analyses show the mud to have been formed between 23,000 and 13,000 B.P. The upper salt layer has an age of from about 10,000 to 6,800 B.P. According to this reading, the salt essentially began to be deposited only with increasing evaporation of the lake water at the beginning of the postglacial warm period. The underlying mud, called "Parting Mud," dates, however, from the maximum of the last cold period. Since very little salt precipitated, and since mud was brought into the basin, the precipitation must have been appreciably higher than at present. The sedimentation rate between 24,000 and 19,000 B.P. comprised 3.7 centimeters per century. This is considerably higher than the rate between 19,000 and 14,000 B.P., during which time only 1.2 centimeters per century accumulated. It must be recalled in this connection that the peak of the last cold period in central Europe and North America was reached immediately after 25,000 B.P.

Below the Parting Mud lies the lower salt layer, which accumulated between 32,700 and 24,200 B.P. At that time the climate was again warm and dry, so that the already extant lake slowly desiccated. This time corresponds exactly with the Stillfried B interstadial, and it is very noticeable that the phase of evaporation was broken several times by shorter wet periods of which the most important occurred at 27,800 B.P. It will be recalled (above, p. 67) that the Gravette cultural horizon at the Moldovo V paleolithic station on the Dniester occurred in a soil whose age was determined as 23,000 to 23,700 B.P. [360], separated from a somewhat lower fossil soil which probably belongs to the true Stillfried B interstadial by a thin loess horizon which seemed to represent a cold dry climate. It is possible that this short interval of cold climate within the Stillfried B (*sensu lato*) may be seen as the equivalent of the above-mentioned phase of mud accumulation at around 27,800 B.P. in Searle's Lake. That would show that this minor climatic fluctuation was also of worldwide significance, and is a further good example of the worldwide, almost simultaneous occurrence of climatic fluctuations.

Below the lower salt layer, which is split up by several mud horizons, there is a further very homogeneous salt layer, whose age can be estimated by extrapolation to be about 48,000 B.P. Since the dating is not exact, the possible equivalents of this dry period in other regions need not be discussed [723, 633].

In the Llanos Estacados a borehole was sunk in an old lake basin. At present a steppe flora of grasses, *Chenopodiaceae,* and many shrubs prevail [325]. About 33,500 years ago there was a vegetation with pine and spruce trees as well as steppe. The climate was cold and wet. Between 33,500 and 22,500 B.P. this vegetation was replaced by a shrub and grass steppe in which oak (*Quercus*) occurred locally. This phase of warm dry climate clearly represents the Stillfried B interstadial, or the time at which the lower salt layer of Searle's Lake was formed. Between 22,500 and 14,000 B.P. the forest vegetation again occupied the region and now dominated it. The most important tree was the pine. The climate once more was cold and moist, but it was more humid than during the first cold phase. It was the time of origin of the Parting Mud in Searle's Lake and corresponds to the maximum of the Weichselian cold period in Europe. From 14,000 to 10,000 B.P.—that is, during the European late glacial—the woodland was gradually replaced by steppe. This characterizes the present region. At the same time that the cold, wet climatic phase set in—a pluvial climate in this southern basin—the upper tree limit in the Chuska Mountains of New Mexico sank at least 850 meters below its present level, and an alpine *Artemisia* community (wormwood) with pine and spruce thickets occurred in place of the present woodland, which is formed of *Pinus ponderosa* [42], a clear indication of strong cooling. The preceding observations from the southwestern part of North America are a much clearer indication of the former existence of pluvial periods, which coincided with the cold periods or, better, with phases of particularly cold climate in the Pleistocene cold periods. Yet these results should not be exaggerated, for here as elsewhere an increase in rainfall and a lower evaporation caused by the lower temperatures might have operated simultaneously.

These interconnections can quickly be seen in an example from New Mexico reported by Leopold [642]. The climatic snow line during the last cold period was about 1,500 meters lower than it is today. With a vertical temperature gradient of 0.6°C/100 meters this snow-line depression corresponds to a decrease of 9°C. In the same region there is the basin of a now dried-up lake, Lake Estancia. Its size during the last cold period can be accurately determined, and in addition, the drainage basin of the lake is also known. Leopold assumed that during the cold period the temperature decline was particularly marked in the summer months, since at this time the cooling influence of the northern ice sheet must have been especially noticeable. If on the basis of these observations and hypotheses a calculation is made of the evaporation from the former water surface, we find that the fall in temperature is insufficient to permit the lake to rise to its established level. Leopold concluded that in New Mexico at the peak of the last glaciation not only was the temperature lower than at present, but the precipitation must have been higher; so in fact the climate was pluvial.

The figure of 9°C which Leopold used for the cold period temperature fall may be too small (above, pp. 142 ff.). In addition there is nothing to suggest that only the summers were colder than today, with winter temperatures about that of the present. If, instead, the figures for the center of North America (on pp. 142 and 144 above) are used, i.e. a fall in the July temperature of about 12°C and in the mean annual temperature of 13 to 15°, and the rest of the calculation is left unchanged, then the important result obtained is that the precipitation was not higher but to the contrary must have been about 15 percent lower than at present to obtain a comparable water budget of the lake levels during cold periods. This discussion shows that the distinction of the extent to which the cold-period climates in the arid regions of North America can be considered as pluvials and how much the increase in the available precipitation in comparison with current conditions must have been, despite the much clearer paleobotanic and sedimentological data, still remains highly uncertain.

In my opinion, at least in certain regions in the American Southwest, true pluvials prevailed, even if the increase in precipitation was not great. The basis for this assumption lies, on the one hand, in the very convincing observations reported and, on the other, in the fact that the calculated reduction in total precipitation for Lake Estancia is relatively small. For if it is further considered that the major part of the rain probably fell in winter [518], and that the hottest part of the year was thus poor in rain, then the possibility of a still smaller reduction or even a small increase in rainfall must be considered.

Conclusion

The reported observations may suffice to show the basic characteristics and problems of the Pleistocene climatic fluctuations in the arid regions. At least for the last cold period, which is easier to review, it can be established that the northern margin of the arid zone in the Northern Hemisphere was better watered than it is at present. There are clear signs of increased precipitation in the arid zones of the southwestern parts of North America, but the same does not show up with the same clarity in North Africa—at least not so far as we can tell given our present knowledge. From this the impression arises that local differences played a significant role. Among them position, trend, and elevation of mountains were probably very important. This difference also appears to have affected the age of the period of maximum precipitation during the last cold period. Many observations in the northern Sahara and in southern Russia indicate that the time immediately following the Brørup interstadial (*sensu lato*) was characterized by a relatively wet climate, but the research of Hafsten in the Llanos Estacados shows that the wettest period was reached only after the Stillfried B interstadial. This is certainly the consequence of local differences in the atmospheric circulation in regions with different orographic form.

How far the pluvial periods at the southern margin of the Northern Hemisphere's arid zone correspond to those along the northern margin is not known. The spread of the savanna vegetation in British Guiana and in the Llanos Orientales of Venezuela show that at least regionally the pluvial along the northern margin of the arid zone corresponded to a dry period along the southern. It would be hazardous, however, to draw a general conclusion from this at the present time without awaiting further observational data and without a consideration of the peculiarities of each individual region.

In spite of these difficulties, it is clear that the change in position of the most important belt of rainfall of the earth during the cold periods did not convert the present deserts to woodland or bushlands, although simultaneously in the high geographic latitudes the woodlands so characteristic there during the warm periods were replaced by cold steppe or semidesert.

In the introduction to this chapter the question of the nature of the pluvials and their relationship to the glacials of higher geographic latitudes was raised. It must nevertheless be stated that the recognition of pluvial periods causes extraordinarily great difficulties, provided that under this term are understood periods of increased precipitation outside the former periglacial region. Over wide areas of the present arid zone the water budget was more favorable than at present, but this circumstance seems much more due to the great fall in temperature than to an increased precipitation.

It is understandable that a general, appreciable cooling of the earth's surface must lead to a significant reduction in the amount of precipitation, for with lower temperatures less water is evaporated than at higher temperatures. This explains why the majority of climates during the cold periods had appreciably lower rainfall than at present. Yet it does not necessarily follow from this that in all climatic regions a reduction in rainfall must be found. It would be perfectly conceivable that independently of the general decrease in average rainfall in most regions the precipitation in certain zones might have increased. This is particularly true when it is considered that the polar front was displaced toward the Equator. What makes true pluvials harder to understand is that reliable indications of their existence are known only in rare cases and then only observed at certain parts of the earth. Should the term "pluvial" be dropped? It seems to me that this is not justified. On the one side, there are, as has been noted, at least in places many indications that speak in favor of a pluvial climate; on the other, it is one of the most important paleoclimatic tasks to find the cause of the formerly better water-budget conditions of many of the dry regions of the earth. If the term "pluvial" were rejected, then the discussion of this very difficult problem would become one-sided and would rapidly lead to false conclusions. In addition to "pluvial" for a period of increased rainfall it seems well to apply also "hygroclinic period"—that is, a period that was moister—without associating this with increased precipitation. In the majority of cases hygroclinic changes are caused by a reduction in evaporation. In general, even today, it may be remarked that neither pluvials nor hygroclinic periods correspond to the full duration of a cold period in higher latitudes, but essentially only to the segments of coldest climate. Which of these phases within a cold period had the wettest climate in the present arid zone may change considerably both regionally and during the course of the individual cold periods.

In view of the uncertainties in knowledge of the climatic character of the individual pluvial periods, it is to be recommended that the problem of climatic changes within the present arid region should be deferred, particularly as Büdel [102] has recently discussed the case of the Sahara.

5. Climatic Fluctuations During the Last Cold Period in the Northern Hemisphere, Excluding Present-Day Dry Regions

Certain of the important problems of climatic fluctuations and changes during the Ice Age have been considered in the preceding chapters, and they provided essential criteria in establishing the most important steps in the development of climatic fluctuations. In particular, the comparison of the climatic optimum of a warm period with the peak of a cold period shows clearly the unexpectedly great magnitude of the fluctuation. Yet these observations are insufficient for a complete understanding of the history of climatic development, especially as they are not available in the same number or quality from all parts of the earth. It becomes much more necessary to consider precisely the course of individual cold periods, the better to understand the transition from a warm to a cold period, and to study intensively the cold period itself. For this purpose the last cold period is undoubtedly the most inviting, for we are much better informed about it than about the earlier ones. Consequently, in what follows the course of climatic development during this cold period will be presented in detail for Europe, Siberia, and North America. The results so obtained may be evaluated as an indication of what climatic fluctuations may have occurred during earlier cold periods about which we are less well informed. This should not be taken to mean that the climatic fluctuations of the early cold periods closely follow the same pattern. Reconstruction of the various phases in the climatic history of the last two warm periods has shown how dangerous it is to erect generalities too quickly, for with all the similarity in the gross patterns of climatic fluctuation there repeatedly occurred new variations and episodes, which, although seemingly unimportant with respect to the climatic history, may be of fundamental importance for developments in the plant and animal kingdoms. The same is certainly true of the cold periods.

Temperature Changes

Earlier, (p. 119) the remarkable fact was discussed that at the end of the last and next to last warm periods the bogs forming up to those times were repeatedly flooded, so that the normal succession from open water, through silting up, to the formation of a land surface could repeat itself. Yet the process was soon broken off and eliminated by erosion processes or by abiotic deposition. It was shown that this wet phase, at least in the case of the last warm period, was not the consequence of increased precipitation, a point which Klute [408] earlier made clear, for it began earlier in continental central Russia than in the oceanic regions of central and western Europe, where the effect of precipitation should have been more marked. This leads to the thought that the wetness may have been the consequence of the already considerable fall in temperature which must have become apparent earlier in continental regimes than in oceanic. The correctness of such an assumption could be confirmed by the temperatures estimated during the first half of the spruce-fir phase, closing the first half of the last warm period. This next raises the question of the further progress of temperature changes during the last cold period. At first only the temperatures during the different phases of a cold climate will be considered. The climatic character of interstadials will be the subject of discussion below (p. 219).

The fossilized evidence of formerly frozen soils forms a useful thermometer. In the last few years a large number of such forms have been described in Europe [cf. the maps in 391, 601, 771]. Yet the number of those forms which can, within limits, be assigned to closely defined stratigraphical horizons is appreciably smaller. The distribution of these permafrost forms is illustrated in Figure 98. From the cold climate phase between the end of the last warm period and the beginning of the Amersfoort interstadial (for the divisions of the last cold period see above, Table 4, p. 65), very few occurrences of permafrost forms have been recognized. They were described in Denmark by Andersen [11]. Dücker (lecture in Cologne, July 1965) reported corresponding appearances in Schleswig-Holstein, and Velichko [769, 770] described important solifluction processes which occurred between the end of the last warm period but prior to the accumulation of loess on the middle reaches of the Desna. As this had begun within the relatively oceanic climates of Czechoslovakia and lower Austria before the beginning of the Amersfoort interstadial, the same must have occurred but to a greater degree in the middle reaches of the Desna, so that, with reservations, the permafrost forms noted should be assigned to the very beginning of the last cold period. However, we are dealing here only with solifluction forms, as in Denmark and Schleswig-Holstein—that is, soil movements produced by frost action. They belong to those frost-action criteria which are climatically ambiguous and which need not necessarily be associated with the permafrost zone. It may be assumed from this that neither Denmark nor Schleswig-Holstein nor central Russia lay in the permanently frozen region, but rather that the southern limit of the permafrost region was not far distant.

Supporting this view is the simultaneous retreat of the forests over wide regions of Europe north of the Alps, and the replacement of woodlands in the eastern half of central Europe by steppe. Possibly the mean annual temperature in Denmark and on the Desna River was about 0°C or even lower. Thus the climate in Denmark and on the middle Desna must have been about 8°C colder than at present. According to Andersen's determinations [11], the mean July temperature of Denmark was less than 10°C (Zone W2b of Andersen). Today it is about 15° to 17° in that area. The changes in the mean annual and July temperatures further indicate that the mean January temperature must have fallen appreciably, indeed to an even greater extent than the July temperature, for otherwise the low mean annual temperature cannot be explained. The continentality of the climate must have greatly increased with respect to present-day conditions even in such a near oceanic region as Denmark and Schleswig-Holstein: In place of the Atlantic climate which reigned there at the end of the last warm period, a cold continental climate was already established immediately at the beginning of the last cold period.

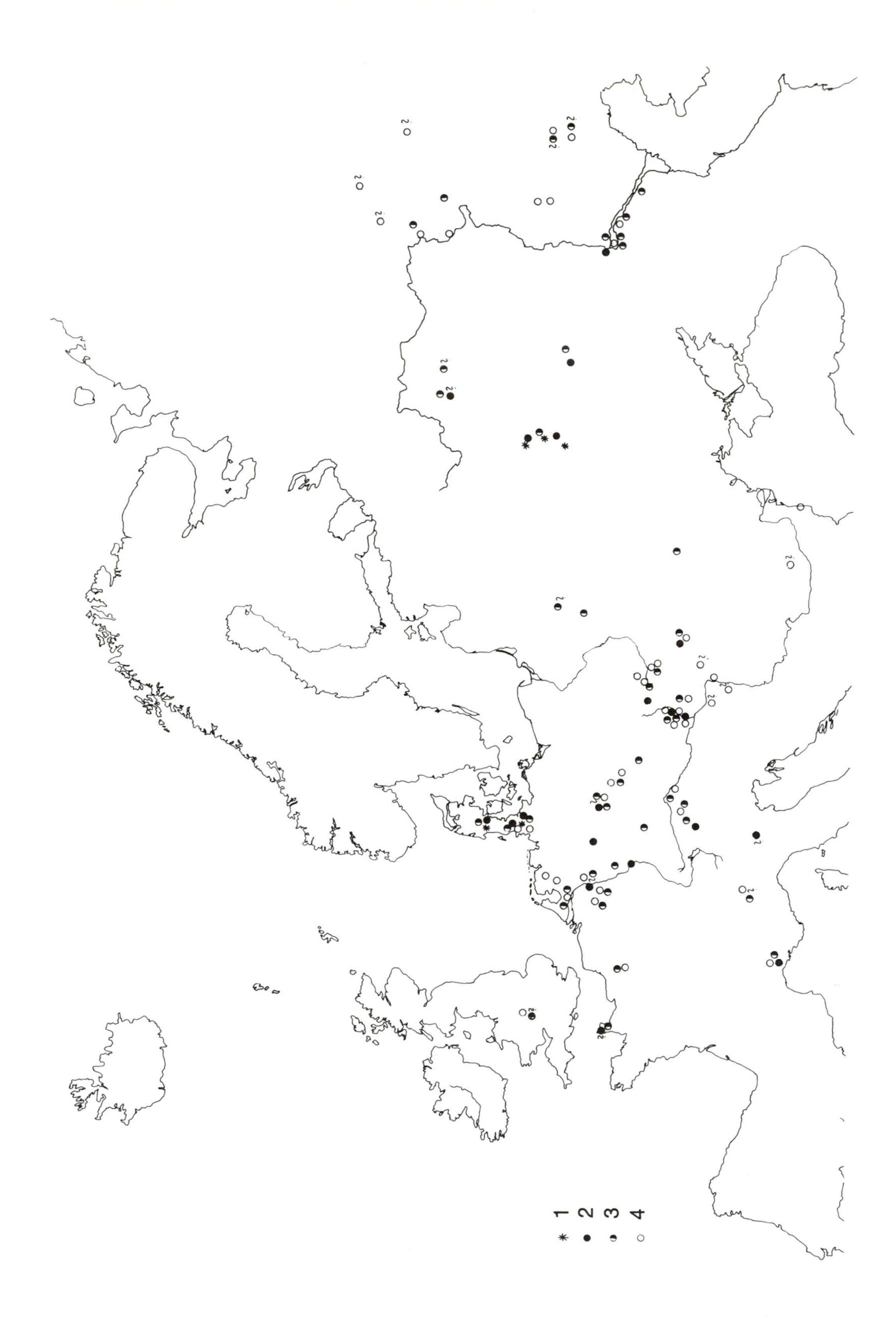

1
2
3
4

Many fossil frozen soil forms have been observed which formed during the phase of cold climate between the Amersfoort and Brørup interstadials. Among them reliable indications of permafrost are to be found; the area of permanently frozen ground had for the first time in Europe extended far to the west and south. Its southern limit seems, from the evidence of the permafrost forms, to run from the southern margin of the north German plain, through the Moravian Gate and northern Hungary into the northern Ukraine, finally reaching the middle Volga. If the southern limit of the permafrost region agrees approximately with the −2.0°C isotherm, an assumption which is probably too convenient, then the former mean annual temperature in northern Germany, in central Germany, and in the Moravian Gate must have been at least 11°C lower, and in the northern Ukraine certainly 7.0°C lower than at present. The German Mittelgebirge region, from which many observations of permafrost forms of the same age have been recorded, was deliberately excluded from this consideration because local climatic conditions could considerably distort the picture. It can be presumed nonetheless that the former permafrost region actually extended much farther to the south than has been indicated here. Yet in the framework of present research we are concerned not so much with the maximum temperature change as with the minimum, which should be determined in order to obtain a much more reliable knowledge than is possible by deceptive estimates of maximum figures.

Andersen [11] also estimated a mean July temperature of less than 10°C during this phase in Denmark. The greatest number of traces of frost action are found first immediately after the Brørup interstadial and then after the much later Stillfried B interstadial. It follows from the distribution of these forms immediately after the Brørup interstadial (Fig. 98) that the mean annual temperature in northern France was 13°C, in Bavaria and Moravia about 11°C, in northern Hungary from 12° to 13°C, and in the lower Volga and Ural River at least 9°C lower than at present.

The temperatures sank to their lowest immediately after the Stillfried B interstadial. The relative amounts have been calculated (above, p. 141) from the former distribution of pingos. The combination of these figures with those deduced from the maximum extent of permafrost forms (Fig. 98) results in a mean annual temperature in southern England and Wales at least 12°C below the present level; Kopp [429] gave a reduction in temperature of 12° to 13°C in the east Cantabrian Mountains in northern Spain; in southern France the temperature fall must have been 15° to 16°C; comparable reductions were reached in Hungary; and on the middle and lower Volga the annual temperature was at least 9° to 10°C lower than today. It should be recalled in this connection that earlier (above, p. 154) a fall of mean July temperature of about 10°C in comparison with present levels in west, central, and east Europe was calculated from the position of the climatic snow line and the polar tree limit (where this can be definitely ascertained, as even at the peak of the last cold period and even under the most favorable con-

FIGURE 98. Age of the permafrost forms of the last cold period in Europe: 1. between the end of the last warm period and the beginning of the Amersfoort interstadial; 2. between the Amersfoort and the Brørup interstadials; 3. immediately after the Brørup interstadial; 4. after the Stillfried B interstadial.

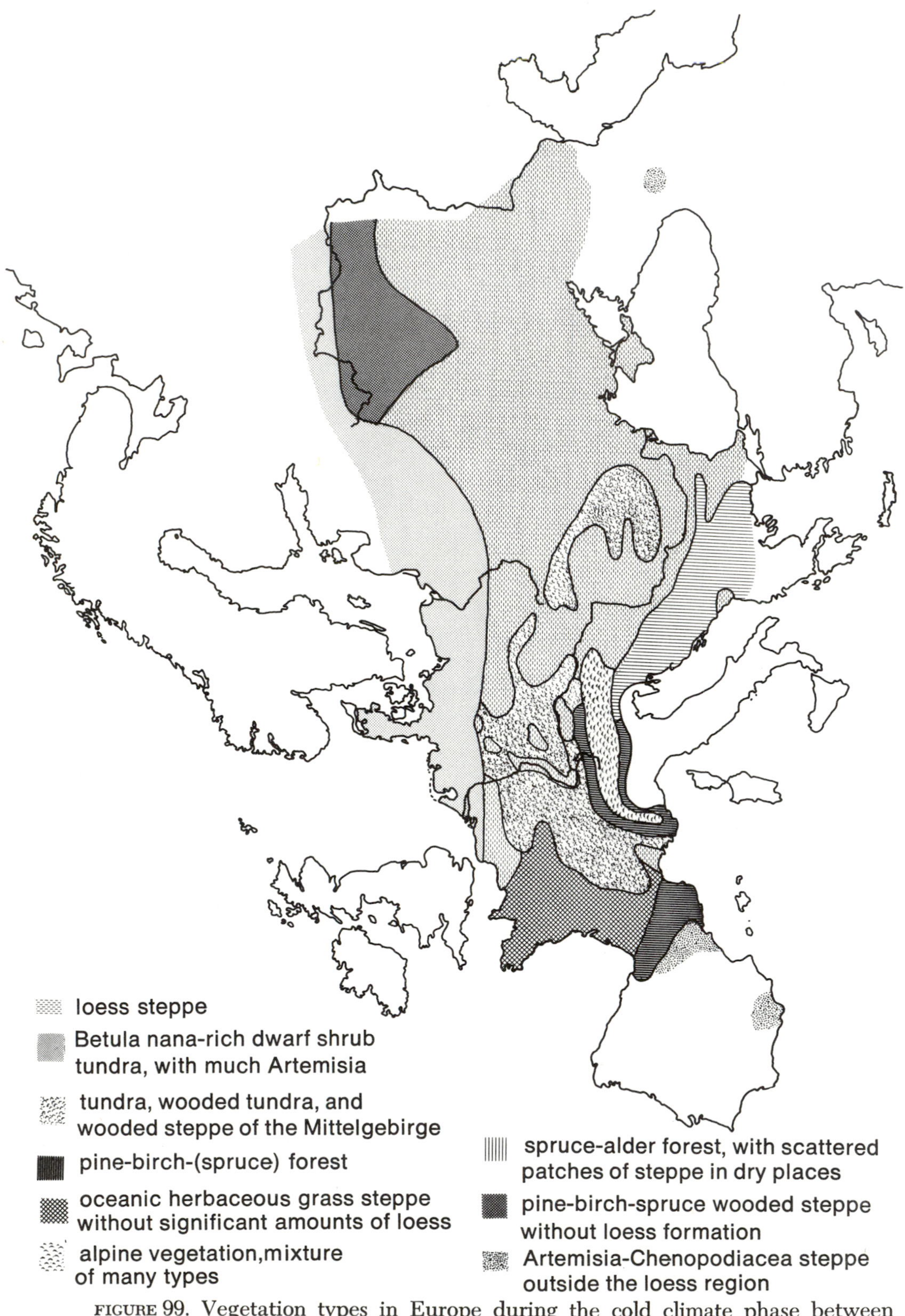

FIGURE 99. Vegetation types in Europe during the cold climate phase between the Amersfoort and Brørup interstadial [260].

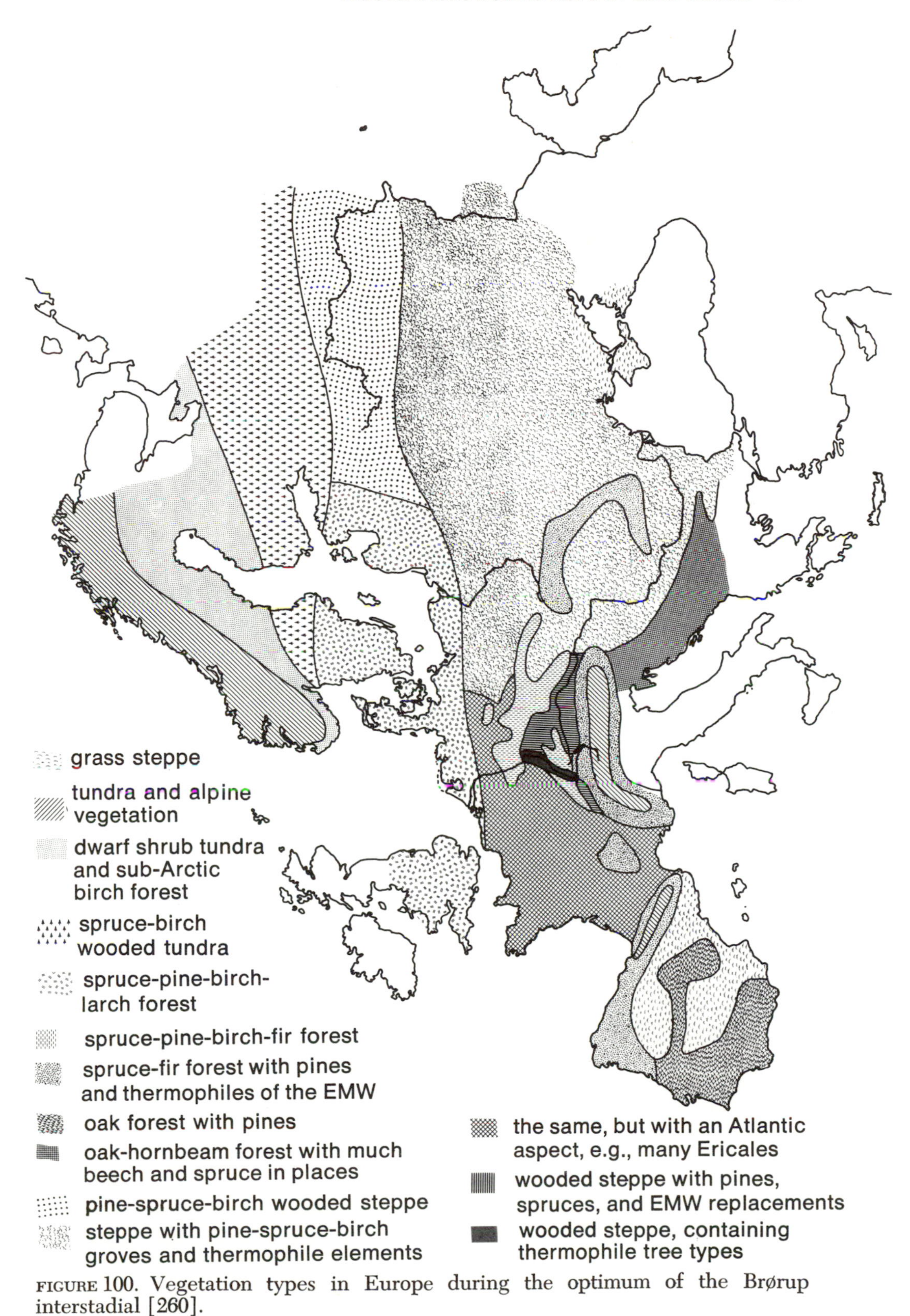

FIGURE 100. Vegetation types in Europe during the optimum of the Brørup interstadial [260].

ditions, isolated thickets could occur within the steppes). Despite all the uncertainties of the data, it can be concluded that mean temperature of the coldest month must have fallen considerably more, since in other cases the comparably small decrease in the mean July temperature is not consistent with such low mean annual temperatures. The changes in temperature alone during the last cold period thus enable one to recognize the appreciable increase in continentality during the cold climate phases. This process must be reflected too in the water budget of the region.

The Water Budget

Figures 99 and 100 give some impression of the most characteristic of the physical-geographical conditions at the beginning of the last cold period. They depict the state of the vegetation during the phase of cold climate between the Amersfoort and Brørup interstadials in the one and during the Brørup interstadial in the other. The details, in particular the evidence supporting these attempts at reconstruction, must be sought in another work [260]. In evaluating the maps it must be understood that only the basic features of the former types of vegetation are shown, for it is not possible to be absolutely certain of the ages of all the formations concerned. However, in general the Amersfoort interstadial was less well defined than the Brørup, which comprised several subphases, and the phase of cold climate prior to the Amersfoort was not characterized by such extreme conditions as that intervening between the Amersfoort and Brørup. Even if formations from a somewhat older phase (one which instead of being from the Brørup is from the Amersfoort, etc.) are erroneously included in one or the other of the diagrams, this does not fundamentally change the map.

The most important result which arises out of the data assembled in Figures 99 and 100 consists of the recognition that during the cold climate phase between the Amersfoort and the Brørup interstadials and also during the Brørup interstadial itself, central and eastern Europe was covered by steppe and wooded steppeland and only in the rarest instances by forest. During the phases of cold climate, just as during the warm-climate phases, the water budget of most European regions, even at the beginning of the last cold period, must have been extraordinarily unfavorable. This pattern can be confirmed with respect to the Brørup interstadial from examination of the distribution of the more important types of soil [323, 260], just as, in regions now under woodland where at the present time brown earths or parabrown earths, pseudogley, or associated soil types are widespread, steppe black earths predominated. The existence of steppe or at best wooded steppe must be inferred from this, especially when it is considered that most of the occurrences are observed on slopes or in basins where the local water supply conditions are much more favorable than on the neighboring watersheds and plains. It must be admitted that during the Brørup interstadial in the mountainous regions, as in the Rheinischen Schiefergebirge, the Mittelgebirge of central Germany and the western Carpathians, there were indeed locally developed types of soil which can best be described as brown earths or even as parabrowns in the most favorable cases. They signify a better water supply and heavy forestation. Their distribution indicates, however, that often they are only isolated formations developed because of particularly favorable local conditions. By and large this connection between the more mature soil types and the rain-capturing mountains shows that during the Brørup interstadial also the amount of available precipitation was very low and much less than at present.

If the large number of loess profiles which have been subject to careful sedimentological investigation are excluded, there are, regrettably, few detailed observations available to estimate the water budget of the later segments of the last cold period. Regrettably, the sedimentology of the layers of loess allows less reliable indications to be drawn concerning the water budget than the biological remains. Of particular interest are the pollen analyses of Bastin [39] and the written communications of Professor Mullenders and Dr. Bastin concerning the Rocourt loess profile near Liège, Belgium, the mollusk faunas in the loess profile of Dolní Věstonice (Wisternitz) in southern Moravia [407], and the pollen analysis on material from the Stillfried an der March loess profile in lower Austria [258 and Frenzel, unpubl.]. At the localities mentioned thick loess sequences were formed during the last cold period. They all rest upon a soil formed during a warm period, which is referred to in Belgium as "Rocourt soil" ("sol de Rocourt"), in Czechoslovakia as the interglacial soil of "PK III," and in lower Austria as the basal soil of the "Stillfried complex." Above the warm-period forest soil several black earths follow relatively thin layers of loess, the former developing during the first warm fluctuations of the last cold period—that is, in the Amersfoort and Brørup interstadials. A thick layer of loess separates these black earths from a still younger soil horizon, namely the Kesselt soil in Belgium, the "PK I" in Czechoslovakia, and the "Stillfried B" soil (Paudorf interstadial) in lower Austria. Above it follows further loess, upon which the modern soil is formed. Up to the present time no pollen analyses, excluding the provisional first analyses of Schütrumpf [68], have been carried out in Dolní Věstonice, in contrast to the other two localities. Instead, there is the model analysis by Ložek of the molluscan fauna in the approximately 20-meter profile of Dolní Věstonice, the results of which, given in Table 20, can be directly compared with the results of the pollen analyses from Stillfried an der March only some 57 kilometers distant.

Although in these investigations very different climatic indicators (mollusks and pollen) have been used, the results agree remarkably well. This impression is strengthened by the observations of Bastin (insofar as the preceding analysis permits such a conclusion) according to which during the time between the origin of the Rocourt soil and the uppermost black earth below the Kesselt soil—that is, within the "Stillfried complex"—a grass steppe prevailed. As in lower Austria, there occurred in Belgium, during the interval in which the thick loess horizon between the Stillfried complex and the Stillfried B soil horizon developed, an herbaceous steppe in which the importance of grasses was appreciably reduced. A corresponding change between an initial grass steppe, approximately from the end of the last warm period to the Brørup interstadial and a subsequent time of widespread steppe vegetation characterized by herbaceous plants after the Brørup interstadial but prior to the Stillfried B interstadial, is recognized in the pollen analyses on material from near Oberfellabrunn in western lower Austria [258]. It may be assumed from this, with some reservations, that from the beginning of the last cold period in Europe during the cold climate phases a predominantly grass steppe prevailed, but after the Brørup interstadial it was replaced by a herbaceous

TABLE 20. COMPARISON OF THE ECOLOGICAL AND PHYTOGEOGRAPHICAL CONDITIONS IN DOLNÍ VĚSTONICE AND STILLFRIED AN DER MARCH DURING THE LAST COLD PERIOD

GEOLOGICAL HORIZON	DOLNÍ VĚSTONICE (LOWER WISTERNITZ) (MOLLUSCAN FAUNA)	STILLFRIED (POLLEN FLORA)
Loess about 50 cm. above the "Stillfried B" soil	Cold tundra with wet localities	Herbaceous steppe relatively rich in tundra elements
"Stillfried B" soil	Grass steppe, rich fauna	Spruce (fir) woods and thermophilous deciduous trees along the rivers, grass steppe herbs on the loess plateaux
Loess between the "Stillfried complex" and the "Stillfried B soil"	Grass steppe	Herbaceous steppe rich in grasses
"Stillfried complex" Uppermost humus horizon	Xerothermic wooded steppe	Thickly wooded steppe with hornbeam thickets
Loess horizon	Loess steppe	Grass steppe
Middle humus horizon	Xerothermic steppe	Less thickly wooded steppe; a higher proportion of grassland elements
Loess horizon and lowermost humus horizon	Relatively poor steppe community	Grass steppe without thermophilous trees
Soil of warm period	No mollusks; upon the soil, reworked interglacial mollusks occur	Probably warm period woodland vegetation of a hornbeam phase
Loess	Very cold loess steppe	Herbaceaus steppe rich in *Artemisia* (wormwood) and grasses

steppe. This provides a basis for the attempted reconstruction of the former hydrological conditions.

Within the grass steppe zone, but beyond the permafrost region today are Saratove, Akmolinsk, and Omsk. Tomsk is in the transition zone between steppe grasslands and birch-poplar (*Populus tremula*) woodland. The figures for the mean annual temperature and total annual precipitation are:

LOCALITY	MEAN ANNUAL TEMPERATURE °C	MEAN ANNUAL PRECIPITATION (mm.)
Saratove	+5.3	360
Akmolinsk	+0.6	300
Omsk	−1.1	340
Tomsk	−1.7	460

It can be assumed that the mean annual temperature during the first loess phase of the last cold period before the Amersfoort interstadial in Belgium, southern Moravia, and lower Austria was somewhat above 0°C. This follows from the former distribution of solifluction phenomena (above, p. 207). At that time the July temperatures even in relatively oceanic Denmark were at least 6° to 7°C below the present day. The same can certainly be assumed for regions in central Europe; their July temperatures lay at least 3°C to 4°C below those of the four south Russian and western Siberian localities which were given for comparison. Thus during the first phase of cold climate considered, the evaporation was certainly less in central Europe than in the given comparable localities today. If in spite of this the vegetation was sparse, it can scarcely be assumed that the average precipitation in Moravia, lower Austria, and even in Belgium was greater than 350 to 400 millimeters per year, particularly when it is recalled that in lower Austria and Moravia, at least, loess was accumulating. Thus the annual precipitation during the first phase of cold climate prior to the Amersfoort interstadial was at least 200 to 250 millimeters less than at present.

During the cold-climate phase between the Amersfoort and Brørup interstadials the same vegetation prevailed as during the intervals already discussed, although temperature fell still further.

It follows from this that the precipitation certainly did not increase, but was at least as great as during the first phase of cold climate or was even somewhat less.

With the end of the Brørup interstadial phytogeographical conditions became established which were until then unknown. A herbaceous steppe rich in grasses now reigned. The temperatures and plant communities resembled those now to be found in Outer Mongolia in the region of Ulan Uda and Kjachta. With a mean annual temperature of about −3.0° to −3.5°C the mean annual rainfall there is from 205 to 295 millimeters (cf. above, p. 158), so that the average precipitation appears to have dropped still further during the loess phase following the Brørup interstadial, and in lower Austria and Moravia was about 250 to 300 millimeters—that is, about 100 millimeters lower than during the first two cold phases of the last period.

After the Stillfried B interstadial still more extreme conditions became established (Fig. 101). The flora and fauna indicate not only wet locations (probably above the permafrost layer) but also herbaceous steppe in which grasses are scarce. The amount of precipitation at that time was probably no greater than during the preceding cold phase; possibly there may have been less, or at least it was unfavorably distributed over the year.

From this follows the paleoclimatically important conclusion that in central and eastern Europe since the end of the last warm period (Eemian warm period) and during the last cold period no time is known in which the precipitation was higher than or even equal to the present day. Rather, all the indications are that the climate immediately at the beginning of the last cold period was very dry, and this aridity increased with the passage of time.

The numerous data concerning an early glacial spread of tundra fauna appear to contradict these observations. Much later, after the main thickness of loess had accumulated, this fauna was replaced by a steppe fauna [564]. A further argument opposing the conclusions established above is given by the character and scale of frozen soil forms during the various phases of the last cold period. For only at the beginning of the very cold phases were solifluction and cryoturbation (the mixing of a highly mobile soil layer due to frost action) important processes in generating surface forms. Later, the soil forms so produced were

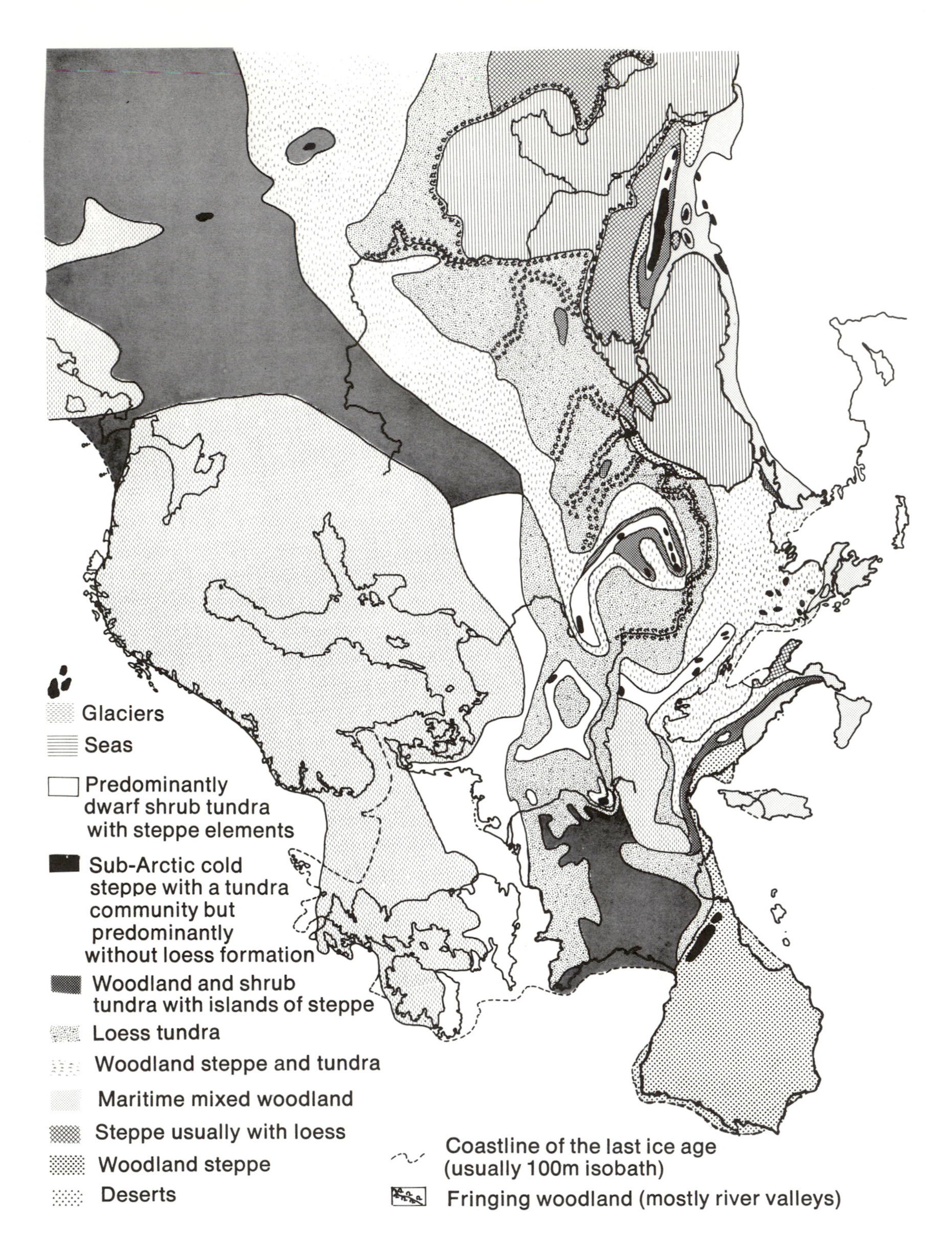

FIGURE 101. Vegetation types in Europe at the time of maximum glaciation after Stillfried B interstadial, Pleniglacial B [260].

FIGURE 102. Strong solifluction at the transition of a warm period to the following cold period. After the formation of the now fossil soil during a warm period (not too clearly shown on the lower two steps in the lower right of the picture), the soil horizons at the beginning of the succeeding cold period were intermixed (wave and tonguelike deformation of the white layer) by effects of solifluction. Subsequently, everything was covered by loess in which only rare traces of solifluction or associated frost working are to be seen. Loess profile from Hrubieszów on the Bug (Poland).

covered by loess (Fig. 102), still within the same cold phase; simultaneously in the place of solifluction and cryoturbation, nets of ice wedges developed—that is, large rectangles with 20 to 30-meter sides whose fissures were ice-filled. A wide distribution of solifluction and cryoturbation presupposes that the soil is well provided with water. This condition can be fulfilled if the precipitation is relatively great, or when as a result of the strong vapor-pressure gradient at the surface of slowly developing ground ice, water is drawn up from lower horizons. If solifluction was always particularly active at the beginning of very cold phases, there would seem to have been a higher precipitation. In any discussion of this problem, however, it must be remembered that the soil at the beginning of the cold phase may be very well provided with water, since formerly the annually thawing layer was in all probability extensive. During the interstadials the soil does not in general seem to have contained permafrost. It is true that Shotton and Strachan [685] described a block of dead ice dating from the beginning of the last cold period which persisted through the Stillfried B interstadial and began to melt only around 10,670 ± 130 years B.P. near Rodbaston (Staffordshire). From this it would seem that permafrost persisted, at least locally during the Stillfried B in the soils of Europe. However, it is uncertain whether this occurrence was actually the remains of glacier ice or whether it is the relict of a fossil pingo such as have been recently discovered at various localities in southern England. They were formed outside the glacial region, and upon melting left surface structures behind which appear to be dead ice holes (Sölle, kettle holes). Thus before this block of "dead ice" is used as evidence of cold climate during the Stillfried B it must be reinvestigated from this point of view. I thus assume that the strong solifluction and cryoturbation immediately at the beginning of phases of severe cold climate is essentially dependent upon the much thicker annual layer of thaw than occurred later. Furthermore, at the beginning of a cold period the amount of precipitation available may have been somewhat higher than at later times. Yet this more favorable water budget should not be taken as an indication of a pluvial beginning to a phase of cold climate.

It is true that the observations of Brunnacker [92] and Reich [622] on the occurrence of slaty coal at Grossweil and at Pfefferbichl near Füssen (northern margin of the Bavarian Alps) seem to indicate a higher rainfall in central Europe at the beginning of the last cold period than assumed above. Here, following the end of the last warm period, a pluvial phase set in, during the course of which a lake formed. The lake was partially silted up during the first warm phase of the last cold period (phase 11 of Reich); the silting up was reversed as a renewed cold climate set in, and the lake once more flooded the region, until during a further period of climatic improvement (phase 13 of Reich) silting up recommenced. Brunnacker [92] discovered evidence of cryoturbation in the sediments which accumulated during the second phase of lake development. Thus ground frost was a contributing cause of the flooding. Yet this does not explain the process. On the other hand, the question must be asked whether it should be generally concluded that formerly the rainfall was higher over wide areas. A fall in temperature would permit a rapid increase in precipitation in a mountainous rain-catchment area. It also follows that the amount of plant-transpired water would be appreciably reduced, and simultaneously the ground water and run-off would increase, once the tree cover was destroyed by the low temperatures. Around the margins of the mountains the effects of such an increased run-off would be particularly marked, so that this process would lead to an increase in the lake level. Thus the pluvial and lake-forming phase noted above appears to be important only in mountainous regions; it cannot be used as an indication for an all-round improvement in the water budget at the beginning of the last cold period.

This example of an early glacial phase of lake formation along the northern margin of the Bavarian Alps shows once more that the effects of relief on climate should not be neglected in an investigation of Pleistocene climatic history. This modifying factor was apparent (above, p. 204) with respect to the expression of the pluvials. Further indications can also be obtained from Brunnacker's [89] observations in southern Bavaria and Mainfranken as well as from Fink's [217] on the eastern margin of the Alps. Then in the wet regions of western Mainfranken, at the northern margin of the Bavarian Alps, in southeastern Burgenland, and in Steiermark—all of which are also wet at the present day—the loesses of the last cold period which accumulated in a relatively wet climate of a steppe area with interspersed woodlands were converted to loams, so that they only occur there now in the form of blanket loam: *Decklehme* (Brunnacker) or *Staublehme* (Fink). In this connection there is an important distinction between the wet and dry loess regions of Austria to be noted [216], which clearly reflects climatic differences dominant at the time of deposition. Nevertheless, even in the regions of Deck- and Staublehme a steppe climate prevailed, although conditions were somewhat moister than in many other parts of Europe.

The Climatic Character of Interstadials

As has been noted, the last cold period was broken up by several interstadials of which the most important were the Amersfoort, Brørup, Stillfried B, and, probably, Lascaux-Ula. At these times woodlands expanded northward along the rivers, and the picture of a wide steppe woodland developed (Fig. 100). The woodland migration was certainly dependent upon a climatic fluctuation. The following changes may be considered: (1) it could have become warmer; (2) it might have remained as cold as previously but with the moisture increased; (3) both factors may be superposed.

No definite indications of notable frost action in the soil are known during the interstadials of the last cold period in central Europe; there is no trace of perpetually frozen ground. Thus the second possible explanation for the interstadials, namely continuing low temperatures with increased rainfall, can be excluded, especially as in this case particularly strong frost action would have to be assumed. It seems, then, that the distinctive feature of the interstadials was a warming trend. Unfortunately, even today every attempt to determine the climate of the interstadials of the last cold period in Europe encounters considerable difficulties.

The cause is to be found in the numerous sources of error which leave the picture uncertain: it follows from Figure 100 that steppe woodland was widespread during the Brørup interstadial, the most important interstadial of the last cold period in central and eastern Europe. The same is probably also true for the other interstadials. Woodland then, at most, occupied only relatively isolated localities, and as a result the amount of tree pollen in the pollen flora was small. Among the species of trees, pine, birch, and others with wide ecological ranges were particularly favored, so the amounts of pollen of more exacting species occur only in extraordinarily low amounts. Under these conditions it is very difficult to distinguish whether the small amount of hornbeam, oak, linden, and elm pollen in the interstadial sediments consists of older reworked material or was derived from trees growing in the vicinity or whether it was transported great distances by wind. According to the way that question is answered, the paleoclimatic consequences deviate widely.

The second uncertainty is related to the plants' ability to migrate. Zagwijn [817] supposed that the end of the last warm period occurred some 70,000 years ago. A radiocarbon age for the first part of the Amersfoort interstadial gives about 64,000 years, while the strong afforestation of the Netherlands associated with the following Brørup interstadial began some 59,000 B.P., ending with the interstadial about 53,000 B.P. It would seem from this that the vegetation had ample time for migration; within 5,000 to 6,000 years the more exacting types of vegetation could have retreated at the end of the last warm period and subsequently readvanced. The time between the Amersfoort and the Brørup interstadials probably lasted some 3,000 years, while the Brørup had a duration of about 6,000 years. This time is as long as that between the beginning of the Alleröd and the optimum of the postglacial warm period. During this latter interval the vegetation, which at the end of the last cold period consisted of open tundra and steppe in central Europe, developed through several progressive and regressive steps into the rich mixed oak forest of the postglacial. It would therefore seem that the same amount of time should have been sufficient at the beginning of the last cold period for the migration of flora conformable to the prevailing conditions. The former floral distribution must therefore correspond exactly to the optimal conditions of climate prevalent at the time. However, it can be argued against this view that the time elapsed during the late and postglacial, while sufficient for the woodland to replace the open vegetation, was not enough to permit the completion of all plant migrations. On the contrary, in central Europe the migration of the beech from its southern refuge came much later [223, 46, 440]; the woodland flora of the northern and southern Alpine margins changed appreciably after a phase of mixed oak forest without human intervention [829]. In the Mediterranean region, too, natural changes in vegetation were continuing until significantly interfered with by man [53, 54]. A time of from 5,000 to 6,000 years is thus still insufficient to repair the damage caused by the Pleistocene cold periods, to permit the reestablishment of the optimum limits of distribution of very different species under the new climatic conditions.

In the case of the interstadials of the last cold period a further difficulty arises from the fact that at least the Brørup interstadial and probably other warm fluctuations too were made up of several phases with quite different climates. This is characterized for the Brørup interstadial by Andersen [11], whose data may be presented as in Table 21. Apart from the introductory phase W2e, the Brørup interstadial consisted of four phases of warm climate (W3a; c; e, W4, W5a; and c) and three of a cooler climate (W3b; d; and W5b). A comparable yet less detailed stratigraphy can frequently be observed in other profiles in central Europe (especially in Dolní Věstonice; see [407]). Table 20 (above, p. 214) lists the principal changes in vegetation during these phases (upper two humus horizons of the Stillfried complex at Stillfried and Dolní Věstonice as well as the intervening layer of loess); it demonstrates that the climatic deterioration within the Brørup interstadial, whose exact correlation with the phases in Denmark cannot be determined at the present time, must have been appreciable, for it permitted a renewal of steppe conditions where woodland had previously been established. The climate thus had become much drier.

It is therefore recognized that the Brørup interstadial at least was composed of several minor fluctuations during which quite different climates became established. Such climatic fluctuations within an interstadial must have considerably retarded the establishment of the optimal vegetation for the climatic conditions of the interstadial. It is possible that during the Brørup the woodland vegetation never reached the maximum extent climatically possible. Should that be the case, then the values of some climatic factors deduced from the former limits of distribution of the most important plant types will appear less favorable than was actually the case during the warm phases of this interstadial. Obviously the same is valid for the Stillfried B, which can be divided into at least two phases of favorable climate separated by a phase of unfavorable steppe climate; as has already been described. Earlier, Leroi-Gourhan [463, 464] had suggested that this interstadial was made up of the older Arcy interstadial and the younger Stillfried B interstadial (*sensu strenuo*).

Probably the terminology proposed by Leroi-Gourhan was not very happily chosen, for the more important Arcy interstadial is said to have occurred about 32,000 to 30,000 B.P., while the Stillfried B interstadial is said to date from about 26,000 B.P. It may be supposed that what Leroi-Gourhan terms the Arcy interstadial in France is the Stillfried B of Austrian terminology. Yet the Stillfried B in the sense of Leroi-Gourhan may comprise the minor warm fluctuation noted many times which occurs shortly after the true Stillfried B. However, this difficult problem goes beyond the scope of the present book and will not be discussed further.

The important consequences of the preceding is that because of the difficulties discussed with respect to the Brørup interstadial, but which are perhaps also valid for the other interstadials of the last cold period, only the general trend of climatic development can be given. How close the attempts at a reconstruction approach the actual conditions remains far from clear.

TABLE 21. DIVISIONS OF THE BRØRUP INTERSTADIAL IN DENMARK

ZONE ACCORDING TO ANDERSEN	VEGETATION TYPE	SOIL CONDITIONS	MEAN JULY TEMP. (°C)
W5	c. Pine woods (*Pinus*)	important layers of acid humus	13 to 15
	b. Spruce (*Picea*) ? birch (*Betula*	very important	ca 13?
	a. *pubescens*), larch (*Larix*), alder buckthorn (*Rhamus frangula*), *Calluna* (*ling*) *Empetrum* (crowberry)		13 to 15
W4	woods of *Betula pubescens, Picea, Larix, Rhamnus frangula.* Spruce of increasing importance	growth of acid humus layer still accumulating	13 to 15
	e. woods of *Betula pubescens*, poplar (*Populus*) and alder buckthorn. Dwarf birch (*Betula nana*) Juniper (*Juniperus*) and shrubs widespread. *Calluna* more important	increasing development of acid humus layer	13 to 15
	d. Open woodland	slight erosion	ca. 12
W3	c. woods of *Betula pubescens* and poplar (*Populus*), *Betula nana, Juniperus*, shrubs important	humus formation	13 to 15
	b. more open woodland	slight erosion	ca. 12
	a. woods of *Betula pubescens, Populus. Betula nana* and shrubs important	humus formation	ca. 13
W2	e. *Betula nana* and shrubs predominant. *Betula pubescens* and poplar occur	virgin soil beginning of humus formation	ca. 12, increasing

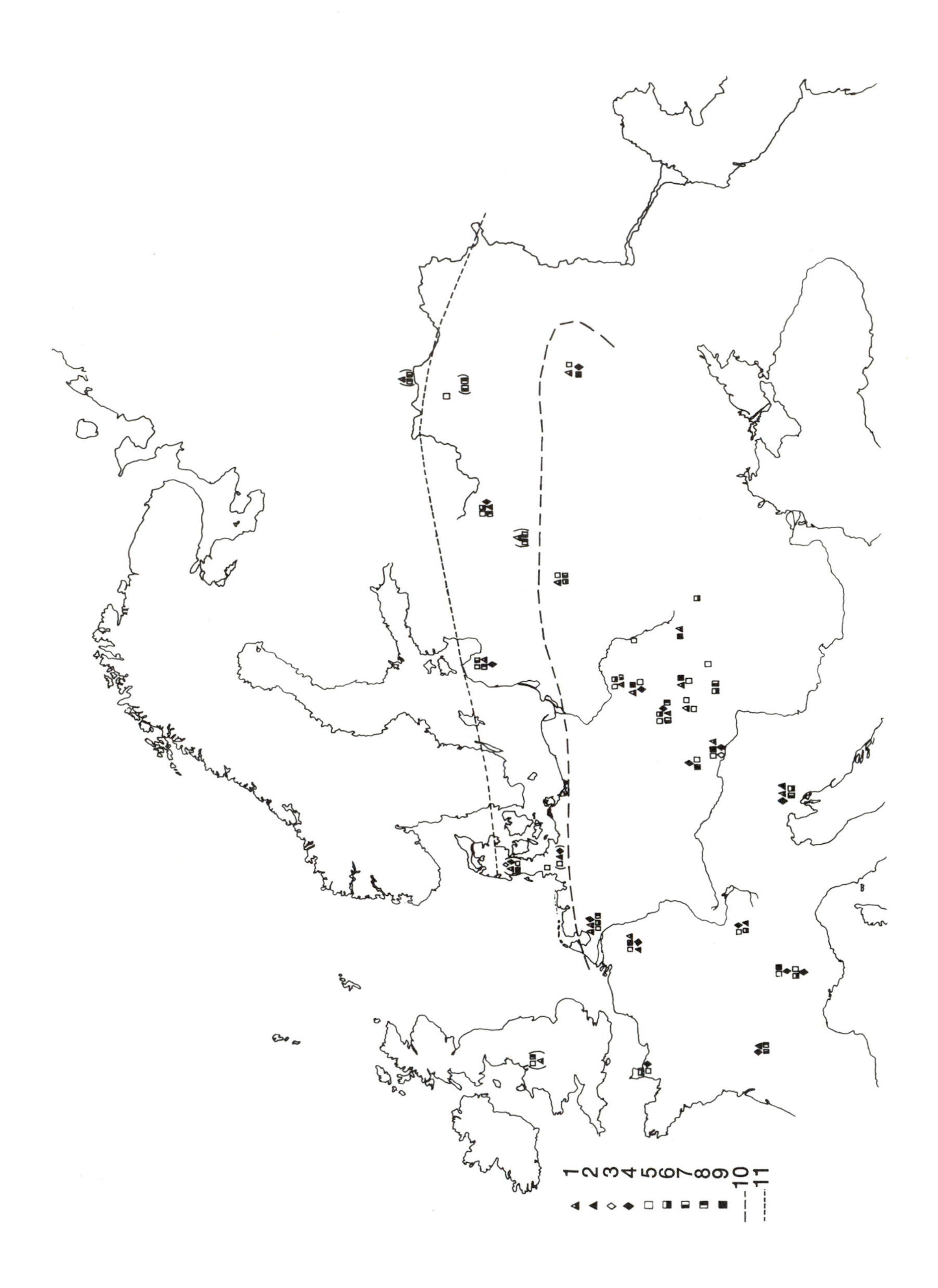
1
2
3
4
5
6
7
8
9
10
11

Figure 103 shows the distribution of the hornbeam (*Carpinus betulus*), oak (*Quercus*, probably mostly *Q. robur*), linden (usually *Tilia cordata*), elm (*Ulmus* sp.), ash (*Fraxinus excelsior*), beech (*Fagus silvatica*), and hazel (*Corylus avellana*) as determined by pollen analysis at the time of optimal conditions during the Brørup interstadial. The broken line shows the approximate position of the northern boundary of the above-mentioned thermophilous species, excluding all problematic occurrences. The dotted line shows the distribution limit of the alder. These distribution data are taken as a basis for the attempt to calculate the mean temperature at the optimum of the Brørup interstadial. The time of the climatic optimum was formerly displaced considerably toward the end of the interstadial. In northern Germany and adjacent regions the mean January temperature was apparently about 2°C lower than today. The July temperature was about 1°C lower than now, and the annual mean temperature about 2°C below present levels. In central Russia the corresponding figures are a 2° to 4°C lower January temperature and a 2°C lower mean annual temperature. The July temperature appears to have equaled that of today. Zagwijn [817] determined that the July temperature during the Brørup interstadial in the Netherlands corresponded approximately to the present; Andersen [11], however, assumed a temperature 2° to 3°C lower in Denmark. The figures agree very well, considering the difference in geographic location and the difficulties in the determination of former temperatures. The calculations show that the former summer temperatures were not appreciably different from the present; at most they were only a few degrees lower.

On the whole the January temperatures were somewhat lower than today. However, the correctness of the estimated figure of less than −3°C for the Netherlands [Zagwijn, 817] or the −10° to −15°C for outcrops of the same age from near Manchester [689] must be questioned, for they are based on a climatic evaluation of the former distribution of the spruce. Such an aid, however, is far from reliable (p. 46).

Uncertain though the individual determinations may be, they do serve to show that temperatures in central and eastern Europe during the climatic optimum of the Brørup interstadial in general cannot have been appreciably different from the present day.

The mean precipitation can scarcely be reliably determined in the same way, as up to the present time the pollen analyses have been carried out without exception on material from lake beds, basins, or rivers. The water budget of such locations in consequence of local conditions was appreciably more favorable than that of the surrounding areas. Nevertheless, it can be assumed that

FIGURE 103. The occurrence of macro- and microfloral remains of thermophilous trees in the sediments of the Brørup interstadial in Europe: 1. hornbeam (*Carpinus*); 2. beech (*Fagus*); 3. ash (*Fraxinus*); 4. hazel (*Corylus*); 5. alder (*Alnus*); 6. oak (*Quercus*); 7. linden (*Tilia*); 8. elm (*Ulmus*); 9. mixed oak forest (*Quercus* + *Tilia* + *Ulmus*); 10. presumed northern limit of *Carpinus*, *Corylus*, *Tilia*, *Quercus*, and *Ulmus*; 11. presumed northern limit of alder. In parentheses: very low pollen counts thought to be due to long transportation or to the reworking of older materials. Source: Appendix 7, p. 251.

the total precipitation increased in comparison with the preceding phase of cold climate, particularly in locally favorable climatic localities or near the sea, and was high enough to make possible thick woodland growth beyond the small wet locations. Yet it still lay considerably below present values (above, p. 212). Unfortunately, more precise statements are not yet possible.

During the preceding Amersfoort interstadial the vegetation seems to have been more sparse than during the Brørup. Andersen [11] recorded only heathland in Denmark made up of dwarf birch and juniper; along the upper Don were stands of pine and spruce (paleolithic station Kostenki XVII [215]), and near Stillfried an der March a grassland steppe prevailed [Frenzel, unpubl.], which in the groves of the western part of lower Austria near Oberfellabrunn passed into wooded steppe in which species of the subalpine pine and spruce forests were important [258]. There were also isolated trees of more exacting species, such as hazel, hornbeam, pine, and linden on the northern slopes of the Bavarian Alps [622]; elm, alder, and hazel near Stillfried [Frenzel, unpubl.]; as well as oak, ash, and elm near Amersfoort in the Netherlands [817]. Zagwijn concluded from this that the July temperature in the Netherlands averaged about 14° to 15°C—that is, about 3°C below the present value. Reich gives somewhat higher figures for the temperature depression; however, it scarcely seems wise to take this further in view of the scarcity of material available. Generally, of course, it may be assumed that the climate during the Amersfoort interstadial was colder and possibly also drier than during the Brørup.

Our knowledge of the climatic changes within the Stillfried B are equally uncertain. Near Arcy-sur-Cure in eastern France, oak, linden, hornbeam, and alder were widespread; on the north slopes of the Pyrenees near Isturitz, hazel, alder, and elm were to be found [463, 464]; in southern Moravia and in eastern lower Austria subalpine spruce woods spread along the river courses, in which in addition to spruce, pine, and stone pine (*Pinus cembra*) there were occasional larch [410], fir, oak, linden, elm, alder, hornbeam, ash, maple, and hazel [258]. The outcrops of Česky Těšin (near Ostrava), which Kneblová [413, 416] placed in the Göttweig interstadial, may perhaps date from this time. The poverty of this flora and that of Bialka Tatrzańska in the western Carpathians [700] suggest referral to the Stillfried B. At that time in those areas spruce and pine woods occurred in which there were possibly isolated hazel and alder. The contemporaneous flora of the Laibach moors in Slovenia was appreciably richer; there stands and forests of pine, oak, hornbeam, elm, and beech as well as bushes of hazel are found [680]. Possibly of the same age is a poorly developed soil resting on, and covered by, loess in western Podolia. Upon its surface is found charcoal of spruce (or larch) and of pine [432]. Near Moldovo on the middle Dniester, oak charcoal occurs, together with countless bones of the horse, European caribou (*Rangifer tarandus*), mammoth (*Mammonteus primigenius*), woolly rhinoceros (*Coelodonta antiquitatis*), and bison (*Bison priscus*) [360]. This was a cool, wooded steppe in which stands of timber, of the species noted, developed. Finally, the coeval spruce-forest steppe of the Belaya in the western foothills of the Urals must be recalled (above, p. 189). The total number of useful observations is still too small to be able to determine the former climate from them. Furthermore, consideration must be given to the fact that the immediately preceding phase of the last cold period brought with it a severe cooling, so that the former woodland vegetation was pushed far to the south. Probably it was during this first significant and long phase of cold climate that various steppe species could populate the old woodlands of the northern Mediterranean. The loess outcrops described by Fränzle [253], Mancini [507], and Markovič-Marjanovič [514] on the southern slopes of the Alps and in Montenegro favor this interpretation, as do contemporaneous steppes developed at the feet of the Sierra Nevada in southern Spain [528, 529]. It would therefore require a very long time before the forest flora of central Europe could reestablish itself, particularly so when the Stillfried B appears to have comprised two warm waves. It is thus hardly possible to draw certain conclusions about the former climate, for in all probability the vegetation did not have time enough to adapt fully to all the climatic possibilities. This result is important, for at the same time a high sea stand has been observed near Götcborg [87], which leads one to think of temperatures higher than would appear from the plant distributions described. Yet in the Caribbean Sea the surface water temperature lay about 3°C above the values characteristic of the coldest intervals, but still 3°C below present-day values [199].

The youngest important interstadial was the Lascaux-Ula. At that time in favorable locations in Lithuania stands of spruce, pine, and birch developed, with alder and willow along the river courses [320, 426]. At the same time, oak, elm, and hazel were to be found on the western slopes of the Massif Central of France [365, 463, 464], so that the climate must have been quite similar to the interstadials of the late glacial phase of Europe—that is, to the Bölling and Alleröd interstadials Firbas [222] estimated the July temperature for the more important Alleröd interstadial. According to this, it appears that the Baltic region during July was some 6°C colder, but the southern parts of central Europe only 4°C colder than today. At the same time, according to Emiliani [194], the equatorial Atlantic surface water temperatures were only half a degree below the present values. In the present scanty state of knowledge, these data serve as an indication that the climate during the interstadials of the last cold period, and most of all during the Brørup and Stillfried B (*sensu lato*), was astonishingly warm, so that these warm climate phases do represent distinct pauses in the course of the cold period. In contrast, at no time does the precipitation appear to have increased on a comparable scale. So cold dry stadials seem to have alternated with warm dry interstadials, which were themselves interrupted by cooler phases. The indication of dryness of the climate of the interstadials concerns only the basic characteristic of the climate. It is understandable that locally, as well as during the course of time, deviations occurred. The fact that certain interstadials were composed of different successive time intervals with quite different climates makes this a reasonable assumption. Furthermore, Selle [671] showed in the Orel and Nedden-Averbergen interstadials of northwestern Germany that many of these warm fluctuations of the last cold period had a cold, dry initial phase. At such times birch woods established themselves as an essential floral element. Later, pine and spruce migrated in; *Ericaceae* heath became widespread, and *Sphagnum* moors again became important. This process, on the one hand, shows the impoverishment of the soil during the course of the interstadial, and, on the other, permits one to recognize that the climate during the second phase became wetter. The amount of precipitation, unfortunately, cannot be estimated, even in these examples.

The Göttweig Interstadial and Its Analogues

Up to the present only the climatic changes during the Amersfoort, Brørup, Stillfried B, and Lascaux-Ula interstadials have been described. Frequently, however, a further interstadial is differentiated, which is said to have lasted from about 50,000 to 34,000 years B.P. Gross [315, 316, with an exhaustive bibliography] named this time interval, earlier called the Göttweig interstadial, the Würm Interpleniglacial. It was a time, according to Gross's data [316], of cold climate interrupted by several warm fluctuations. At various places in central, western, and eastern Europe observations made in the last decade seem to have clearly demonstrated one or more warm fluctuations during the last cold period. Generally, these warm fluctuations have been referred to simply as the Göttweig interstadial, without any real assurance that they date from that time.

The interstadial takes its name from the Göttweig monastery near Krems on the Danube, in the neighborhood where a well-developed fossil soil is found within a thick loess section. This soil is said to have been formed during the Göttweig interstadial. In other loess profiles north of the Danube immediately above the equivalents of this soil, several other humus horizons are found, so that it might be presumed that the Göttweig interstadial spanned a long time during which several climatic fluctuations of different magnitude occurred. Fink [216, 218], on the basis of field observations, much earlier was convinced that the fossil soils described, occurring in the thick loess beds of southern Germany, Moravia, and lower and eastern Austria, had very different ages and were formed during the last warm period as well as during the first warm fluctuations of the last cold period that can be assigned to the Amersfoort and Brørup interstadials. The correctness of this conclusion was later confirmed by both the snail fauna of the loess layers and by pollen analysis of the fossil soils. Thus the original assumption that there was a further important interstadial during the last cold period consisting of three to four warm fluctuations separated by periods of cold, dry loess climate can no longer be upheld [315, 316, 407, 258]. Yet this statement should not be construed as implying that the climate between the Brørup and Stillfried B was one of unchanging dryness and cold. In this connection, Gross [316] has recently collected data which forcefully demonstrate the need for caution. For in this space of time there occurred several intervals during which the growth of dense vegetation was possible. Thus in Dithmarschen (northern Germany), after the end of the Brørup interstadial but prior to the beginning of the Stillfried B, there was an important phase of forest advance [29]. This was named the Odderade interstadial, which according to Dücker [lecture in Cologne, July 1965], had an age of about 45,000 to 46,000 years B.P. In the loesses in Hessen, western Germany, above the fossil soils formed immediately at the beginning of the last cold period, there occur several poorly developed soils which developed in that time interval, between the Brørup and Stillfried B [657]. It must finally be remembered that according to the evidence of the molluscan fauna in the Czech loesses the climate during this part of the last cold period was not as unfavorable as that following the Stillfried B. This also seems to be indicated by the lower Chvalyn transgression of the Caspian (above, pp. 187 ff.), which in turn appears to show that between the end of the Brørup and the beginning of the Stillfried B there was a series of fluctuations of warm and moistness. Yet in my opinion the significance of these observations should not be exaggerated. Much too little is known about that period. As a result, no certain distinction can be made whether these periods with a better water balance were minor fluctuations of general significance, or, during a time of not particularly severe climatic conditions, the local water budget improved at various places at various times. For this, very different causes might be responsible. At the present time, in view of the small number of reliable radiocarbon dates and field geological investigations, many of the warm fluctuations can only be guessed. In spite of all the uncertainties, I incline to the first interpretation. Yet to regard the interval as an interstadial does not appear justified, as it was established above (p. 215) that the climate was in general dry and cold. Short-term alterations in temperature and water budget change the picture very little, even if they were of significance for man and the plant and animal world.

It has been shown that repeated fluctuations in temperature occurred during the last cold period in central and western Europe which occasionally reached appreciable amounts. Yet these conditions were never sufficient to support a renewed migration of a rich forest flora; the available moisture was insufficient. In view of the facts, it is very surprising that recently, numerous observations have been reported from Russia from which it appears that during a significant interstadial a vegetation comparable to the present time occupied the cold steppes. Consequently, in Russia reference is often made to a younger interglacial which interrupted the last cold period of central and western European terminology. Following a suggestion of Moskvitin [543], this was called the Mologo-Sheksna interglacial in central Russia but was named the Karginsk interglacial in Siberia. It is important that in the Karginsk interglacial the forest flora is said to have migrated further north than it does at present. The Mologo-Sheksna interglacial, according to the radiocarbon dating, should be equivalent to the Göttweig interglacial [551], but the Karginsk interglacial corresponds to the Stillfried B of central European terminology [400, 401]. If these assumptions prove correct, the greater climatic fluctuations of the last cold period in northern Eurasia were indeed simultaneous, but their magnitude must have increased eastward in such a way that the strongest climatic fluctuation becomes progressively younger, the further east the region considered lies. Furthermore, it is important that in both central Russia and Siberia two glacial advances have been described which delimit this significant warm fluctuation. In eastern Europe the Mologo-Sheksna interglacial was preceded by the important "Kalinin" glaciation, and followed by the less pronounced "Ostashkov" glaciation. In Siberia the older glaciation is generally called the "Zyrjanka" and the younger and relatively unimportant one the "Sartan." Thus the maximum glaciation during the last cold period in eastern Europe is held to have occurred some 50,000 years ago, but in western Siberia, shortly before 32,000 years B.P. In contrast, in central Europe, if the still uncertain Stettin stadial is excluded, only a single glaciation of the last cold period is known, the outermost limit of which is referred to as the "Brandenburg stadial." This maximum advance appears to have occurred after the Stillfried B interstadial—that is, some 23,000 years B.P. Thus the division of the last cold period and the scale of the climatic fluctuation appear to show that these were very different in different regions of northern Eurasia. This becomes clear from a comparison of the climate during the optimal phases within the last cold period. In central and western Europe the climate during the Brørup interstadial was dry and in general somewhat cooler than today; in eastern Europe (the Mologo-Sheksna interglacial) and Siberia (the Karginsk interglacial) it was moister and warmer than today. An explanation for such climatic deviations is hard to find, and it seems appropriate to inquire whether the dating is reliable everywhere.

At all localities in eastern Europe where a subsequent pollen analysis test was possible, it appears that under the designation Mologo-Sheksna formations of very different ages have been included. In places there are sediments belonging to the last warm period (Eemian, Mikulino warm period); as, for example, the deposits in the Rybinsk reservoir on the upper Volga, where the pollen flora permits the recognition of the forest history of the last warm period [126] although in part distorted by an over-representation of alder which grew near the river [290]. In other cases these were postglacial sediments subsequently overlain by detrital material, thus giving in the field the impression of a much earlier age. In this respect the pollen analysis of Grichuk [302] of the Balanza and of the Kasplja regions in the Smolensk district are important. They indicated a luxuriant deciduous forest whose composition differed a litte from both that of the last warm period and of the postglacial age. The stratigraphy of the sediments favors a postglacial age. Nonetheless Grichuk [302] placed the formation in the Mologo-Sheksna interglacial because of the somewhat different forest history and regarded it as a clear indication of the existence of this interglacial. Subsequent radiocarbon dating has shown, however, that the formation in question has an age in the range between 5120 ± 200 and 2760 ± 100 (1310 B.P.), so that it actually comes from the younger part of the postglacial and not from some interstadial or even interglacial of the last cold period.

Insofar as I can review the problem, there is at the present time no incontrovertible evidence for the existence of the Mologo-Sheksna interglacial [260]. Probably this interglacial and its equivalent, the Göttweig interstadial, never existed. The fact, however, that among many others, so experienced a geologist as Moskvitin has repeatedly asserted that the climate of the last cold period passed through this warm fluctuation is probably related to the more intense and longer-lasting periglacial planation and transportation around the margin of the northern ice sheet in the continental climate of central Russia in comparison with central Europe, and that the local relief prior to this was much lower and less differentiated, so that Quaternary geological research is made extraordinarily difficult. Another reason may be found in that, just as in central Europe, so in central Russia the radiocarbon dating of samples with a relatively high absolute age were presented too early and applied as a reliable criterion of age, although the geological findings opposed the apparently exact and reliable radiocarbon dating. Unfortunately, it has to be admitted that this dating technique, so invaluable for dating younger materials, has led to large errors extremely difficult to evaluate in samples with a relatively high absolute age (above, p. 17 ff.).

In the event that the Mologo-Sheksna interglacial in fact did not exist, the question arises of the significance of the still older glacial advance of the last glaciation in central Russia—that is, the Kalinin glaciation. Chebotareva, Nedoshivina, and Stoljarova [129, 130] as well as Gerasimov, Serebrjannyj, and Chebotareva [273] dispute the existence of this older glaciation, and Chebotareva, Mal'gina, and Nedoshivina [128] have shown that the youngest moraine in the region of Lake Ilmen, often referred to as the Ostashkov glaciation—that is, the one which first occurs after the Mologo-Sheksna interglacial—is separated from older horizons only by beds of the last warm period (Eemian, Mikulino) and not by sediments of an interstadial of the last cold period, so that they can only represent sediments laid down during the maximum advance of the last ice age in the terminology of central Europe. Because in central Europe, too, no continental glaciation prior to the Stillfried B interstadial is known, the reality of the Kalinin glaciation is open to doubt. However, in North America a significant glacial advance is known immediately at the beginning of the last cold period, appreciably earlier than the period of maximum glacial advance in Europe. Before a final conclusion can be reached, further observations are advisable.

Only some radiocarbon dates from the lower Yenisey favor the correlation of the Karginsk interglacial of Siberia with the Stillfried B of Europe. The river, below the mouth of the Nishnya Tunguska, cut a valley into the glacial sediments which accumulated there during some unknown phase of the last cold period. This erosion valley does not appear to have again been covered by ice, for moraine deposits are absent from the 20 to 25-meter terrace of the lower Yenisey, which has been called the Karginsk Terrace. Radiocarbon analysis of wood from this terrace gave ages of 37,000 ± 600 years B.P., 26,800 ± 400 B.P., and 21,350 ± 750 B.P. They thereby indicate a time of origin of the sediment which in part corresponds to the Würm interpleniglacial, in part to the Stillfried B interstadial, or even younger warm fluctuations. The next still younger terrace, at 10 to 14 meters, is composed of fluviatile material overlain by one or two fossil soils. Wood from below these soils had radiocarbon ages of 12,940 ± 270, 14,320 ± 330, and 15,460 ± 320 years B.P. [134]. These clearly relate to a late glacial formation, so that the fossil soils probably belong in the Alleröd interstadial, or perhaps may even be placed in the Bölling interstadial; for above the soils are loam and clay which were deposited during the spring highwater of the late glacial under a very cold climatic regime [285]. This latter cold-climate phase should correspond to the younger tundra period of Europe. The climatic history of Siberia during the latter stages of the last cold period is thus comparable with that of Europe, and in no case was the climate more favorable. In spite of this, organic material found in the Karginsk Terrace appears to show a basically different development of the older climatic fluctuations of the last cold period. The reliability of the radiocarbon ages of the wood of the so-called Karginsk Terrace must therefore be questioned. This terrace is seemingly a typical erosion terrace built up of heterogeneous material [401]. Obviously the river during its breakthrough to the north laid down material of various ages, in part eroded and transported material to be later redeposited in other localities. In my opinion the radiocarbon dates quoted do not have great reliability, for they indicate a period appreciably longer than the Stillfried B of central Europe, and they do not accord very well with the geological results [260]. In addition, the careful research of Ravskj, Aleksandrova, Vangengejm, Gerbova, and Golubeva [621] on material from the Tunka Basin south of Lake Baikal gives no indication of the existence of the Karginsk interglacial, although the authors repeatedly refer to this important warm period. In the region under consideration there occurs near Irkut in the Mondinsk Basin a young river terrace that is cut into the moraines of the last cold period and stands some 6 meters above the river and about 10 meters below the upper surface of the terrace sediments on the river islands. Its pollen flora is extremely poor. It is characterized only by birch with, occasionally, pollen of larch and spruce. Above this horizon, nonarboreal pollens predominate. The author assigns the birch phase to the Karginsk warm period but notes that this does not prove the existence of a Karginsk interglacial. Of much greater significance are the peats formed on the terrace surface which are intersected by ice wedges. This fissure system is regarded as having been formed during the closing phase of the last cold period (that is, during the Sartan stadial), so that the peat must date from the Karginsk warm period. This conclusion is not too convincing, for today in the same region perpetually frozen ground is forming, either as a natural consequence of a change in the environment or as a result of human interference with the vegetation cover. Insofar as there is no more conclusive investigation, it must be assumed that there was no Karginsk interglacial, at least not in the form in which it is many times asserted. Probably the effects of the warm fluctuations of the last cold period also made themselves felt in Siberia, so that many formations which were assigned to the Karginsk warm period actually belong to one or another of these interstadials. At the present time there is no reliable indication that the climate during any phase of the last cold period in Siberia was ever relatively more favorable than in Europe. From this it follows that during the last cold period no warm fluctuation occurred that in any way confirms the assumption of a "Göttweig interstadial," a "Mologo-Sheksna interglacial," or a "Karginsk interglacial." It is my opinion that careful investigations must be repeated in those regions of the earth where formations assigned to this age occur to permit an evaluation of their true age.

The Problem of Prevailing Wind Direction

Up to the present, various aspects of the climatic fluctuations during the last cold period of Europe have been discussed. Particular attention has been given to fluctuations in temperature and to the water budget. The question of the former prevailing wind direction has been left to one side.

It is conceivable that one consequence of the strong cooling over the continental ice sheets in the north was the development of a region of stable high pressure. This must have influenced the climate of Europe through northerly winds, if it was really of great significance. Klute [408], at any rate, questions it.

The resultant circulation system may be valid for all the Pleistocene glaciations in Europe as well as, in modified form, for northern Asia and North America. This view has been energetically defended by Solger [841], particularly on the grounds of the structure of the late glacial dunes of northern Germany. Subsequently, the correctness of this hypothesis for late glacial times has been questioned from many sides with good reason, for often two genetically different dune types, the barchan and the longitudinal dune, have been confused with each other. All available observations show, to the contrary, that during the last glacial in Europe the prevailing wind direction was the same as at the present time [602–04 and Fig. 104].

However, the prevailing wind direction during the earlier phases of the last cold period, or even of the preceding cold periods, is unknown. It is particularly

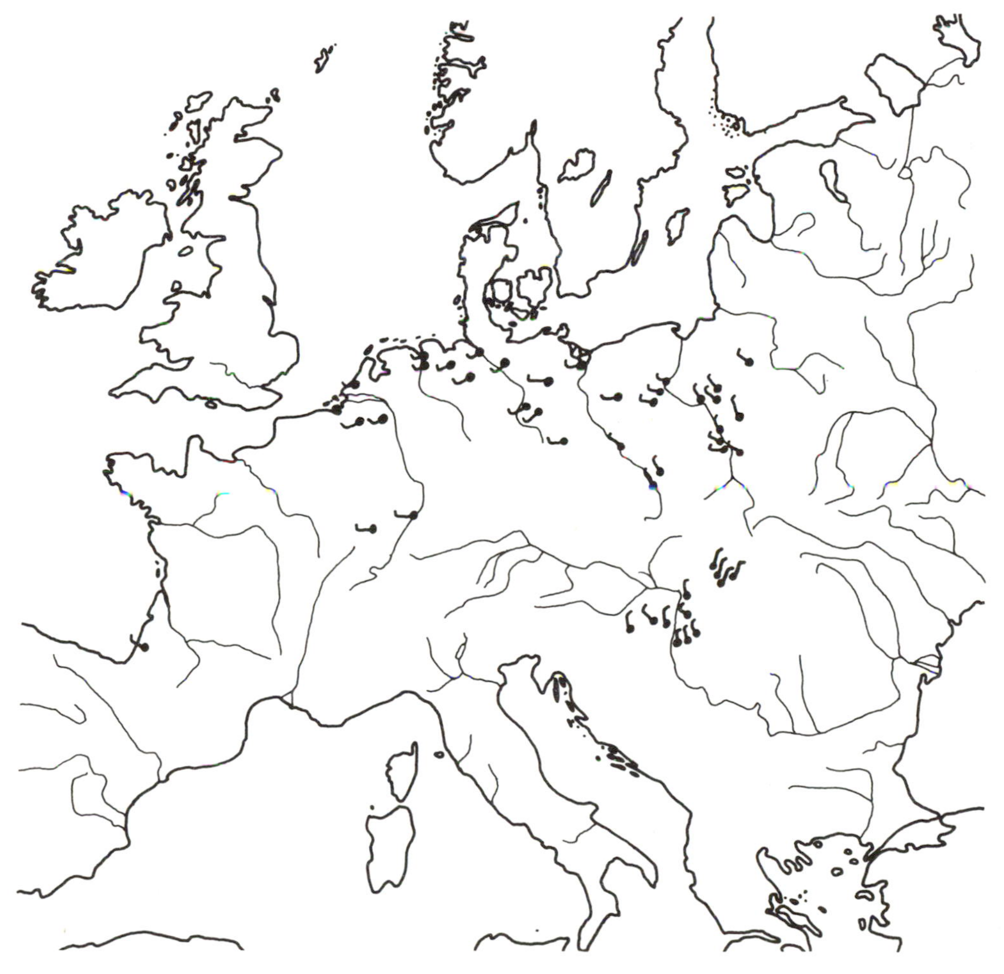

FIGURE 104. Wind directions during the late glacial phase in central Europe, deduced from the structure of the migrating dunes formed at that time. Points indicate the actual locations, and wind directions are shown by the symbols. The length of the symbol has no relation to former wind strengths. Simplified after Poser [604].

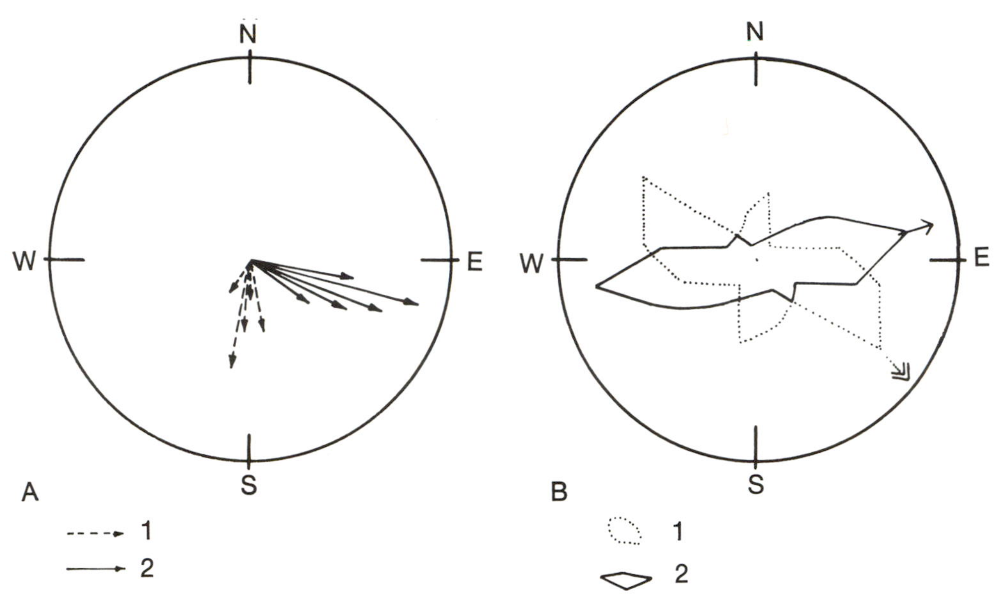

FIGURE 105. Prevailing wind directions during the last cold period in the Netherlands: A. Gelderse Vallei; B. Central Netherlands. The length of the arrow or the wind rise is a measure of the frequency of observations. A: 1. end of the main Pleniglacial B, transition to the oldest tundra period; 2. older tundra period. B: 1. older tundra period; 2. younger tundra period. After Maarleveld and van der Schans [498].

disappointing in this context that even today the scale of glaciation of Europe during the various segments of the last cold period is still unknown, yet the radiation and reflection conditions must have been very different, depending upon whether wide continental areas were covered by snow and ice or were ice free. In the one case, radiation and reflection favor the establishment of an anticyclone; in the other, they do not. The literature, consequently, is full of contradictory results [cf. detailed review in 604], although during the cold-climate phases an aeolian sediment, loess, was laid down over wide areas. However, the bedding of the loess, and with it the possibility of determining wind direction, is far less clearly defined than in the dunes. Possibly much can be deduced from the limits of the late glacial loess and their relation to dunes of the same age. Late and high glacial loesses, however, were probably accumulated from winds from different directions, and seasonal changes in wind directions may have occurred, complicating the picture still more. Maarleveld and van der Schans [498] as well as van der Sluys and Maarleveld [691] carried out careful inclination measurements—that is, the direction of greatest dip of sand horizons which were wind deposited during the peak of the last cold period (Pleniglacial B in Dutch terminology), in the older and during the younger tundra periods in the Gelders Valley and in the Zeeuws Vlaanderen (Fig. 105). During the period of maximum ice cover in the last cold period the prevailing wind in the Netherlands blew out of the north; with the retreat of the ice the wind backed to northwest and finally to west southwest. On the eastern slopes of the Sudeten Mountains the high glacial loess was blown in from the northeast

[231], and from the mineralogical researches of Rjabchenkov [627] on the loess of the Ukraine a wind direction from the northern quadrant can be recognized during the principal loess phase of the last cold period. Lomonović [476] gave this latter direction for the time of loess accumulation on the northern slopes of the Trans-Ili-Alatau in southeastern Kazakstan. These directions are not readily reconciled with the picture of winter atmospheric circulation sketched by Poser [601] during the Weichselian cold period. Yet they agree well with the data of Schönhals [656], according to whom the loess formed during the last cold period in northern Bohemia was transported by easterly winds; for the number and thickness of the sand layers intercalated in the loess derived from fluvial deposits decreases westward. It must be realized, however, that sand transport may have occurred readily during the winter when the plant cover was sparse. The north and east winds derived from the cross-bedding in the sands is thus not surprising for winter conditions. The lighter loess was probably retained only by the thicker vegetation during the summer, so that from its bedding conditions the former summer-wind direction may be deduced. At least in central and western Europe, westerly winds during the summer should have been more important than during the winter. Such seasonal differences in wind directions certainly must be considered everywhere. As an example, the conditions of the North American loess can be cited. In Illinois they indicate generally a prevailing westerly wind. Occasionally an easterly wind direction can also be deduced which blew with regularity at certain periods [231, 690]. This seasonal difference became particularly marked at the time of a strong ice advance. The pollen flora of sediments formed during a cold period in the mountains of the central Sahara make it obvious that the former spring and summer winds blew from the north (above, p. 196). Unfortunately, from the northern limits of the Mediterranean region, apart from observations on the height of the former climatic snow line (Fig. 84), scarcely any other worthwhile indication of wind direction is known, for the data of Butzer [111] and of Butzer and Cuerda [114] on the cross-bedding directions of fossil dunes from the southern point of Mallorca are so widely scattered, despite the small size of the area investigated, that it is better not to use them in this context. In every approximately meridional-striking mountain range the climatic snow line rose from west to east, so that it may be supposed that the snow-bearing wind blew from a westerly direction. How the prevailing wind direction changed during the course of the year and during the course of various segments of the last cold period there, as in other areas of the Northern Hemisphere, is not known, for the preliminary data are far too sparse to reconstruct any reliable wind regime for the last cold period.

Glaciation in North America

In considering the problem of the pluvials, it was shown that the climatic fluctuations in southwestern North America were simultaneous with those in Europe, but that the climatic development in North America ran a different course in many respects compared with that in Europe. The same impression is given by the very different history of glaciation during the last cold period when the two land masses are compared. It would appear that the climatic character of North America was in general more favorable for the accumulation of con-

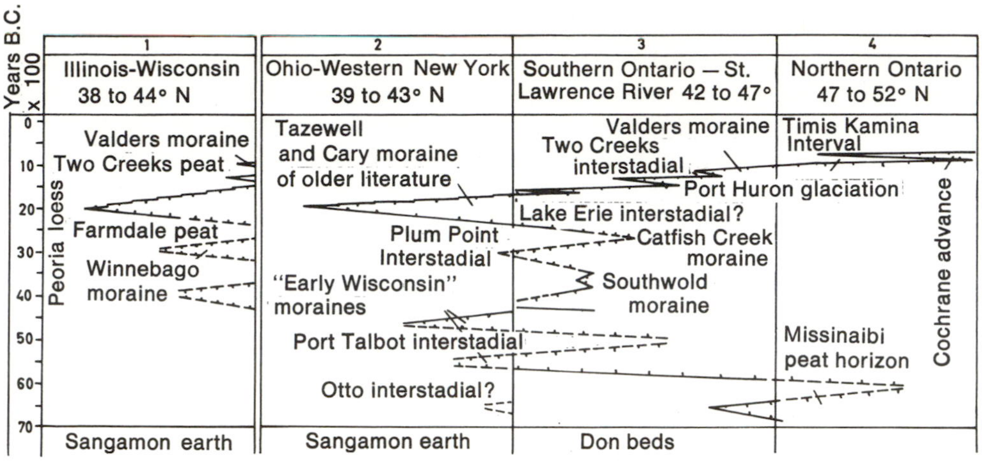

FIGURE 106. History of glaciation of North America. The apparent movement of the ice advancing from the north (right) is indicated by the line. The time scale at the left is in thousands of years B.P. Simplified after Flint [233].

tinental ice sheets than that of northern Eurasia. This must have as a consequence that the climatic fluctuations were differently composed compared to those in Europe. Accordingly, an attempt will be made to investigate this, step by step, for stadials and interstadials.

The maximum ice advance in Europe, as previously noted, succeeded the Stillfried B interstadial—that is, it began about 25,000 years B.P. It was assumed at first that the same was true of North America. The last cold period in North America, the Wisconsin, appeared to be subdivided, from oldest to youngest, into the Iowan, Tazewell, Cary, and Mankato stadials. The oldest glacial sediments of these advances date from about 25,000 years B.P.; their age thus corresponds approximately with the maximum advance of the Weichselian glaciation. In between, the picture is considerably altered (Fig. 106), for an advance of the North American ice sheet far to the south in the Great Lakes region has been recognized immediately at the commencement of the last cold period, although the maximum ice advance of the last cold period ("classical Wisconsin") occurred about 25,000 to 20,000 B.P. [230, 233, 736, 282, 174, 175, 396, 779, 234, 286, 392, 559].

The continental ice sheet immediately at the beginning of the last cold period had already advanced into the St. Lawrence basin and had reached the southwestern corner of present-day New York State, and its moraines are to be found in southeastern Indiana. In addition, Heusser [342] has described moraines from the Olympic Peninsula in the western part of Washington State which are probably of the same "early Wisconsin glaciation," whose scale appears to be greater than the succeeding glaciations, including the "classical Wisconsin." This oldest and manifestly important continental and valley glaciation apparently preceded the St. Pierre interstadial, which corresponds to the Amersfoort interstadial of Europe [234]. It covered wide areas of North America at a time when, even in the vicinity of the Alps, no strong glaciation has been recognized.

The results of the pollen analyses of Muller [559] on the loams and peats formed near Otto in southwestern New York State during the St. Pierre interstadial indicate the existence of pine and spruce woods which were presumably composed largely of *Pinus banksiana* and *P. strobus* as well as of *Picea glauca* and *P. mariana*. Instead of this, the present-day vegetation is largely made up of a varied deciduous flora in which various oak, hickory (*Carya*) and maple species are particularly common. Fir, larch, and the hemlock fir (*Tsuga*) appear to have been almost wholly absent from the former forest. The pollen of oak, elm, maple, walnut (*Juglans*), hickory, and linden is, however, extremely rare in these sediments and must be considered as either redeposited or as far-traveled wind-blown material. Muller compared this vegetation with the forest type which is widespread north of North Bay in Canada at the present time. The January temperature there averages about −11.1°C today, with the mean annual and July temperatures being about +3.9°C and 19.4°C respectively. From this it follows that the mean January temperature during the St. Pierre interstadial may be estimated to have been about 8°C below present levels, with the July and mean annual differences being −2°C and −5°C. These estimates are wholly consistent with those determined by Zagwijn [817] for the Amersfoort interstadial in the Netherlands upon sediments whose age corresponded to those of Otto at 63,900 ± 17000 years B.P. Terasmaë [736] described a peat horizon from between Pierreville and Donnacona on the St. Lawrence, which lies between two moraines and banded clays. In the peat a change from a subarctic or an arctic vegetation to a woodland phase and a final return to subarctic vegetation can be recognized. During the optimal climatic phase in the woodland, pine (apparently *Pinus banksiana*) and spruce (probably *Picea mariana* and *P. glauca*) predominated; during the first part of the warm fluctuation there also occurred in minor amounts the Weymouth pine (*Pinus strobus*) as well as oak, elm, maple, ash, hickory, hornbeam, and linden. The hemlock fir was absent. Presumably the time available for its migration was too short [736]. A similar woodland flora was widespread at the time of formation of the Scarborough beds near Toronto. Teresmaë [737] placed this woodland phase also in the St. Pierre interstadial. The former mean annual temperature there should have been about 4.5°C lower than at present. It seems that the flora mentioned was richer than that described by Muller in southwestern New York State. Its floral character suggests temperature conditions which may be more closely compared with those of the present day in the same region than with those estimated for the St. Pierre interstadial near Otto. It is therefore reasonable to question whether this interstadial and that near Otto are really of the same age, or whether that in the St. Lawrence basin corresponds to the Brørup interstadial should the Otto interstadial be of the same age as the Amersfoort. In this event, the St. Pierre interstadial, which appears to unite the two outcrops, must be further subdivided.

Younger than those just discussed is the Port Talbot interstadial, which takes its name from a particularly good outcrop on the north shore of Lake Erie. At first an age of more than 39,000 years [174] was assigned to this interstadial, which occurs between two significant ice advances, so that it might be correlated with the warm fluctuation known as the Gottweig interstadial [175].

Subsequent repeat dating gave ages of 47,000 ± 2500 and 47,500 ± 2500 B.P., while wood in the ground moraine above the interstadial had an age of 44,200 ± 1500 years B.P. [779]. According to the macro- and micro-fossil evidence, initially pine-spruce forest was widespread and later replaced by spruce woods. In addition, larch (*Larix*) and birch occurred. The pollen of *Abies*, *Ulmus*, and *Alnus* was probably blown in from regions further to the south or derived from the reworking of older material. Dreimanis [174] compared the forest type with that which at present is found in the headwaters of the St. Maurice River in Quebec (48.5°N, 74°W). This region has a mean January temperature of −17.8°C, with only 80 frost-free days in the year; nevertheless, in warm summers in mid-July +20°C is reached. If these figures can be applied to the Port Talbot interstadial on the north shores of Lake Erie, it follows that the winters were about 12°C colder than at present, the number of frost-free days some eighty fewer, but with a summer about as warm as the present. Here again is clearly repeated what has been learned with respect not only to the older Otto interstadial of New York but also to the early glacial interstadials of Europe, namely that during these warm fluctuations the summer temperatures corresponded approximately to those of the present day, but the winter temperatures were appreciably lower. It seems that this process was more pronounced in North America than in Europe. It is not easy to suggest to which of the warm fluctuations in Europe the Port Talbot interstadial may have corresponded. Possibly it was the Odderade interstadial of Schleswig-Holstein, for which dates of 44,760 and 46,250 B.P. have been obtained [Dücker, lecture, Cologne, July 1965].

During the early interstadials of the last cold period, the phytogeographical conditions in southeastern Indiana were very similar. Essentially there were pine and spruce forests in which only rare hard woods occurred. Kapp and Gooding [392] assumed from this that the climate warmed but little. The observations, however, are unfortunately insufficient for a more exact paleoclimatic study.

South of the ice front in North America as in northern Eurasia loess accumulated over wide areas during the last cold period (Fig. 80.) The conditions of deposition suggest that there was a prevailing wind from the west (p. 233). This makes it particularly interesting to analyze the history of climatic fluctuations of a region lying on the weather side of the Rocky Mountains. In this context Heusser [342] has valuable results.

In the western part of Washington State lies the Olympic Peninsula. The highest mountain, Mount Olympus, reaches an elevation of 2,420 meters. Even today it carries glaciers up to 5 kilometers long. The high precipitation of about 2,900 millimeters per year, frequent mist, and uniform temperature (January 7.2°C, July 16.7°C) are indications of a high oceanic climate, which permits the growth of a luxuriant forest of Sitka pine (*Picea sitchensis*), Douglas fir (*Pseudotsuga menziesii*), hemlock fir (*Tsuga heterophylla*), arborvitae (*Thuja plicata*), various species of pine (*Pinus contorta* and *P. monticola*), and alder between sea level and about 500 meters. Its prevailing vegetation during the various phases of the last cold period was considerably different (Table 22).

It is important to note that on the whole near the profile considered (Hump-

TABLE 22. VEGETATIONAL AND CLIMATIC HISTORY OF THE WESTERN SIDE OF THE OLYMPIC PENINSULA, WASHINGTON (ISL = INTERSTADIAL OR RETREAT PHASE; GL = STADIAL OR PHASE OF ICE ADVANCE)

ZONE		VEGETATION
GL 5		Rich grass flora; *Tsuga heterophylla* not significant; increase in the importance of *Tsuga mertensiana* and *Abies*. This phase probably corresponds to the "classical Wisconsin." Mean July temperature 5–6° lower than at the present day.
1SL 5		*Tsuga heterophylla* more important; common *Alnus* and *Sphagnum*; *Pinus contorta* common; *Myrica* very common. July about 2°C colder than at the present day.
GL 4		*Tsuga heterophylla* almost wholly absent; consequently grasses and *Tsuga mertensiana* of greater significance. Age of a sample from the upper surface of GL 4: 27,400 ± 2,000 B.P.
	1SL 4	Spread of *Pinus contorta*
GL 3		Vegetation as in GL 4. Age of one sample from the upper surface of GL 3: more than 30,000 B.P.
	1SL 3	Spread of *Pinus contorta*
GL 2		Maximum spread of grasses; the importance of Sitka spruce (*Picea sitchensis*), *Tsuga mertensiana* and firs much diminished. Simultaneously maximum spread of the middle Wisconsin glaciation. Older than 30,000 B.P. Mean July temperature about 6°C below present levels.
	1SL 2	Grasses less significant; increasing density of tree cover.
GL 1		Spread of the Graminaceae community. Simultaneous glacial advance not known.
	1SL 1	At first *Pinus contorta* very common, later increasing significance of alder. This process can be considered as the succession of plant communities on a surface newly freed from ice.

NOTE: The interstadial was apparently very marked. It separates the time of maximum glacial advance on the Olympic Peninsula from a period of appreciably less intense glaciation. From this, Heusser considers the section from GL 1 to the end of 1SL 4 as middle Wisconsin, 1SL 1 and the still older glacial advances of the Wisconsin cold period as early Wisconsin. Glaciation of the early Wisconsin: the strongest glaciation of the last cold period.

tulips, Gray's Harbor County) during the phase of warm climate, conditions were never found which can be compared with the postglacial. This means that the sedimentary succession does not contain beds of the last warm period, so that the beds examined belong exclusively to the last cold period, unless some erosional discordance has been overlooked. This interesting observation must be confirmed and amplified, yet it should now be clear that even in westernmost North America the divisions of the last cold period are identical with the European. The figures estimated for the reduction of the mean July temperatures, considering the strong contrast in climatic character (extreme oceanic climate during the cold period on the Olympic Peninsula; continental climate in central Europe) compare very well with those in Europe. Identity is not to be expected. Of special paleoclimatic value is a further item of information which arises out of the observations detailed: even where a heavy rainfall has to be considered, the precipitation during the stadials of the last cold period, from the floral evidence, was appreciably less than at present, and even during the interstadials the amount of precipitation remained appreciably less than today. Also here, in the immediate vicinity of the Pacific Ocean, the increasing dryness of climate, so clear in Europe and in central North America on the basis of the spread of loess, steppes had set in. Observations once more lead to reflections on the nature of the atmospheric circulation compared with conditions in the continental interior. Goldthwait [283] assumed that the wind blew out of the same quarter as it now does, but that the climatic zones south of the ice front were compressed and the wind strength increased. This obvious assumption of an intensified atmospheric circulation in middle and low latitudes has been presented many times from other aspects. Yet the extreme continentality of the climate during the cold periods in western Europe as well as the low precipitation on the western slopes of the Olympic Peninsula make this assumption questionable. Unfortunately, the present state of knowledge, in my opinion, precludes the possibility of examining this important matter in detail. The clearly recognizable strong-wind action, represented by the thick loess and dune fields, cannot be used as evidence of increased atmospheric circulation, for even if the wind were no stronger than at present, the far-reaching degradation of the vegetative cover must have permitted strong deflation. Only in regions that lie some 8° to 10° further south in North America does the amount of precipitation appear to have somewhat increased.

As has already been seen, the climatic fluctuations of the last cold period in North America essentially occurred simultaneously with those in Europe, and the scale of the climatic fluctuations with respect to present-day conditions during both stadials and interstadials in both regions was comparable. Differences can be recognized in that the decrease of winter temperatures during the interstadials in North America was probably greater than that occurring simultaneously in Europe. Particularly striking are the important glacier advances with the stadials in North America, for which there is nothing comparable during the first part of the last cold period in Europe. The glaciers of the continental ice sheet in North America advanced very rapidly (unfortunately no comparable observations can be made in Europe in the absence of well-datable "forest beds"). During the maximum advance of the classical Wisconsin, the ice front south of the Great Lakes advanced about 52 meters per year [229], about 115 meters per year in northern Ohio, but only 27 meters [282] in southwestern Ohio. Burns [104], however, obtained much smaller amounts in the same region (e.g. instead of 35 meters only about 12 meters per year, etc.) in determinations based upon dendrochronology. It is understandable that these values reflect local conditions and those of the ice flow and hence give values that may be quite different. Nevertheless, they are astonishingly high and may indeed be correct if the values taken from Flint [229] from a wide area are used as a basis.

It might be assumed that the reason for the notable difference between the glacial histories of Europe and North America is that although in both land areas the ice sheets advanced simultaneously to the south at the beginning of the last cold period, some time had to elapse in Europe before the Baltic was filled with ice, whereas in America there is nothing comparable. This view, however, is opposed by the results of pollen analysis and radiocarbon dating recently carried out in northern Sweden. At Ale, near Lulea [264], at Porsi in Lapland [495], and near Gallejaure in northern Sweden [503] at more than 24,000 years, more than 40,000 years, and more than 35,000 years respectively, tundra or birch

TABLE 23. MEAN ANNUAL TEMPERATURE OF LOCALITIES ON APPROXIMATELY THE SAME GEOGRAPHIC LATITUDE (ABOUT 60°N) IN NORTH AMERICA AND NORTH EUROPE

NORTH AMERICA		NORTH EUROPE	
Ungava, Payne River	−8.0°C	Lerwick	+6.7°C
Fort McKenzie	−8.2°C	Oslo	+5.1°C
Fort Harrison	−11.2°C	Stockholm	+5.9°C
Churchill	−10.7°C	Leningrad	+4.3°C

tundra prevailed. In the opinion of the authors cited, they do not date from the last interglacial but are probably younger. Fromm, Kolbe, Persson [264] and Lundqvist [495] regarded them as of the Göttweig interstadial, whose existence must be questioned (above, pp. 226 ff.). In contrast, Magnusson [503] refers to an early interstadial of the last cold period, and the same could be true for the other two outcrops. In this event a wide zone around these localities in northern Sweden must have been ice free, so that the woodland flora could migrate. If now it is assumed that the ice during early stadials of the last cold period in northern Europe, as in North America, had already advanced far to the south, then with the beginning of the interstadials discussed, at least in northern Scandinavia, the ice masses must have vanished very rapidly. It can be concluded that such a hypothesis is nearly impossible. Instead of this there is another, more significant, state of affairs. The mean annual temperature at present is much lower at the same latitude in North America than in northern Europe. See Table 23.

The mean annual temperature in Berlin (52° 27′N) in central Europe is +8.8°C; in Detroit in North America (42° 20′N), it is only 0.4°C higher. As the precipitation in northeastern Canada at about 400 to 800 millimeters per year is relatively high with a relatively small negative temperature anomaly during the summer months, a large, firm snow field could rapidly collect. Figure 107 gives a good impression of this. It is a photo of a region of the Payne River in Ungava, North Labrador. Dr. Rodewald wrote as follows, "I personally observed in northern Ungava between 60° and 61° on the 8 August 1956 a large quantity of old snow patches which reached to sea level. This was after a May about 4°C colder than normal and June about 3°C cooler than normal, a July about 0.5°C warmer than normal. As the first snowfall of the new cold season usually occurs in September, I estimate that with a fall in temperature from May to September of from 4° to 6°C, snow could persist and a new ice sheet could gradually establish itself."

The present climatic character of wide areas of Canada is thus significantly more favorable for the formation of a continental ice sheet than is northern Europe or the Alps. Presumably conditions during the Pleistocene cold periods were quite similar. If this were so, then the smaller climatic fluctuations would have had a more marked effect in North America than in Europe. This appears actually to have been the case, and the late glacial history of the two regions offers good examples of it. By way of conclusion and as an indication of the cor-

FIGURE 107. Air photo of northern Ungava, North Labrador, 8 August 1956, after a cold early summer. The widespread snow fields which can be recognized at many points will probably survive the summer. In Europe, Oslo lies in the same latitude. Photo by Dr. Rodewald of the region between the mouth of the Payne River and Cape Horn Advance.

rectness of this interpretation, one postglacial climatic fluctuation, the Cochrane advance, which occurred not long before 6,800 B.C. [233, 806], may be singled out. At that time the ice sheet advanced into northern Ontario. Simultaneously in Europe afforestation was in full swing. Because of this Antevs [19] has questioned whether the radiocarbon dating for the Cochrane advance is really correct. Instead, he places this glacial advance in the younger tundra period and consequently assumes a greater age for all older glacial formations in North America, so that in his view the Two Creeks Forest Bed is no longer the equivalent of the Alleröd interstadial (Table 6, p. 68). New dates, however, have confirmed that the Cochrane advance actually does lie at the beginning of the postglacial. At the same time there is evidence from Europe of a noteworthy glacial advance. This is the Piottino fluctuation [828], which occurred in the southern Alps between 8100 and 7700 B.C. Lang [450], like Antevs, queried the correctness of the dating and placed the Piottino fluctuation in the younger tundra period. Yet the arguments expressed are not convincing, especially because Welten [787] had earlier pointed out that the Aletsch glacier freed the upper Rhone valley for the first time during the European Boreal (about 700 to 5500 B.C.). It appears to me that observations both in North America and in the Alps (after Zoller [828] and Welten [787] and also after Beug [56]) are well founded. This means that in this case, too, an important continental glaciation in North America is paralleled by a relatively modest valley glaciation in the Alps. The present climatic differences between the two regions were thus clearly operative in former times, so that one is justified in tracing back older differences in the glacial histories to this source. The climatic fluctuations of the last cold period and in general terms for the whole Quaternary were accordingly simultaneous, and their scale, in comparable regions of the Northern Hemisphere, was of about the same magnitude. They took place, however, against a background of regionally different climatic character, so that the results are also quite different.

Fluctuations in Sea Level

When large volumes of water are locked up in the form of ice, then sea level falls; with the melting of the ice, the level rises again. Thus along coastlines not affected by tectonic movement, a formerly high sea-level stand corresponds to a warm period, while a regression signifies a cold period. It is thought that during the last warm period two notable terraces were formed, of which the higher is today found at elevations of about 15 to 18 meters and the lower at 5 to 8 meters. It has been observed at different times that between the two transgressions, the so-called Monastir I and II, a regression occurred. Examples of this are found on the coast of Lebanon [811], in Mallorca [112, 114], and on the coast of southeastern Virginia [509]. The sea level during this time fell back almost to the present level. If these changes in sea level were not caused by local tectonic events, they must be atttributed to climatic causes. These appear to have been relatively strong. They must represent two significant warm periods and an intervening phase of cold climate. According to Wright [811] the actual seawater temperatures appear to have been quite high on the Lebanese coast, not only during Monastir I but also during Monastir II, although the observations of Butzer and Cuerda [114] on the southern tip of Mallorca lead to a different conclusion.

These alterations of sea level pose a problem. Despite the extraordinarily large numbers of pollen analyses carried out in the most diverse regions of northern Eurasia and North America, paleobotanical evidence of a cooling within the last warm period is entirely lacking, and this has been suggested as the cause of the change in sea level. The paleobotanical evidence indicates much more clearly that there was no such cold-climate phase. The cursory indications of Lona [478] need not be considered here. Also, the evidence of Dansgaard, Johnsen, Clausen, and Longway [149b] for a two-fold division of the last warm period does not seem to me as convincing as they present it as being. If, consequently, the last warm period was not broken by the required cold-climate phase, then the two marine terraces, Monastir I and II, cannot form part of the last warm period. On the other hand, there is no evidence whatsoever to support the assumption that the Monastir I terrace could be somewhat older than the last warm period. The sole remaining conclusions are that only the Monastir I terrace dates from the last warm period, and the Monastir II terrace must be younger.

Directly linked with this hypothesis are weighty consequences. It must follow that the climate at the time of formation of the Monastir II terrace was warmer than at present, for the former sea level appears to have been higher than today. There is no evidence for this. In my opinion the Monastir II terrace corresponds approximately with the warm fluctuation known as the Brørup interstadial, since this interstadial represents the only truly significant phase of warm climate since the beginning of the last cold period. The magnitude of the transgression must be explained by a series of factors whose full significance cannot yet be reliably estimated.

Woldstedt [807], as noted (above, p. 182), assumed that during the warm periods the sea level stood at much the same position as today and that the present varied elevation of former strand lines is to be explained by later epeirogenic movements. The following considerations are relevant to such an hypothesis. According to it the sea level at the optimum of the last warm period was about at the present level; during the approximately 80,000 years (as a round figure) since then, the land surface must have risen at least 15 meters, for the former coastline now lies at about that elevation. The optimum of the Brørup interstadial occurred some 55,000 years ago. Assuming that epeirogenic uplift occurs with constant velocity and that the ice had melted to an extent comparable with the present day, it must follow that the coastline of that time should now lie some 10.5 meters above the present sea level. In fact its elevation is from 5 to 8 meters. It then follows that significantly more ice was present than today, particularly so when it is recalled that the calculation founded on the assumption of a 15-meter strand line is too small, for it does not take into consideration that a greater amount of ice was locked up in Antarctica than at the present time (above, p. 181). It is important that the calculation of the minimum sea-level elevation during the Brørup interstadial is valid to a degree if the various elevations of the warm-period marine terraces was due not to the epeirogenic uplift but to a significant accumulation of ice in the Antarctic during the Pleistocene. The correlation of the Monastir II terrace with the Brørup interstadial is thus not as astonishing as might appear; rather, it may be possible in this way to estimate later the minimum extent of glaciation during this interstadial.

Subsequently the sea level changed, but these changes are hard to evaluate paleoclimatically. This is particularly so for the Stillfried B interstadial. Between 26,000 and 30,000 B.P.—that is, during the time of the Stillfried B—the sea level at Göteborg was essentially indistinguishable from that of the present day [87]. According to Coope, Shotton, and Strachan [143] and Donovan [168], the sea level in the Severn Estuary and in the Bristol Channel in southwest England between 45,000 and 24,000 B.P. was approximately the same as it is now. It might be concluded from this that the sea level before the Stillfried B had not fallen by any significant amount. This idea agrees well with the assumption that the maximum glacial advance of the last cold period occurred only after the Stillfried B. Yet as still older advances of the ice sheet are known in North America, there are serious objections to this. It seems impossible at this stage to discuss the problem further, for the number of unknowns is far too great.

6. Concluding Remarks

The climatic fluctuations of the Ice Age have become better known during the last decade, thanks to the detailed research of many workers in many lands, than during the preceding thirty years. Yet the number of problems is by no means diminished; rather, it has become greater. Already the investigation of the climatic fluctuations of the last cold period has thrown up such an abundance of questions that a general review is threatened. In conclusion the most important results will be summarized.

(1) The climatic fluctuations of the Ice Age affected the whole earth simultaneously. This is a valid statement not only for first-order fluctuations but also for second and still higher orders. Cold or warm periods affected the whole earth at the same time, and all attempts to reconstruct the former atmospheric circulation must take cognizance of this fact.

(2) Even the first great climatic fluctuation of the Ice Age, the Pretiglian cold period, markedly affected the present-day cool temperate latitudes and led to a situation where, for the first time, the woodlands that had reigned supreme over wide regions of the Northern Hemisphere up to the end of the Tertiary were replaced by steppe or associated grass and shrub communities. These plant communities characterize all subsequent cold periods in northern Eurasia and to a lesser extent also North America and New Zealand.

(3) Overprinted upon the climatic fluctuations of the Ice Age were slow, progressive climatic changes. They find expression in the increasing coldness and dryness at the peak of succeeding cold periods, although the climate during the optima of the warm periods remained approximately the same thermally. Yet in Europe, at least, the precipitation appears to have decreased.

(4) The most important climatic changes seem to have taken place during the first part of the Pleistocene—that is, from the end of the Pliocene to about the end of the Menapian cold period, in Dutch terminology. Furthermore, it appears that the postglacial was less climatically favorable than the preceding Pleistocene warm periods, with which the postglacial has many features in common.

(5) The climatic fluctuations of the Ice Age were furthermore superposed upon the differing climatic character of various regions, so that despite the similar order of magnitude of the climatic changes, during the course of time appreciable differences find expression in the climatic fluctuations.

(6) The outstanding characteristic of the cold periods was the great dryness of the climate. This extended not only over cold temperate latitudes but also over the warm temperate. Consequently, true pluvials that corresponded to the cold periods of higher latitudes did not occur; rather, the marked fall in tem-

perature reduced evaporation, and this may explain many occurrences which may casually be assessed as evidence of true pluvials.

(7) During the warm periods the climate of the present cool temperate latitudes was milder and moister than at the present time. But many of the differences in the history of development of vegetation are due not to this but to the difference in plant migration from the different cold period refuges or to the first stages of climatic amelioration; they are thus not expressions of essentially different climates.

(8) The transition from a warm period to a cold period was apparently not marked by a rise in precipitation in contrast to the view often cherished, but was marked by a rapid decrease accompanied by a very rapid cooling. Wet and cold or moist-cold initial phases of the cold periods have not been recognized, though the water budget was somewhat more favorable than that at the maximum of the cold periods.

(9) The interstadials of the last cold period in Europe at least developed with a very dry but warm summer and cold winter climate. It must be supposed that the same was true in North America. Furthermore, the interstadials show that they were composed of many smaller climatic fluctuations that must have considerably affected the plant and animal worlds.

(10) It is not clear that the atmospheric circulation was intensified during the cold periods.

The summary once more shows the fullness of the spectrum of Pleistocene climatic fluctuations. In the discussion of certain aspects of it the question of their origin repeatedly arises. This must at first certainly be sought in the particulars of atmospheric circulation, then in those factors which may eventually have caused those circulation changes. In recent years particular interest has been expressed in the former atmospheric circulation. In this respect the works of Grigor'ev [308], Willett [796, 797], Willett and Sanders [798], Flohn [235-39], Brooks [86], and Butzer [105, 106, 109, 110] as well as the symposia of Shapley [682] and Nairn [562] can be cited. Although it is natural to consider present extreme weather situations as models to explain the former atmospheric circulation during cold and warm periods, this question will not be pursued further. The reason is that it lies in a field of research far removed from the botanists, and observations in earlier chapters clearly indicate the need for extreme caution. It seems far better to assemble as much reliable observational data as possible than to develop interesting hypotheses. The same holds true with respect to the origin of Pleistocene climatic fluctuations, especially as such related problems are considered extensively in the symposia mentioned.

The more intensively we become involved in the problems of the Ice Age, the more clear it becomes that we are still at the threshold of a true understanding of them. The prospect can only be improved by intensive research at carefully selected localities and by critical testing of present knowledge.

Appendices

In the following appendices are to be found the sources of the data used in the construction of some of the text figures. The bracketed numbers refer to items in the Bibliography.

APPENDIX 1

Former Expansion of Permafrost or Ground Frozen Many Months of the Year in Northern Eurasia

a) Maximum advance of the Saale-Dneprovsk-Samorov glaciation
   - England [176]
   - Central and western Europe [321, 391, 497, 579, 580, 644]
   - Southern Europe [65]
   - Eastern Europe [361, 545, 552, 762]
   - Western Siberia [133, 291, 466, 506, 567, 681, 820]
   - Central and eastern Siberia [5, 6, 274, 284, 455, 621, 761, 830, 831]

b) Eem-Mikulino-Kazancev warm period
   - Northern Eurasia [260]

c) Last cold period
   - Central and western Europe [600, 601, 391]
   - Eastern Europe [256, 771]
   - Western Siberia [820, 256]
   - Northern Eurasia [256, 756]

d) Postglacial warm period
   - Northern Eurasia [256, 257]

APPENDIX 2

Age of the Alleröd Interstadial (source of data for the diagrams)

- Ireland [312, 280]
- England and Scotland [280, 312, 404]
- Denmark [312]
- Norway and Sweden [225, 325]
- Baltic and central Russia [127, 426, 566]
- Poland [428, 781]
- Netherlands and Germany [314, 225, 312, 313, 30]

APPENDIX 3

Distribution of Certain Tree Species at the End of the Pliocene (about Reuverian B in Dutch terminology) in Northern Eurasia

Fagus

France: Pas de la Mougudo, St. Vincent-La-Sabie, *F. silvatica* [452]; Normandy, La Londe [187, 188].

Germany and Netherlands: Sylt [794]; Reuver, *F. decurrens, F. silvatica* [775]; Elmpt (Waldniel) [81]; Weilerswist [294]; Dernbach, *F. attenuata* [555]; Weckesheim [465]; Frankfurt/M., *F. ferruginea, F. decurrens* [502]; Willershausen [436]; *F. silvatica, F. grandifolia* [721, 741]; Eichenberg [136].

Poland: Kroscienko, *F. decurrens, F. ferruginea, F. silvatica* [731]; Warsaw-Ochota [637, 715].

Roumania: Chiuzbaia, Maramures, *F. pliocaenica, F. attenuata, F. angusta* [278].

Bulgaria: Kurilo, Sofia, *F. orientalis* [719].

Southern Russia: Sal̦-Manyč watershed [300]; Kerch Peninsula, Cape Chengulek [521]; Kama, Rybnaja Sloboda [36].

Caucasus: Cape Pithyusa, *F. orientalis* [617]; Suchum, *F. orientalis* [423].

Central and northwestern Russia: White Russia [494]; east Lithuania, Varis, and Schlawe [320].

Eastern Siberia: estuary of the Kei in the Aldan [394].

Amur-Zeja Depression: ? [533].

Central Japan: [383, 683, 684, 702–4, 814].

Ilex

France: Meximieux [648].

Germany and Netherlands: Sylt [743, 794]; Reuver [775], *I. aquifolium*; Frankfurt/M., *I. aquifolium* [502]; Eichenberg [136].

Poland: Kroscienko, *Ilex* sp. [731]; Warsaw-Ochota [637, 715].

Roumania: Chiuzbaia, Maramures, *Ilex* sp., *I. studeri* [278].

Bulgaria: Kurilo, Sofia, *Ilex aquifolium* [719].

White Russia: [494]; Kameneck-Rayon, Brest district [124].

Eastern Lithuania: Varis and Schlawe [320].

Lower reaches of the Kama: Rybnaja Sloboda, *Ilex* sp. [35].

Caucasus: Suchum, *I. colchica* [423].

Amur-Zeja depression: [533].

Central Japan: [383, 683, 684, 704, 814].

Juglans

France: Normandy, La Londe [187, 188].

Germany and the Netherlands: Reuver, *J. cinerea* [755); Dernbach, *J. globosa* [555]; Weckesheim [465]; Frankfurt/M., *J. costata* [502]; Willershausen, *J. sieboldiana* (?) [721].

Poland: Kroscienko, *J. cinerea* [731].

Roumania: Chiuzbaia, Maramures, *J. cinerea* [278].

Southern part of European Russia: Volodarsk-Volynsk district [475]; Kerch Peninsula, Cape Chengulek [521]; lower reaches of the Kama, Rybnaja Sloboda [35]; *J. cf. acuminata, J.* sp. [36]; Stavropol/Volga [317–19]; White Russia [494].

Caucasus: Cape Pithyusa [617].

Upper reaches of the Ob: near Barnauls, *J. cinerea* [506].

Eastern Siberia: Junction of the Kel in the Aldan [394]; east banks of Penzina Bay, *J. acuminata* [592]; Amur-

Zeja Depression [533].

Carya

France: Meximieux [648]; Pas de la Mougudo, *C. minor* [452]; Normandy, La Londe [187–88].

Germany and the Netherlands: Sylt [743, 794]; Reuver, *C. angulata*; *C.* sp. [775]; Elmpt, (Waldniel) [81]; Weckesheim [465]; Frankfurt/M., *C. angulata*, *C. tomentosa*, *C. globosa*, *C. longicarpa* [502]; Allershausen, Uslar/Solling [436]; Willershausen, *C. glabra* [721, 741]; Eichenberg [136].

Poland: Kroscienko, *C. cf. tomentosa*, *C.* sp. [731]; Warsaw-Ochota [637, 715].

Roumania: Chiuzbaia, Maramures, *C. serraefolia*, *C. minor* [278].

Southern part of European Russia: Kerch Peninsula, Cape Chengulek [521]; Stavropol/Volga [317–19]; Russkaja Bektjaska, Uljanovsk district ? [447]; White Russia [494].

Caucasus: Cape Pithyusa [617].

Far East: Suifun [43, 44]; Amur-Zeja Depression [533].

Pterocarya

France: Pas de la Mougudo, St. Vincent. La Sabie, Cantal, *Pt. caucasica* [452]; Normandy, La Londe [187, 188].

Germany and the Netherlands: Sylt [743, 794]; Reuver, *Pt. denticulata*, *Pt. limburgensis*, *Pt.* sp. [775]; Elmpt (*Waldniel*) [81]; Weckesheim [465]; Frankfurt/M., *Pt. denticulata* [502]; Allershausen, Uslar/Solling [436]; Willershausen, *Pt. fraxinifolia* [721, 741]; Eichenberg [136].

Poland: Kroscienko, *Pt. fraxinifolia* [731]; Warsaw-Ochota [637, 715].

Roumania: Chiuzbaia, Maramures [278].

Southern part of European Russia: Sal-Manyc-Watershed [300]; Kerch Peninsula, Cape Chengulek [521]; lower reaches of the Kama, Rybnaja Sloboda [35, 36]; Stavropol/Volga [317–19]; White Russia [494].

Caucasus: Suchum, *Pt. pterocarpa* [423].

Eastern Siberia: Junction of the Kel in the Aldan [394]; Namcy near Yakutsk [277].

Far East: Suifun [43, 44].

Amur-Zeja Depression [533].

Liriodendron

France: Meximieux [648].

Germany and the Netherlands: Reuver, *L. aptera*, *L. tulipifera* [721]; Frankfurt/M., *L. tulipifera* [502]; Willershausen, *L. tulipifera* [721].

Poland: Kroscienko, *L. tulipifera* [731].

Central Urals: Noviyjj Log, Visim-Rayon, Sverdlovsk district, *L. tulipifera* [169].

Caucasus: Cape Pithyusa [617].

Nyssa

France: Normandy, La Londe [187, 188].

Germany and the Netherlands: Sylt [743]; Reuver, *N. silvatica* [775]; Elmpt (Waldniel) [81]; Weilerswist [294]; Weckesheim [465]; Frankfurt/M., [502]; Willershausen [721], cf. *Nyssa* [741]; Eichenberg [136].

Poland: Kroscienko, *N. silvatica* [731]; Warsaw-Ochota [637, 715].

Eastern Lithuania: Varis and Schlawe [320].

Southern part of European Russia: on the Uborta, northern part of the Zvitomir district [474]; White Russia [494]; Kameneck-Rayon, Brest district [124].

Far East: Suifun [43, 44].

Central Japan: [393, 683, 684, 702–4, 814].

Liquidambar

France: Meximieux [648]; Normandy, La Londe [187, 188].

Germany and the Netherlands: Sylt [743]; Reuver, *L. orientalis, L. europaea* [775]; Elmpt (Waldniel) [81]; Weilerswist [294]; Frankfurt/M., *L. pliocaenica* [502]; Willershausen, *L. orientalis* [721]; Eichenberg [136].

Poland: Kroscienko, *L. europaea* [731].

Bulgaria: Losenetz, *L. europaea* [578].

Roumania: Chiuzbaia, Maramures, *L. europaea* [278].

White Russia: Cape Pithyusa [617].

Far East: Suifun [43, 44]; Amur-Zeja Depression [533].

Tsuga

France: Normandy, La Londe [187, 188].

Germany and the Netherlands: Sylt [743]; *Ts. diversifolia, Ts. canadensis* [794]; *Reuver* [775]; Elmpt (Waldniel) [81]; N. Eifel [294]; Weckesheim [465]; Willershausen, *Ts. diversifolia, Ts. canadensis* [436, 721, 741]; Allershausen, Uslar/Solling, *Ts. diversifolia, Ts. canadensis* [436]; Eichenberg [136].

Poland: Kroscienko, *Ts. caroliniana, Ts. europaea* [731]; Warsaw-Ochota, *Ts. cf. diversifolia* [637, 715].

Roumania: Chiuzbaia, Maramures [278].

Bulgaria: Kurilo, Sofia, *Ts. aff. canadensis* [719].

Southern part of European Russia: Sal-Manyc watershed [300]; Kerch Peninsula, Cape Chengulek [521]; lower reaches of the Kama, Rybnaja Sloboda, *Ts. europaea* [35]; Stavropol/Volga [317–19]; Russkaya Bektjaska, Uljanovsk district [447]; Novyj Log, Nisim-Rayon, Sverdlovsk district [169]; White Russia [494].

Eastern Lithuania: Schlawe [320].

Caucasus: Cape Pithyusa [617]; Suchum, *Ts. europaea* [423].

Western and eastern Siberia: Upper reaches of the Cuja [496, 842]; junction of the Kel in the Aldan, *Ts. canadensis* [394]; Namcy near Yakutsk [277]; Aldan, Mamontova Gora, Tandinsk outcrop [276, 69, 760]; Ilimpeja and Tajmura [132]; junction of the Bystraja in the Angara, Torsk Basin [295, 296]; Tunka Basin [620]; south bank of Lake Baikal between the rivers Anosovka and Dulicha [563]; Nerca Valley between the rivers Chila and Torga [832]; Amur-Zeja Depression [533]; Holy Cross Bay [583]; upper reaches of the Indigirka, Promezutocnyj tributary [492]; upper reaches of the Kolyma on the Beneljocha River [395.]

APPENDIX 4

Distribution of Certain Plants During the Holsteinian Warm Period in Europe

Carya

Neede, Holland [241]; Zydowszcyzna near Grodno [388]; Po Delta [576, 577]; Wlodawa on the Bug [713]; Suszno on the Bug [714].

Vitis

Tornskov, S. Jutland [12]; Cava Nera Molinaria near Rome [66]; Ciechany Krzesimovskie, Lublin uplands [83]; Neede, Holland [241];

Cava Bianca near Rome [249]; Valle dell'Inferno [250]; Riano near Rome [251]; Rakow [433, 434]; Essen-Vogelheim [435]; Olszewice [470]; Hötting breccia near Innsbruck [560]; Wunstorf near Hannover [632]; Syrniki on the Weipoz, Lublin uplands [695, 696]; Olszewice [698]; Hoxne, East Anglia; discovery of Dr. R. G. West at the end of April 1963 in sediments which probably belong to the Hoxnian interglacial.

Azolla
British Isles: [782, 177].
Northern Germany and the Netherlands: [136, 240, 346, 472, 632, 644].
Poland: [281, 696, 697].
European Russia: [170, 171, 317–19, 480, 628].
Western Siberia: [430, 532, 821].

Trapa
Northern Germany: [632].
Poland: [83, 181, 371, 388].
Central Russia: [480, 299, 727, 759].

Buxus
Ireland: [382, 782].
France: [572].
Germany and Czechoslovakia: [202, 414, 415, 435, 632].
Denmark: [12].
Italy: [66, 249].

Fagus
Ireland: [382].
France: [572].
Germany: [47, 335].
Poland: [388, 433, 434, 698].
Alps: [560].
Northern Carpathians and surrounding regions: [414, 415, 427, 501, 500].
Italy: [247, 250, 251].
European Russia: [475, 479, 480, 727, 759].

Taxus
British Isles: [382, 782, 791].
Denmark: [12, 14, 381].
Germany: [435].
Alps: [560].
Italy: [66].
Poland: [709, 730, 60, 172].
Central Russia: [727, 759].

Ilex
British Isles: [382, 782, 177, 591, 720].
Denmark: [12, 14].
Italy: [66].
Poland: [709].
Central Russia: [759, 299].

Tilia platyphyllos
France: [572].
Alps: [560].
Italy: [250].
Poland: [371, 388, also *Tilia tomentosa*, 433, 434, 470, 753].
Central Russia: [289].

APPENDIX 5

Distribution of Certain Plants During the Eemian Warm Period in Europe

Acer tataricum: [172, 299, 590, 706].

Tilia platyphyllos
Northern Germany and Denmark: [28, 58, 59, 309, 381, 614, 676, 660, 661, 742, 783–85].
Poland: [84, 442, 590, 618, 619, 706, 712, 729, 734, 748, 754].
Central Russia: [125, 147, 148, 165, 843, 172, 290, 299, 303, 306, 361, 385, 389, 390, 531].

Tilia tomentosa
(from west to east): [614, 717, 614, 172].

Hedera
British Isles: [348, 705].
Germany and the Netherlands: [28, 41, 844, 530, 614, 622, 676, 742, 817].
Poland: [699, 748].

Taxus
British Isles: [254, 782].
Northern Germany and Denmark: [309, 381, 425, 614, 670, 676, 784, 785, 59].
Poland: [706, 708, 729, 733, 748].
Czechoslovakia: [489, 610].
Lower Austria: [258].
Southern Alps: [85].
Italy: [752].

Buxus
British Isles: [348].
Central Europe: [202, 411, 412].
Italy: [85].

Ilex aquifolium
British Isles: [254, 705, 782].
Northern Germany, the Netherlands, and Denmark: [28, 41, 202, 221, 246, 381, 614, 676, 742, 783, 785, 817].
Alps: [622].
Carpathians: [411, 412].
Poland: [699, 708, 733, 748, 369].
Central Russia: [165].

Fagus
Northern Germany: [41 ?, 202, 670].
Poland: [179, 619, 699, 706, 754].
Eastern margin of the Alps, North-western Carpathians: [405, 258, 259, 489, 581, 610].
Southern Carpathians: [598].
Jugoslavia: [515].
Italy: [510, 511, 576, 752].
European Russia: [164, 385, 389].

Trapa
British Isles: [254].
Northern Germany and Denmark: [41, 309, 381, 717].
Poland: [63, 590, 729, 748].
Central Russia: [125, 163, 172, 531].

Cladium Mariscus
Northern Germany and Denmark: [58, 309, 61, 381, 670, 676, 661, 783].
Poland and Western Russia: [845, 172, 712].

APPENDIX 6

Works That Record Waterlogging of the Peat Layers Formed During the Last Warm Period (in which the beginning of waterlogging can be dated with assurance from the vegetation history)

Lower Rhine: [82, 246, 817].
Northern Germany: [28, 309, 424, 587, 614, 676].
Denmark: [381].
Poland: [63, 75, 590, 699, 712, 748].
Alps: [622].
Central and western Russian: [25, 26, 123, 125, 129, 147, 165, 297, 306, 320, 531, 687].

APPENDIX 7

Works in Which Are Recorded Data on the Occurrence of Macro- or Micro-Remains of Thermophilous Wood Types in Sediments of the Brørup Interstadial or Its Analogues

British Isles: [689].
Denmark: [11].
Lower Rhine: [242, 593, 39].
Northern Germany: [29, 328].
Poland: [62, 74, 616, 619, 701, 732, 275, 615, 369].
Eastern Alps and Foreland: [68, 258, 259, Frenzel unpubl.].
Western Alps: [78, 457, 458, 493].
France: [119, 188].
Carpathians and Hungary: [180, 370, 389, 711, 665].
Northern Jugoslavia: [678–80].
Ukraine and central Russia: [25, 31, 33, 215, 273, 300, 304, 432, 481, 494].
Baltic: [267, 127].

References

The following abbreviations are used:

Ber. D. Bot. Ges.	Berichte der Deutschen Botanischen Gesellschaft
Biull. chetvert.	Biulleten' Komissii po izucheniiu chetvertichnogo perioda
Bull. Geol. Soc.	Geological Society of America, Bulletin
Bot. Zhurn.	Botanicheskij Zhurnal
C.R. Acad. Sci.	Comptes Rendus hebdomadaires des séances de l'Académie des sciences, Paris
DAN	Dokladi Akademii Nauk SSSR
Izv AN	Izvestija Akademii Nauk SSSR
Materialy chetvert.	Materialy vsesojuznogo soveshchaniia po izucheniiu chetvertichnogo perioda
Tezisi	Tezisi dokladov vsesojuznogo mezhduvedomstvennogo soveshchanniia po izucheniiu chetvertichnogo perioda
Trudi Geogr.	Trudi Instituta Geografii, Akademiia Nauk SSSR
Trudi Geol.	Trudi Geologicheskogo Instituta Akademii Nauk SSSR

[1] *Aario, L.*, Waldgrenzen und subrezente Pollenspektren in Petsamo Lappland. Annales Acad. Sci. Fenn., Ser. A, **54** Nr. 8, 1940

[2] ——— Über die pollenanalytischen Methoden zur Untersuchung von Waldgrenzen. Geol. Fören. i Stockh. Förhandl., **66**, 337–54, 1944

[3] *Ahlmann, H. W.*, The Present Climatic Fluctuation. Geogr. Journ., **62**, 165–95, 1949

[4] *Ahorner, L.*, and *K. Kaiser*, Über altpleistozäne Kalt-Klima-Zeugen (Bodenfrosterscheinungen) in der Niederrheinischen Bucht. Decheniana, **116**, 3–19, 1964

[5] *Aleksandrova, L. P., E. A. Vangengejm, V. G. Gerbova, L. V. Golubeva* and *E. I. Ravskii*, Neue Beiträge zur Kenntnis der anthropogenen Sedimente des Berges Tologoj (West-Transbaikalien). Biull. chetvert., **28**, 84–101, 1963

[6] *Alekseev, M. N.*, Stratigraphie kontinentaler neogener und quartärer Sedimente der Viljuij-Senke und des Unterlaufes der Lena. Trudi Geol., **51**, 1961

[7] *Alekseev, M. N., R. E. Giterman, N. P. Kuprina, A. I. Medjancev* and *I. M. Choreva*, Quartäre Sedimente Jakutiens. Voprosy geologii antropogena, 129–40, Moscow 1961

[8] *Alekseev, V. A., I. V. Kazachevskii, V. V. Cherdyncev* and *P. S. Enikeev*, Über die Bestimmung des absoluten Alters quartärer Bildungen (Arbeiten des Laboratoriums für absolute Datierung des G.I.N.A.N. S.S.S.R.). Absoljutnaja geochronologija chetvertichnogo perioda, 22–26, Moscow 1963

[9] *Aleshinskaja, Z. V.*, Zur Stratigraphie mittel- und jungquartärer Sedimente im Nordostteil der Westsibirischen Tiefebene (auf Grund von Diatomeen-Analysen). Paleogeografija chetvert. per. SSSR, 150–59, Moscow 1961; Fr. summary

[10] *Aljavdin, F. Y.* Einige Probleme der Paläogeographie des Quartärs im Nordteil der Westsibirischen Tiefebene. Tezisi; Sekc. Zapadnoj Sibiri i Urala, 29–30, 1957

[11] *Andersen, S. Th.*, Vegetation and its Environment in Denmark in the Early Weichselian Glacial (Last Glacial). Danmarks Geol. Unders., II. raekke, Nr. 75, 1961

[12] ——— Pollen Analysis of the Quaternary Marine Deposits at Tornskov in South Jutland. Danmarks Geol. Unders., IV. raekke, **4**, Nr. 8, 1963

[13] ——— Interglacial Plant Successions in the Light of Environmental Changes. Report of the VIth Internat. Congr. on Quaternary Warsaw 1961, **2**, 359–68, Łódź 1964

[14] ——— Interglacialer og interstadialer i Danmarks kvartaer. Meddelelser fra Dansk Geol. Foren., **15**, 486–506, 1965

[15] *Andersson, G.*, Hasseln i Sverige fordom och nu. Sveriges Geol. Undersök., ser. **Ca**, Nr. 3, 1902

[16] *Antevs, E.*, The Big Tree as a Climatic Measure. Publ. Nr. 352, Carnegie Inst. of Washingt., 115–53, 1925

[17] ——— Rainfall and Tree Growth in the Great Basin. Publ. Nr. 469, Carnegie Inst. of Washingt., 1938

[18] ——— Climate of New Mexico during the last Glaciopluvial. Journ. of Geol., **62**, 182–91, 1954

[19] ——— Transatlantic Climatic Agreement versus C^{14}-Dates. Journ. of Geol., **70**, 194–205, 1962

[20] *Archipov, S. A.*, and *O. V. Matveeva*, Pollenspektren von Prä-Samarov-Sedimenten des Anthropogens in dem ehemals vergletscherten Gebiet des Jenisseij-Flußgebietes der Westsibirischen Tiefebene. DAN, **135**, 1453–56, 1960

[21] ——— Die Präsamarov-Serie des Anthropogens am Südrand der Jenisseij-Senke. Trudi in-ta geol. i geofiz., AN SSSR, sibirskoe otdelenie, **25**, 5–22, 1964

[22] ——— Das Anthropogen am Südrand der Jenisseij-Mulde. Inst. geol. i geofiz., AN SSSR, sibirskoe otdelenie, **29**, Novosibirsk 1964

[23] *Archipov, S. A., E. V. Koreneva* and *Ju. A. Lavrushin*, Stratigraphie der quartären Sedimente im Jenisseij-nahen Gebiet der Westsibirischen Tiefebene (Flußgebiet des Mittellaufes des Jenisseij). Materialy chetvert., **3**, 151–56, 1961

[24] *Argus, G. W.*, and *M. B. Davis*, Macrofossils from a Late Glacial Deposit at Cambridge, Massachusetts. Amer. Midland Naturalist, **67**, 106–17, 1962

[25] *Aseev, A. A.*, Zur Vegetations-Geschichte der Meshchera im oberen Pleistozän. DAN, **115**, 175–78, 1957

[26] ——— Paleogeografiia dolini srednei i nizhnej Oki v chetvertichnyj period. Moscow 1959

[27] *Auer, V.*, The Pleistocene of Fuego-Patagonia. II. The History of the Flora

and Vegetation. Annales Acad. Sci. Fenn., ser. A, III. Geol.-geogr., **50**, 1958
[28] *Averdieck, F. R.*, Das Interglazial von Fahrenkrug in Holstein. Ein Beitrag zur Frage des Buchenvorkommens im Jungpleistozän. Eiszeitalter und Gegenwart, **13**, 5–14, 1962
[29] ——— Frühweichselinterstadiale in Dithmarschen (Schleswig-Holstein). Berichte d. Geobot. Inst. der ETH, Stiftung Rübel, **34**, f. 1962, 58, Bern 1963
[30] *Averdieck, F. R.*, and *H. Döbling*, Das Spätglazial am Niederrhein. Fortschr. Geol. d. Rheinlande u. Westfalens, **4**, 341–62, 1959
[31] *Bader, O. N.*, Die Station Sungiŕ. Ihr Alter und ihre Stellung im Paläolithikum Osteuropas. Trudi Kom. po izuch. chetvert. per., **18**, 122–31, 1961
[32] ——— Kultur-Hinterlassenschaften des Paläolithikum im südlichen Ural und ihre stratigraphische Lage. Antropogen juzhnogo Urala, 239–45, Moscow 1965
[33] *Bader, O. N.*, *V. I. Gromov* and *V. N. Sukachev*, Die oberpaläolithische Station Sungiŕ. Voprosy geologii antropogena, 64–66, Moscow 1961; Engl. summary
[34] *Balout, L.*, Pluviaux interglaciaires et préhistoire saharienne. Travaux Inst. des Recherches Sahar., **8**, 9–19, 1952
[35] *Baranov, V. I.*, Neue Funde einer pliozänen Flora im Volga-Kama-Gebiet. Bot. Zhurn., **33**, 90–92, 1948
[36] *Baranov, V. I.*, and *L. M. Jatajkin*, Entwicklung der Flora und der Vegetation im Kinel' am Unterlauf der Kama. Problemy Botaniki, **6**, 18–26, Moscow 1962
[37] *Barghoorn, E. S.*, Evidence of Climatic Change in the Geologic Record of Plant Life. In: *Shapley, H.*, Climatic Change, Evidence, Causes, and Effects. 235–48, Cambridge 1953
[38] *Bartley, D. D.*, Pollen Analysis of a Small Peat Deposit at Baños de Tredos, near Viella in the Central Pyrenees. Pollen et Spores. **4**, 105–10, 1962
[39] *Bastin, B.*, Essais d'analyse pollinique des loess en Belgique, selon la méthode de *Frenzel*. Agricultura. **12**, deuxième sér., 703–6, 1964
[40] *Becker, J.*, Étude palynologique des tourbes flandriennes des Alpes françaises. Mém. du Service de la carte géol. d'Alsace et de Lorraine, **11**, 1952
[41] *Behre, K. E.*, Pollen- und Diatomeenanalytische Untersuchungen an letztinterglazialen Kieselgurlagern der Lüneburger Heide (Schwindebeck und Grevenhof im oberen Luhetal). Flora, **152**, 325–70, 1962
[42] *Bent, A. M.*, and *H. E. Wright*, Pollen Analysis of Surface Materials and Lake Sediments from the Chuska Mountains, New Mexiko. Bull. Geol. Soc., **74**, 491–500, 1963
[43] *Bersenev, I. I.*, Stratigraphie der quartären Sedimente des Primor'e. Tezisi; Sekc. Vostochnoi Sibiri i Dal'n. Vostoka, 59–60, 1957
[44] ——— Stratigraphie der quartären Sedimente des Primor'e. Materialy chetvert., **3**, 318–20, 1961
[45] *Bertsch, K.*, Lehrbuch der Pollenanalyse. Stuttgart 1942
[46] ——— Die Einwanderung der Buche in Südwest-Deutschland. Ber. D. Bot. Ges., **68**, 223–26, 1955
[47] *Bertsch, K.*, *A. Steeger* and *U. Steusloff*, Fossilführende Schichten der sogenannten Krefelder Mittelterrasse. Sitzungsber., herausgegeb. vom Naturhist.

Ver d. preuß. Rheinlande u. Westfalens, **1929**, C, 5–22, 1931

[48] *Beschel, R.*, Flechten als Altersmaßstab rezenter Moränen. Zs. f. Gletscherkde. u. Glazialgeol., N.F. **1**, 152–61, 1949/1950

[49] ——— Lichenometrie im Gletschervorfeld. Jahrb. d. Vereins zum Schutze d. Alpenpfl. u. -tiere, **22**, 164–85, 1957

[50] ——— Ricerche lichenometriche sulle morene del gruppo del Gran Paradiso. Nuovo Giorn. Bot. Ital., n.s. **65**, 538–91, 1958

[51] ——— Dating Rock Surfaces by Lichen Growth and its Application to Glaciology and Physiography (Lichenometry). Geology of the Arctic, Univ. of Toronto, 1044–62, 1961

[52] *Beucker, F.*, and *G. Conrad*, L'âge du dernier Pluvial saharien. Essai sur la flore d'un épisode lacustre. C.R. Acad. Sci., **256**, 4465–68, 1963

[53] *Beug, H. J.*, Beiträge zur postglazialen Floren- und Vegetationsgeschichte in Süddalmatien: Der See „Malo Jezero" auf Mljet, Teil I: Vegetationsentwicklung. Flora, **150**, 600–31, 1961

[54] ——— Pollen-analytical Arguments for Plant Migrations in South Europe. Pollen et Spores, **4**, 333–34, 1962

[55] ——— Über die ersten anthropogenen Vegetationsveränderungen in Süddalmatien an Hand eines neuen Pollendiagramms vom „Malo Jezero" auf Mljet. Veröff. d. Geobot. Inst. d. ETH, Stiftung Rübel, **37**, 9–15, Bern 1962

[56] *Beug, H. J.*, Untersuchungen zur spät- und postglazialen Vegetationsgeschichte im Gardaseegebiet unter besonderer Berücksichtigung der mediterranen Arten. Flora, **154**, 401–44, 1964

[57] *Beug, H. J.*, and *F. Firbas*, Ein neues Pollendiagramm vom Monte Baldo. Flora, **150**, 179–84, 1961

[58] *Beyle, M.*, Über ein altes Torflager im hohen Elbufer vor Schulau. Verhandl. d. Ver. f. naturwiss. Unterhaltung zu Hamburg, **11**, 1898–1900, 199–205, 1901

[59] ——— Über einige Ablagerungen fossiler Pflanzen der Hamburger Gegend. Mitt. a. d. Mineral.-Geol. Inst. in Hamburg; Beih. z. Jahrb. d. Hamburg. Wissensch. Anstalten, **36**, f. 1918, 33–47, 1920

[60] ——— Über ein altes Torflager in Stubbenberg bei Burg in Dithmarschen. Abh. d. Naturwiss. Ver. zu Bremen, Sonderheft zum **28**. Band, 43–50, 1931/1932

[61] ——— Über ein altes Torflager in Bramfeld im südlichen Holstein. Mitt. a. d. Mineral.-Geol. Staatsinst. in Hamburg, **14**, 17–22, 1933

[62] *Birkenmajer, K.*, and *A. Środoń*, Interstadial oryniacki w Karpatach. Z badań czwartorzędu w Polsce, **9**, 9–70, 1960

[63] *Bitner, K.*, Flora interglacjalna w Otapach. Z badań czwartorzędu w Polsce, **7**, 61–142, 1956; Engl., Russ. summaries

[64] *Blagoveshchenskii, G. A.*, Waldgeschichte des ehemals vereisten Gebietes im Europäischen Teil der UdSSR, im Zusammenhang mit den Klimaschwankungen des Quartärs. Trudi Geogr., **37**, 267–92, 1946

[65] *Blanc, A. C.*, Ricerche sul Quaternario Laziale. III. Avifauna artica, crioturbazioni e testimonianze di solifluzzi nel pleistocene medio-superiore di Roma e di Torre in Pietra. Il periodo glaciale Nomentano, nel quadro della

serie di glaciazzioni riconosciute nel Lazio. Quaternaria, **2**, 187–200, 1955
[66] *Blanc, A. C., G. Cova, P. Franceschi, F. Lona,* and *F. Settepassi,* Ricerche sul Quaternario Laziale. II. Una torba glaciale, avifauna arctica e malacofauna montana nel pleistocene medio-inferiore dell' Agro Cerite e di Roma. Il periodo glaciale Flaminia. Quaternaria, **2**, 159–86, 1955
[67] *Böcker, T. W.,* Oceanic and Continental Vegetational Complexes in Southwest Greenland. Meddelelser om Grønland, **148**, Nr. 1, 1954
[68] *Bohmers, A.,* Die Ausgrabungen bei Unterwisternitz. Forsch. u. Fortschr., **17**, 21–22, 1941
[69] *Bojarskaia, T. D.,* Über die Entwicklung der Vegetation am Unterlauf des Aldan im Jungtertiär und im Quartär. Vestnik Mosk. Univ., ser. 5 geogr., **1964**, Nr. 2, 90–91, 1964
[70] *Bonatti, E.,* I Sedimenti del lago di Monterosi. Experientia, **17**, 252, 1961
[71] ——— Pollen Sequences from the Sediments of the Lake of Monterosi, Central Italy. hectogr., 13 S., 1962
[72] ——— Late Quaternary Pollen Sequences from Central Italy. Pollen et Spores, **4**, 335–36, 1962
[73] *Bordes, F.,* and *H. J. Müller-Beck,* Zur Chronologie der Lößsedimente in Nordfrankreich und Süddeutschland. Germania, **34**, 199–208, 1956
[74] *Borówko-Dlużakowa, Z.,* Dwa nowe profile interglacjalne z Warszawy w świetle badań paleobotanicznych. Z badań czwartorzędu w Polsce, **9**, 105–30, 1960
[75] *Borówko-Dlużakowa, Z.,* and *B. Halicki,* Interglacjaly suwalszczyzny i terenów sąsiednich. Acta geol. polon., **7**, 361–401, 1957
[76] *Bourdier, F.,* Les dépôts quaternaires et le problème du loess dans la vallée de la Durance méridionale. C.R. Acad. Sci., **210**, 405–8, 1940
[77] ——— Essai de chronologie du Quaternaire moyen et supérieur. C.R. Acad. Sci., **215**, 473–75, 1942
[78] ——— Pliocène et Quaternaire dans le bassin du Rhône. Résumé de leurs subdivisions. Geologica Bavarica, **19**, 114–32, 1953
[78a] *Bray, J. R.,* Solar-climate Relationships in the Post-Pleistocene. Science, **171**, 1242–43, 1971
[79] *Brehme, K.,* Jahrringchronologische und -klimatologische Untersuchungen an Hochgebirgslärchen des Berchtesgadener Landes. Zs. f. Weltforstwirtsch., **14**, 65–80, 1951
[80] *Brelie, G. v. d.,* Transgression und Moorbildung im letzten Interglazial. Mitt. d. Geol. Staatsinst. Hamburg, **23**, 111–18, 1954
[81] *Brelie, G. v. d.,* and *U. Rein,* Die Interglazialbildungen im Niederrheinischen Diluvium. Der Niederrhein, **19**, 63–68, 1952
[82] *Brelie, G. v. d., A. Mückenhausen,* and *U. Rein,* Ein Torf aus dem Eiszeitalter im Untergrund von Weeze. Der Niederrhein, **22**, 80–83, 1955
[83] *Brem, M.,* Flora interglacjalna z Ciechanek Krzesimowskich. Acta Geol. Polon., **3**, 475–80, 1953
[84] *Bremówna, M.,* and *M. Sobolewska,* Wyniki badań botanicznych osadów interglacjalnych w dorzeczu Niemna. Acta Geol. Polon., **1**, 335–64 (1950/1951), 1950
[85] *Brockmann-Jerosch, H.,* Fundstellen von Diluvialfossilien bei Lugano.

Beibl. z. Vierteljahrsschrift d. Naturforsch. Ges. in Zürich, **68**, 1–7, 1923
[86] *Brooks, C. E. P.*, Climate through the Ages. A Study of the Climatic Factors and their Variations. 2. Aufl. London 1950
[87] *Brotzen, F.*, An Interstadial (Radiocarbon Dated) and the Substages of the Last Glaciation in Sweden. Geol. Fören. i Stockh. Förhandl., **83**, 144–50, 1961
[88] *Brouwer, A.*, Pollenanalytisch en geologisch onderzoek van het Onder- en Midden- Pleistoceen van Noord-Nederland. Leidse Geol. Mededel., **14B**, 258–346, 1949
[89] *Brunnacker, K.*, Regionale Bodendifferenzierungen während der Würmeiszeit. Eiszeitalter und Gegenwart, **7**, 43–48, 1956
[90] ——— Die Geschichte der Böden im jüngeren Pleistozän in Bayern. Geologica Bavarica, **34**, 1957
[91] ——— Bemerkungen zur Feinstgliederung und zum Kalkgehalt des Lösses. Eiszeitalter und Gegenwart, **8**, 107–15, 1957
[92] ——— Das Schieferkohlenlager vom Pfefferbichl bei Füssen. Jahresber. u. Mitt. d. Oberrhein. Geol. Ver., N.F., **44**, 43–60, 1962
[93] ——— Reliktböden im östlichen Mittelfranken. Geol. Blätter für NO-Bayern, **12**, 183–90, 1962
[94] ——— Über Ablauf und Altersstellung altquartärer Verschüttungen im Maintal und nächst dem Donautal bei Regensburg. Eiszeitalter und Gegenwart, **15**, 72–80, 1964
[95] ——— Böden des älteren Pleistozäns bei Regensburg. Geologica Bavarica, **53**, 148–60, 1964
[96] ——— Schätzungen über die Dauer des Quartärs, insbesondere auf der Grundlage seiner Paläoböden. Geol. Rundschau, **54**, 415–28, 1964
[97] *Brunnacker, M.*, and *K. Brunnacker*, Weitere Funde pleistozäner Molluskenfaunen bei München. Eiszeitalter und Gegenwart, **13**, 129–37, 1962
[98] *Brunnschweiler, D.*, Der pleistozäne Periglazialbereich in Nord-Amerika. Zs. f. Geomorph., N.F. **8**, 223–31, 1964
[99] *Büdel, J.*, Die räumliche und zeitliche Gliederung des Eiszeitklimas. Naturwiss., **36**, 105–12, 133–39, 1949
[100] ——— Die „periglazial"-morphologischen Wirkungen des Eiszeitklimas auf der ganzen Erde (Beiträge zur Geomorphologie der Klimazonen und Vorzeitklimate IX). Erdkunde, **7**, 249–66, 1953
[101] ——— Reliefgenerationen und plio-pleistozäner Klimawandel im Hoggar-Gebirge (Zentrale Sahara). Erdkunde, **9**, 100–15, 1955
[102] ——— Die pliozänen und quartären Pluvialzeiten der Sahara. Eiszeitalter und Gegenwart, **14**, 161–87, 1963
[103] *Burck, H. D. M.*, Quaternary; Pleistocene; Eemian. In: *Pannekoek, A. J.*, Geological History of the Netherlands. 92–96, 's-Gravenhage 1956
[104] *Burns, G. W.*, Wisconsin Age Forests in Western Ohio. II. Vegetation and Burial Conditions. Ohio Journ. of Sci., **58**, 220–30, 1958
[105] *Butzer, K. W.*, Mediterranean Pluvials and the General Circulation of the Pleistocene. Geografiska Annaler, **39**, 48–53, 1957
[106] ——— Russian Climate and the Hydrological Budget of the Caspian Sea. Revue canad. de géogr., **12**, 129–39, 1958

[107] ——— Quaternary Stratigraphy and Climate in the Near East. Bonner Geogr. Abh., **24**, 1958

[108] ——— Studien zum vor- und frühgeschichtlichen Landschaftswandel der Sahara. II. Das ökologische Problem der neolithischen Felsbilder der östlichen Sahara. Akad. d. Wiss. u. d. Lit., Abh. d. math.-nat. Kl., 20–49, 1958

[109] ——— Current Research on Pleistocene General Circulation. Public Lecture, given to the Departm. of Geol. and Geogr., Univ. of Minnesota, 1960

[110] ——— Climatic Change in Arid Regions since the Pliocene. In: *Stamp, L. D.*, History of Land-Use in Arid Lands. 31–56, Paris 1961

[111] ——— Palaeoclimatic Implications of Pleistocene Stratigraphy in the Mediterranean Area. Ann. N.Y. Acad. of Sci., **95**, 449–56, 1961

[112] ——— Coastal Geomorphology of Majorca. Ann. of the Assoc. of Amer. Geographers, **52**, 191–212, 1962

[113] ——— Jungpleistozäne und holozäne Landschaftsgeschichte des nördlichen Eurasiens. Erdkunde, **18**, 67–68, 1964

[114] *Butzer, K. W.*, and *J. Cuerda*, Coastal Stratigraphy of Southern Mallorca and its Implications for the Pleistocene Chronology of the Mediterranean Sea. Journ. of Geol., **70**, 398–416, 1962

[115] *Bykov, G. E.*, Quartäre Sedimente der Atbasar- und Esil'-Gebiete der Kazakischen ASSR. Izv. Gos. Geogr. Obšč., **65**, 241–49, 1933; Engl. summary

[116] *Campo, M. van*, Analyse pollinique des dépôts wurmiens d'El Guettar (Tunisie). Veröff. d. Geobot. Inst. Rübel in Zürich, **34**, 133–35, 1958

[117] ——— Quelques pollens pleistocènes nouveaux pour le Hoggar. C.R. Acad. Sci., **258**, 1297–99, 1964

[118] ——— Représentation graphique de spectres polliniques des régions sahariennes. C.R. Acad. Sci., **258**, 1873–76, 1964

[119] *Campo, M. van*, and *J. Bouchud*, Flore accompagnant le squelette d'enfant moustérien découvert au Roc de Mersal, commune du Bugue (Dordogne) et première étude de la faune du gisement. C.R. Acad. Sci., **254**, 897–99, 1962

[120] *Campo, M. van*, and *R. Coque*, Palynologie et géomorphologie dans le Sud Tunesien. Pollen et Spores, **2**, 275–84, 1960

[121] *Campo, M. van, G. Aymonin, Ph. Guinet* and *P. Rognon*, Contribution à l'étude du peuplement végétal quaternaire des montagnes sahariennes: L'Atakor. Pollen et Spores, **6**, 169–94, 1964

[122] *Campo, M. van, J. Cohen, Ph. Guinet* and *P. Rognon*, Contribution à l'étude du peuplement végétal quaternaire des montagnes sahariennes. II. Flore contemporaine d'un gisement de mammifères tropicaux dans l'Atakor. Pollen et Spores, **7**, 361–71, 1965

[123] *Capenko, M. M.*, and *N. A. Machnach*, Antropogenovye otlozheniia Belorusii. Minsk 1959

[124] ——— Einige Angaben zur Kenntnis des Pliozäns und des frühen Anthropogens Weißrußlands. Trudy Kom. po izuch. chetvert. per., **20**, 85–91, 1962

[125] *Chebotareva, N. S.*, Die Stratigraphie der pleistozänen Sedimente des Zentrums des Russischen Flachlandes (zwischen der Grenze der Valdai-Vereisung und des Moskauer Stadiums der Dneprovsk-Vereisung). In: *Markov, K. K.*, and *A. I. Popov*, Lednikovyi, period na territorii evropeiskoi

chasti SSSR i Sibiri. 116–47, Moscow, 1959
[126] ——— Einige Probleme der Paläogeographie der Valdai-Vereisung im Nordwesten des Russischen Flachlandes. Paleogeografiia i chronologiia verchnego pleistocena i golocena po dannym radiouglerodnogo metoda. 7–22, Moscow, 1965
[127] *Chebotareva, N. S.*, and *E. A. Mal'gina*, Über das absolute Alter einiger jungpleistozäner Interstadiale. Paleogeografiia i chronologiia verchnego pleistocena i golocena po dannym radiouglerodnogo metoda. 27–37, Moscow, 1965
[128] *Chebotareva, N. S., E. A. Mal'gina* and *M. A. Nedoshivina*, Das Alter der oberen Moräne der Ilmensee-Niederung. Paleogeografiia i chronologiia verchnego pleistocena i golocena po dannym radiouglerodnogo metoda. 37–42, Moscow 1965
[129] *Chebotareva, N. S., M. A. Nedoshivina* and *T. I. Stoljarova*, Moskau-Valdai-(Mikulino-)-interglaziale Sedimente im Flußgebiet des Oberlaufes der Wolga und ihre Bedeutung für die Paläogeographie. Biull. chetvert., **26**, 35–49, 1961
[130] ——— Neue Profile Moskau-Valdai-(Mikulino-)interglazialer Sedimente bei Sosnovatka an der B. Dubenka in der Kalinin-Oblast'. Izv AN, ser. geogr., 1961, Nr. 1, 124–27, 1961
[131] *Chebotareva, N. S., L. R. Serebrjannyj, A. L. Devirc* and *E. I. Dobkina*, Das absolute Alter der tiefen Flußterrassen im Zentrum des Russischen Flachlandes. Izv AN, Ser. geogr., 1962, Nr. 4, 70–74, 1962
[132] *Ceitlin, S. M.*, Das Eopleistozän im Flußgebiet der Unteren Tunguska. DAN, **133**, 1183–86, 1960
[133] ——— Parallelisierung quartärer Sedimente im glazialen und periglazialen Bereich Mittel- Sibiriens (Flußgebiet der Unteren Tunguska). Trudi Geol., **100**, 1964
[134] ——— Über die Gliederung der Letzten Eiszeit in Sibirien. Chetvertichnyi period i ego istoriia. 175–82, Moscow 1965; Engl. summary
[134a] *Cepek, A. G.*, Quartär. In: Grundriß der Geologie der Deutschen Demokratischen Republik, Bd. 1, Akademie-Verlag, Berlin, 385–420, 1968
[135] *Cherdyncev, V. V., N. S. Strashnikov, T. I. Borisenko* and *L. M. Poliakova*, Bestimmung des absoluten Alters fossiler quartärer Knochen. Materialy chetvert., **1**, 266–72, 1961
[136] *Chanda*, S., Untersuchungen zur pliozänen und pleistozänen Floren- und Vegetationsgeschichte im Leinetal und im südwestlichen Harzvorland (Untereichsfeld). Geol. Jahrb., **79**, 783–844, 1962
[137] *Chigurjaeva, A. A.*, and *V. L. Yachimovich*, Über die Flora und die Vegetation des südlichen Ural-Vorlandes, vom mittleren Akchagyl bis zum Holozän. In: Antropogen juzhnogo Urala, 164–87, Moscow 1965
[138] *Clisby, K. H.*, An Imperceptible Plio-pleistocene Boundary. Pollen et Spores, **4**, 339, 1962
[139] *Clisby, K. H., F. Foreman,* and *P. B. Sears*, Pleistocene Climatic Change in New Mexico, USA. Veröff. d. Geobot. Inst. Rübel in Zürich, **34**, 21–26, 1958
[140] *Coetzee, J. A.*, Evidence for a Considerable Depression of the Vegetation

Belts during the Upper Pleistocene on the East African Mountains. Nature (L.), **204**, 564–66, 1964

[141] *Colbert, E. H.*, The Record of Climatic Changes as Revealed by Vertebrate Paleoecology. In: *Shapley, H.*, Climatic Change. Evidence, Causes, and Effects. 249–71, Cambridge 1953

[142] *Conrad, G.*, Synchronisme du dernier pluvial dans le Sahara septentrional et le Sahara méridional. C.R. Acad. Sci., **257**, 2506–09, 1963

[143] *Coope, G. R., F. W. Shotton,* and *I. Strachan,* A Late Pleistocene Fauna and Flora from Upton Warren, Worcestershire. Philos. Transact. Roy. Soc. of London, ser. B, Biol. Sci., **244**, Nr. 714, 379–21, 1961

[144] *Couper, R. A.,* and *W. F. Harris,* Pliocene and Pleistocene Plant Microfossils from Drillholes near Frankton, N. Z., New Zealand Journ. of Geol. and Geophys., **3**, 1, 15–21, 1960

[144a] *Cox, A.*, Geomagnetic Reversals. Science, **163**, 237–45, 1969

[145] *Cranwell, L. M.,* and *L. v. Post,* Pollen Diagrams from New Zealand. Geografiska Annaler, **18**, 308–47, 1936

[146] *Cushing, E. J.*, Redeposited Pollen in Late-Wisconsin Pollen Spectra from East-central Minnesota. Amer. J. Sci., **262**, 1075–88, 1964

[147] *Danilova, I. A.*, Pleistozäne Sedimente und das Relief der Umgebung der geographischen Station Krasnovidovo der Moskauer Staats-Universität. In: *Markov, K. K.,* and *A. I. Popov,* Lednikovyi period na territorii evropeiskoi chasti SSSR i Sibiri, 64–115, Moscow 1959

[148] ——— Lakustrin-sumpfige Sedimente oberhalb der Moräne im Oberlaufgebiet des Moskau-Flusses. Paleogeografiia chetvert. per. SSSR, 83–92, Moscow 1961; Fr. summary

[149] *Danilova, V. V.,* and *M. N. Alekseev,* Bestimmung des relativen geologischen Alters fossiler Knochen mit Hilfe ihres Fluor-Gehaltes. DAN, **119**, 1020–23, 1958

[149a] *Dansgaard, W.,* and *H. Tauber,* Glacier Oxygen-18 Content and Pleistocene Ocean Temperatures. Science, **166**, 499–502, 1969

[149b] *Dansgaard, W., S. J. Johnsen, H. B. Clausen,* and *C. C. Longway, Jr.*, Ice Cores and Paleoclimatology. In: *Olsson, I. U.*, see [573a], 337–48, 1970

[150] *Davis, M. B.*, Three Pollen Diagrams from Central Massachusetts. Amer. J. Sci., **256**, 540–70, 1958

[151] ——— On the Theory of Pollen Analysis. Amer. J. Sci., **261**, 897–912, 1963

[151a] *Davis, R. B.*, Pollen Studies of Near-surface Sediments in Maine Lakes. In: *Cushing, E. J.,* and *H. E. Wright,* Quaternary Paleoecology, **7**, Proc. VII Congr. INQUA, Yale Univ. Press, 143–73, 1967

[152] *Deevey, E. S.*, Biogeography of the Pleistocene. Bull. Geol. Soc. Amer., **60**, 1315–1416, 1948

[153] ——— Late-glacial and Postglacial Pollendiagrams from Maine. Amer. J. Sci., **249**, 177–207, 1951

[154] ——— Palaeolimnology and Climate. In: *Shapley, H.*, Climatic Change. Evidence, Causes, and Effects. 273–318, Cambridge 1953

[155] *Deviatova, E. I.*, Zur Geschichte des Quartärs im Norden im Lichte neuer Befunde im Flußgebiet der Onega. Tezisi; Sekc. Severn. i Zapadn. chasti Russk. Ravn. 6–8, 1957

[156] ——— Marine interglaziale Sedimente im Flußgebiet der Onega. DAN, **125**, 162–65, 1959

[157] ——— Zur Geschichte des Quartärs im Norden, im Lichte neuen Materials, aus dem Flußgebiet der Onega. Materialy chetvert., **2**, 24–31, 1961

[158] ——— Stratigrafiia chetvertichnych otlozhenij i paleogeografiia chetvertichnogo perioda v basseine reki Onegi. Moscow 1961

[159] *Devjatova, E. I.*, and *F. I. Loseva*, Stratigrafiia i paleogeografiia chetvertichnogo perioda v basseine reki Mezeni. Leningrad 1964

[160] *Dobzhansky, Th.*, Adaptive Changes Induced by Natural Selections in Wild Populations of *Drosophila*. Evolution, **1**, 1–16, 1947

[161] ——— Evolution in the Tropics. Amer. Scientist, **38**, 209–21, 1950

[162] *Dobzhansky, Th.*, and *C. Epling*, Contributions to the Genetics, Taxonomy, and Ecology of *Drosophila pseudoobscura* and Its Relatives. Publ. Nr. 554, Carnegie Inst. of Washingt., 1944

[163] *Dokturowskii, W. S.*, Die interglaziale Flora in Rußland. Geol. Fören. i Stockh. Förhandl., **51**, 389–410, 1929

[164] ——— Neue Angaben über inter- und postglaziale Sedimente der UdSSR. Prioda, **20**, 704–7, 1931

[165] ——— Neue Angaben über die interglaziale Flora in der UdSSR. Biull. Mosk. Obshch. Ispyt. Prirodi, otd. geol., **9**, 1–2, 214–29, 1931

[166] *Donner, J. J.*, Pleistocene Geology of Eastern Long Island, New York. Amer. J. Sci., **262**, 355–76, 1964

[167] *Donner, J. J.*, and *B. Kurtén*, The Floral and Faunal Succession of "Cueva del Toll," Spain. Eiszeitalter und Gegenwart, **9**, 72–82, 1958

[168] *Donovan, D. T.*, Sea Levels of the Last Glaciation. Bull. Geol. Soc., **73**, 1297–98, 1962

[169] *Dorofeev, P. I.*, Pliozäne Pflanzen des Urals. Bot. Zhurn., **37**, 850–56, 1952

[170] ——— Über eine frühpleistozäne Flora von Fat'ianovka an der Oka. Bot. Zhurn., **43**, 1034–39, 1958

[171] ——— Über die Entdeckung von *Azolla interglacialica* NIKIT. in pleistozänen Sedimenten an der Oka. Biull. chetvert., **23**, 87–91, 1959

[172] ——— Neue Beiträge zur Kenntnis pleistozäner Floren Weißrußlands und der Smolensker Oblast'. Materialy po istorii flory i rastitel'nosti SSSR, **4**, 5–180, Moscow-Leningrad 1963

[173] *Dorogostaiskaia, E. V.*, Versuch einer Charakterisierung der Ruderal- und Segetalflora Vorkutas und seiner Umgebung. Bot. Zhurn. **48**, 1015–21, 1963

[174] *Dreimanis, A.*, Wisconsin Stratigraphy at Port Talbot on the North Shore of Lake Erie, Ontario. Ohio J. of Sci., **58**, 65–84, 1958

[175] ——— The Early Wisconsin in the Eastern Great Lake Region, North America. Abh. d. Dtsch. Akad. d. Wiss., Kl. III, H. 1, 196–205, Berlin 1960

[176] *Duigan, S. L.*, Plant Remains from the Gravels of the Summertown-Radley Terrace near Dorchester, Oxfordshire. Quat. J. Geol. Soc. London, **111**, f. 1955, 225–38, 1956

[177] ——— Pollen Analysis of the Nechells Interglacial Deposits, Birmingham. Quat. J. Geol. Soc. London, **112**, f. 1956, 373–91, 1957

[178] *Duigan, S. L.*, and *B. W. Sparks*, Pollen Analysis of the Cromer Forest Bed Series in East Anglia. Philos. Transact. of the Roy. Soc. London, ser. B, **246**,

Nr. 729, 149–202, 1963
[179] *Dyakowska, J.*, Interglacjał w Poniemuniu pod Grodnem. Starunia, **14**, Cracow, 1936
[180] ——— Interglacjał w Kątach koło Sromowiec Wyżnich (Pieniny). Starunia, **23**, Cracow, 1947
[181] ——— Plejstoceński profil z Wylezina. Z badań czwartorzędu w Polsce, **7**, 193–216, 1956; Russ., Engl. summaries
[182] *Eberl, B.*, Die Eiszeitenfolge im nördlichen Alpenvorlande. Ihr Ablauf, ihre Chronologie auf Grund der Aufnahmen im Bereich des Lech- und Illergletschers, Augsburg 1930
[182a] *Eckstein, D.*, and *J. Bauch*, Beitrag zur Rationalisierung eines dendrochronologischen Verfahrens und zur Analyse seiner Aussageischerheit. Forstwiss. Centralbl., **88**, 230–50, 1969
[183] *Eissmann, L.*, Entwicklung und Verlauf der Saale während des Alt- und Frühpleistozäns in der südwestlichen Leipziger Tieflandsbucht. Geologie, **11**, 41–50, 1962
[184] ——— Riß- und mindelglaziale Eisrandlagen und Flußterrassen zwischen Mulde und Pleiße im Leipziger Raume. Exkursionsführer, Herbsttagung 1962 der Geologischen Ges. der DDR
[185] *Eklund, B.*, Skogsforskningsinstitutets årsringmätningsmaskiner, deras tillkomst, konstruktion och användning. Meddedel. från Statens Skogsforskningsinst., **38**, Nr. 5, 1949
[186] ——— Om granens årsringsvariationer inom mellersta Norrland och deras samband med klimatet. Meddedel. från Statens Skogsforskningsinst., **47**, Nr. 1
[187] *Elhai, M. H.*, Eléments d'interprétation du relief entre la Dives et la Seine, Normandie. Bull. de l'Assoc. de géogr. franç., 143–57, 1961
[188] ——— La Normandie occidentale entre la Seine et le Golfe Normand-Breton. Etude morphologique. Bordeaux 1963
[189] *Ellenberg, H.*, Kausale Pflanzensoziologie auf physiologischer Grundlage Ber. D. Bot. Ges., **63**, 24–31, 1950
[190] ——— Physiologisches und ökologisches Verhalten derselben Pflanzenarten. Ber. D. Bot. Ges., **65**, 350–61, 1952
[191] *Elsasser, W., E. P. Ney*, and *I. R. Winkler*, Cosmic-ray Intensity and Geomagnetism. Nature (L.), **178**, 1226–27, 1956
[192] *Emiliani, C.*, Pleistocene Temperature Variations in the Mediterranean. Quaternaria, **2**, 87–98, 1955
[193] ——— Pleistocene Temperatures. Journ. of Geol., **63**, 538–78, 1955
[194] ——— Note on Absolute Chronology of Human Evolution. Science, **123**, 924–26, 1956
[195] ——— Oligocene and Miocene Temperatures of the Equatorial and Subtropical Atlantic Ocean. Journ. of Geol., **64**, 281–88, 1956
[196] ——— Paleotemperature Analysis of Core 280 and Pleistocene Correlations. Journ. of Geol., **66**, 264–75, 1958
[197] ——— Cenozoic Climatic Change as Indicated by the Stratigraphy and Chronology of Deep-sea Cores of *Globigerina*-ooze Facies. Ann. N. Y. Acad. Sci., **95**, Art. 1, 521–36, 1961

[198] ——— The Temperature Decrease of Surface Sea-water in High Latitudes and of Abyssalhadal Water in Open Oceanic Basins during the past 75 Million Years. Deep-sea Research, 8, London 1961

[199] ——— Paleotemperature Analysis of the Caribbean Cores A 254-Br-C and CP-28. Bull. Geol. Soc., **75**, 129–44, 1964

[199a] *Emiliani, C.*, The Last Interglacial: Paleotemperatures and Chronology. Science, **171**, 571–73, 1971

[200] *Emiliani, C. T. Mayeda,* and *R. Selli,* Paleotemperature Analysis of the Plio-Pleistocene Section at La Castella, Calabria, Southern Italy. Bull. Geol. Soc., **72**, 679–88, 1961

[201] *Enquist, F.*, Sambandet mellan klimat och växtgränser. Geol. Fören. i Stockh. Förhandl., **46**, 202–11, 1924

[202] *Erd, K.*, Vegetationsentwicklung und Feuchtigkeitsschwankungen während der Eem- und Holstein-Warmzeit in Brandenburg. Ber. d. Geol. Ges. in der DDR, **7**, 259–61, 1962

[203] *Erdtman, G.*, An Introduction to Pollen Analysis. Waltham, Massachusetts, 1954

[204] *Erdtman, G.*, and *G. Nordborg,* Über Möglichkeiten, die Geschichte verschiedener Chromosomenzahlrassen von *Sanguisorba officinalis* und *S. minor* pollenanalytisch zu beleuchten. Bot. Notiser., **114**, 19–21, 1961

[205] *Ericson, D. B., M. Ewing,* and *G. Wollin,* Pliocene and Pleistocene Boundary in Deep-sea Sediments. Science, **139**, 727–37, 1963

[206] ——— The Pleistocene Epoch in Deep-sea Sediments. Science, **146**, 723–32, 1964

[207] *Ericson, D. B., M. Ewing, G. Wollin,* and *B. C. Heezen,* Atlantic Deep-sea Sediment Cores. Bull. Geol. Soc. **72**, 193–286, 1961

[208] *Erlandsson, St.*, Dendro-chronological Studies. Data fr. Stockh. Högsk. geokronol. inst., **23**, 1936

[209] *Faegri, K.*, and *J. Iversen,* Textbook of Modern Pollen-Analysis. Copenhagen 1950

[210] ——— Textbook of Pollen Analysis. Copenhagen 1964

[211] *Fairbridge, R. W.*, Recent and Pleistocene Coral Reefs of Australia. Journ. of Geol., **58**, 330–401, 1950

[212] *Farnham, R. S., J. H. McAndrews,* and *H. E. Wright,* A Late Wisconsin Buried Soil near Aitkin, Minnesota, and its Paleobotanical Setting. Amer. J. Sci., **262**, 393–412, 1964

[213] *Fedorov, P. V.*, Stratigraphie pleistozäner Sedimente und die Entwicklungsgeschichte des Kaspischen Meeres. Trudi Geol., **10**, 1957

[214] *Fedorova, R. V.*, Paläobotanische Erforschung der Sedimente der Limane in der Kaspi-Niederung. Trudi Geogr., **50**, 75–90, 1951

[215] ——— Umweltverhältnisse während der Siedlungszzeiten des jungpaläolithischen Menschen im Gebiet von Kostenki, Voronezh Oblast. In: *Boriskowskii, P. I.*, Skizze des Paläolithikum im Flußgebiet des Don. Materiali i Issledovaniia po archeologii SSSR, Nr. 121, 220–29, 1963

[215a] *Ferguson, C. W.*, Dendrochronology of Bristlecone Pine, Pinus aristata. Establishment of a 7484-year Chronology in the White Mountains of Eastern-central California, USA. In: *Olsson, I. U.*, see [573a], 237–59, 1970

[216] *Fink, J.*, Zur Korrelation der Terrassen und Lösse in Österreich. Eiszeitalter und Gegenwart, **7**, 49–77, 1956

[217] ——— Die Südabdachung der Alpen. Mitt. d. Österr. Bodenkundl. Ges., **1961**, 123–83, 1961

[218] ——— Die Gliederung des Jungpleistozäns in Österreich. Mitt. der Geol. Ges. Vienna, **54**, 1–25, 1961

[219] ——— Studien zur absoluten und relativen Chronologie der fossilen Böden in Österreich. II. Wetzleinsdorf und Stillfried. Archaeol. Austr., **31**, 1–18, 1962

[220] *Firbas, F.*, Über die Flora und das interglaziale Alter des Helgoländer Süßwassertöcks. Senckenbergiana **10**, 185–95, 1928

[221] ——— Über die Bestimmung der Walddichte und der Vegetation waldloser Gebiete mit Hilfe der Pollenanalyse. Planta, **22**, 109–45, 1934

[222] ——— Über die späteiszeitlichen Verschiebungen der Waldgrenze. Naturwiss. **34**, 114–18, 1947

[223] ——— Die postglaziale Waldgeschichte Mitteleuropas nördlich der Alpen. Bd. **1**, Jena 1949

[224] *Firbas, F.*, and *I. Firbas*, Über die Anzahl der Keimporen der Pollenkörner von *Carpinus Betulus* L. Veröff. d. Geobot. Inst. Rübel in Zürich, **34**, 45–52, 1958

[225] *Firbas, F.*, and *B. Frenzel*, Floren- und Vegetationsgeschichte seit dem Ende des Tertiärs. Fortschr. d. Bot., **22**, 87–111, 1960

[226] *Firbas, F.*, and *P. Zangheri*, Eine glaziale Flora von Forli, südlich Ravennas. Veröff. d. Geobot. Inst. Rübel in Zürich, **12**, 24–36, 1936

[227] ——— Über neue Funde pflanzenführender Ablagerungen in der südlichen Po-Ebene bei Forli. Nachr. d. Akad. d. Wiss. Göttingen, math.-phys. Kl., biol. phys., chem. Abt., 11–18, 1954

[228] *Fleming, C. A.*, The Geology of Wanganui Subdivision. New Zealand Geol. Survey, Bull., n. s. 52, Wellington 1953

[229] *Flint, R. F.*, Rates of Advance and Retreat of the Margin of the Late-Wisconsin Ice Sheet. Amer. J. Sci., **253**, 249–55, 1955

[230] ——— New Radiocarbon Dates and Late-pleistocene Stratigraphy. Amer. J. Sci., **254**, 265–87, 1956

[231] ——— Glacial and Pleistocene Geology. New York, London 1957

[232] ——— Pleistocene Climate in Low Latitudes. Geogr. Review, **53**, 123–29, 1963

[233] ——— Status of the Pleistocene Wisconsin Stage in Central North America. Science, **139**, 402–4, 1963

[234] *Flint, R. F.*, and *F. Brandtner*, Climatic Change since the Last Interglacial. Amer. J. Sci., **259**, 321–28, 1961

[235] *Flohn, H.*, Atmosphärische Zirkulation und Paläoklimatologie. Geol. Rundsch., **40**, 153–78, 1952

[236] —— Studien über die atmosphärische Zirkulation in der Letzten Eiszeit. Erdkunde, **7**, 266–75, 1953

[237] ——— Kontinental-Verschiebungen, Polwanderungen und Vorzeitklima im Lichte paläomagnetischer Meßergebnisse. Naturwiss. Rundsch., **1959**, 375–84, 1959

[238] ——— Zur meteorologischen Interpretation der pleistozänen Klimaschwankungen. Eiszeitalter und Gegenwart, **14**, 153–60, 1963

[239] ——— Probleme der theoretischen Klimatologie. Naturwiss. Rundsch., **18**, 385–92, 1965

[240] *Florschütz, F.*, *Azolla filiculoïdes* LAM. uit de *Paludina*-Klei van Berlijn. Proceedings, Nederl. Akad. van Wetensch., Sect. of Sci., **44**, 339–41, 1941

[241] ——— The Subdivisions of the Middle and Young Pleistocene up to the Late-glacial in the Netherlands, England and Germany, mainly Based on the Results of Paleobotanical Investigations. Geol. en Mijnb., n. s. **19**, 245–49, 1957

[242] ——— The Flora of the Eemian and the Tubantian. Verhandel. Kon. Nederl. Geol.- Mijnbouwk. Genootsch., geol. ser., **17**, 119–20, 1957

[243] ——— Steppen- und Salzsumpfelemente aus den Floren der Letzten und Vorletzten Eiszeit in den Niederlanden. Flora, **146**, 489–92, 1958

[244] *Florschütz, F.*, and *J. Ménendez Amor,* Beitrag zur Kenntnis der quartären Vegetationsgeschichte Nord-Spaniens. Veröff. d. Geobot. Inst. der ETH, Stift. Rübel, in Zürich, **37**, 68–73, 1962

[245] ——— Sur les éléments steppiques de la végétation quaternaire espagnole. Ber. d. Geobot. Inst. der ETH, Stiftg. Rübel, **34**, f. 1962, 59, 1963

[246] *Florschütz, F.*, and *A. M. H. Anker van Someren,* De resultaten van het palynologisch onderzoek. Mededel. van de Geol. Stichting, n. s. **10**, 55–65, 1956

[247] *Follieri, M.*, Elementi originali per la storia della vegetazione del Lazio. Nuov. Giorn. Bot. Ital., n. s. **66**, 707–8, 1959

[248] ——— Interpretazione cronologica preliminare della diatomite a *Pterocarya* di Riano Romano. Quaternaria, **5**, 261–63, 1958–61

[249] ——— Nuovi elementi botanici nel tufo grigio della Cava Bianca (Via Flaminia), confermanti la glaciazione Flaminia. Quaternaria, **5**, 265–69, 1958–61

[250] ——— Correlazione paleobotanica fra i due bacini diatomeiferi di Valle dell'Inferno e di Valle Pianaperina presso Riano (Roma). Annali di Botanica, **26**, fasc. 3, 1–10, 1960

[251] ——— La foresta colchica fossile di Riano Romano. II. Analisi polliniche. Annali di Botanica, **27**, 245–80, 1962

[252] ——— Persistence de végétaux tertiaires dans les dépôts quaternaires au nord de Rome. Report VI Internat. Congr. Quaternary Warsaw 1961, **2**, 383–87, Łódź 1964

[253] *Fränzle, O.*, Interstadiale Bodenbildungen in ober-italienischen Würm-Lössen. Eiszeitalter und Gegenwart, **11**, 196–205, 1960

[254] *Franks, J. W.*, Interglacial Deposits at Trafalgar Square, London. New Phytologist, **59**, 145–52, 1960

[255] *Frechen, J.*, and *H. J. Lippolt,* Kalium-Argon-Daten zum Alter des Laacher Vulkanismus, der Rheinterrassen und der Eiszeiten. Eiszeitalter und Gegenwart, **16**, 5–30, 1965

[256] *Frenzel, B.*, Die Vegetations- und Landschaftszonen Nord-Eurasiens während der Letzten Eiszeit und während der postglazialen Wärmezeit. I. Teil: Allgemeine Grundlagen. Abh. d. Akad. d. Wiss. u. d. Lit., math.-nat.

Kl., Jahrg. **1959**, Nr. 13, 937–1099, Mainz 1960

[257] ——— II. Teil: Rekonstruktionsversuch der letzteiszeitlichen und wärmezeitlichen Vegetation Nord-Eurasiens. Abh. d. Akad. d. Wiss. u. d. Lit., math.-nat. Kl., Jahrg. **1960**, Nr. 6, 291–453, Mainz 1960

[258] ——— Zur Pollenanalyse von Lössen. Untersuchungen der Lößprofile von Oberfellabrunn und Stillfried (Niederösterreich). Eiszeitalter und Gegenwart, **15**, 5–39, 1964

[259] ——— Über die offene Vegetation der Letzten Eiszeit am Ostrande der Alpen. Verhandl. d. Zool.-Bot. Ges. Wien, **103**, **104**, 110–43, 1964

[260] ——— Grundzüge der pleistozänen Vegetationsgeschichte Nord-Eurasiens. Wiesbaden 1967 (im Druck)

[261] *Fries, M.*, Pollen Profiles of Late Pleistocene and Recent Sediments from Weber Lake, Northeastern Minnesota. Ecology, **43**, 295–308, 1962

[262] *Fries, M., H. E. Wright,* and *Meyer Rubin,* A Late Wisconsin Buried Peat at North Branch. Amer. J. Sci., **259**, 679–93, 1961

[263] *Fristrup, B.*, Climate and Glaciology of Peary Land, North Greenland. Union Géodés. et géophys. internat., Assoc. Internat. d'hydrol. scient., Assembl. générale de Bruxelles, 1951, **1**, 185–93, 1951

[263a] *Fritts, H. C.*, Tree-ring Evidence for Climatic Changes in Western North America. Monthly Weather Rev., **93**, 421–43, 1965

[263b] *Fritts, H. C.*, Growth Rings of Trees: Their Correlation with Climate. Science, **154**, 973–79, 1966

[263c] *Fritts, H. C., D. G. Smith,* and *M. A. Stokes,* The Biological Model for Paleoclimatic Interpretation of Mesa Verde Tree Ring Series. Amer. Antiquity, **31**, No. 2, Pt. 2, 101–21, 1965

[263d] *Fritts, H. C., D. G. Smith, C. A. Budelsky,* and *J. W. Cardis,* The Variability of Ring Characteristics within Trees as Shown by a Reanalysis of Four Ponderosa Pine. Tree Ring Bull., **27**, Nos. 1–2, 1965

[263e] *Fritts, H. C., D. G. Smith, J. W. Cardis,* and *C. A. Budelsky,* Tree-ring Characteristics Along a Vegetation Gradient in Northern Arizona. Ecology, **46**, 393–401, 1965

[263f] *Fritts, H. C., J. E. Mosimann,* and *Ch. P. Bottorff,* A Revised Computer Program for Standardising Tree-ring Series. Tree Ring Bull., **29**, Nos. 1–2, 15–20, 1969

[264] *Fromm, E., R. W. Kolbe* and *H. Persson,* An Interglacial Peat at Ale near Luleå, Northern Sweden. Sveriges Geol. Undersökn., ser. C., Avhandl. och upssats., Nr. 574, **54**, Nr. 5, 1960

[265] *Gabrielian, A. A.*, Paleogen i neogen Armianskoi SSR, stratigrafiia, tektonika, istoriia geologicheskogo razvitiia. Erevan 1964

[266] *Gakkel', Y. Y.* and *E. S. Korotkevich,* Severnaia Jakutiia. Trudi arkt. i antarkt. nauchnoissledov. in-ta glavn. upravl. severn. morsk. puti, **236**, 48–60, Leningrad 1962

[267] *Galenieks, P.*, Interglacial Peat-bed at Dēsele, Kurzemē (Latvia). Acta Univ. Latviensis, **12**, 565–80, Riga 1925; Engl. summary

[268] *Gassner, G.*, and *F. Christiansen-Weniger,* Dendroklimatologische Untersuchungen über die Jahresringentwicklung der Kiefern in Anatolien. Nova Acta Leopoldina, N. F., **12**, Nr. 80, 1942

[269] *Geer, G. de,* Geochronologica Suecica, principles. K. Svensk. Vet. Akad. Handl. (3), 18, Nr. 6, 1940
[270] *Genieser, K.,* Neue Daten zur Flußgeschichte der Elbe. Eiszeitalter und Gegenwart, **13**, 141–56, 1962
[271] *Gerasimov, I. P.,* and *N. S. Chebotareva,* Das absolute Alter der letzten (Valdai-)Vereisung im Nordwestteil des Russischen Flachlandes. Izv. AN, ser. geogr., **1963**, Nr. 5, 36–44, 1963
[272] *Gerasimov, I. P.,* and *K. K. Markov,* Die Eiszeit auf dem Territorium der UdSSR. Trudi Geogr., **33**, 1939
[273] *Gerasimov, I. P., L. R. Serebriannyi* and *N. S. Chebotareva,* Stratigraphische Komponenten des Pleistozäns Nordeuropas und ihre Korrelierung. Antropogen Russk. ravnini i ego stratigraficheskie komponenti. 5–60, Moscow 1963
[274] *Gerbova, V. G.,* and *E. I. Ravskii,* Zum Problem der Stratigraphie der quartären (anthropogenen) Sedimente im westlichen Transbaikalien. Materialy chetvert., **3**, 283–92, 1961
[275] *Gilewska, S.,* and *L. Stuchlik,* Przedwarciański interstadiał z Brzozowicy koło Będzina. Monographiae Botanicae, **7**, 69–93, Cracow 1958
[276] *Giterman, R. E.,* Pollenspektren quartärer Sedimente im Süden und Osten des sibirischen Hochplateaus. Trudi Geol., **31**, 64–84, 1960
[277] ——— Etappen in der Entwicklung der quartären Vegetation Jakutiens und ihre Bedeutung für die Stratigraphie. Trudi Geol., **78**, 1963
[278] *Givulescu, R., V. Ghiurcă* and *B. Diaconeasa,* Vorläufige Mitteilung über die pannonische Flora von Chiuzbaia (Bez. Maramures, Rumänien). Neues Jahrb. f. Geol. u. Paläontol., Mh., **1964**, 25–30, 1964
[279] *Glock, W. S.,* Principles and Methods of Tree-ring Analysis. Publ. Nr. 486, Carnegie Inst. of Washingt., 1937
[280] *Godwin, H.,* and *E. H. Willis,* Radiocarbon Dating of the Late-glacial Period in Britain. Proc. Roy. Soc., B, **150**, 199–215, 1959
[281] *Gołąbowa, M.,* Roślinność interglacjalna z Makowa Mazowieckogo. Inst. Geol., Biul., **118**, 91–107, Warsaw 1957
[282] *Goldthwait, R. P.,* Wisconsin Age Forests in Western Ohio. I. Age and Glacial Events. Ohio J. of Sci., **58**, 209–19, 1958
[283] ——— Scenes in Ohio during the Last Ice Age. Ohio J. of Sci., **59**, 193–216, 1959
[284] *Golubeva, L. V.,* and *E. I. Ravskii,* Das Anthropogen der Tunka-Senke. Trudi Kom. po izuch. chetvert. per., **19**, 240–59, 1962
[285] ——— Über die Klimaphasen während der Zyrjanka-Vereisung Ost-Sibiriens. Biull. chetvert., **29**, 132–48, 1964
[286] *Gooding, A. M.,* Illinoian and Wisconsin Glaciations in the Whitewater Basin, Southeastern Indiana, and Adjacent Areas. Journ. of Geol., **71**, 665–82, 1963
[287] *Goreckii, G. I.,* Über die Grenze zwischen dem Neogen und dem Anthropogen. Mezhdunarodn. geol. kongr., 21. sess. 1960, dokladi sovetsk. geol., 19–26, 1960; Engl. summary
[288] ——— Zur Lösung des Problems der Untergrenze des Anthropogens. Trudi Kom. po izuch. chetvert. per., **20**, 25–46, 1962

[289] *Gorlova, R. N., E. P. Metel'ceva, A. K. Nedoseeva* and *V. N. Sukachev,* Über interglaziale Sedimente mit einer fossilen Flora bei Tutaev an der Volga. Biull. Mosk. Obshch. Ispyt. Prirodi, otdel biol., **67**, Nr. 1, 59–82, 1962

[290] *Gorlova, R. N., E. P. Metel'ceva, V. A. Novskij* and *V. N. Sukachev,* Über interglaziale Sedimente in der Umgebung von Rybinsk in der Jaroslavl-Oblast'. DAN, **140**, 1427–30, 1961

[291] *Gorodeckaia, M. E.,* Spuren einer ehemaligen ewigen Gefrornis in der Pavlodar'-Oblast'. Izv AN, ser. geogr., **1958**, Nr. 5, 65–72, 1958

[292] *Gradmann, R.,* Das Pflanzenleben der Schwäbischen Alb. Stuttgart 1898

[293] *Graul, H.,* and *K. Brunnacker,* Eine Revision der pleistozänen Stratigraphie des schwäbischen Alpenvorlandes. Petermanns Geogr. Mitteil., **1962**, 253–71, 1962

[294] *Grebe, H.,* Die Mikro- und Megaflora der pliozänen Ton- und Tongyttjalinse in den Kieseloolithschichten vom Swisterberg (Weilerswist, Blatt Sechtem) und die Altersstellung der Ablagerung im Tertiär der Niederrheinischen Bucht. Geol. Jahrb., **70**, 535–74, 1955

[295] *Grichuk, M. P.,* Zur Vegetationsgeschichte im Flußgebiet der Angara. DAN, **102**, 335–38, 1955

[296] ——— Ergebnisse paläobotanischer Untersuchungen pleistozäner Sedimente im Angaragebiet. In: *Markov, K. K.,* and *A. I. Popov,* Lednikovyi period na territorii evropeiskoi chasti SSSR i Sibiri. 442–97, Moscow 1959

[297] *Grichuk, M. P.,* and *V. P. Grichuk,* Über die periglaziale Vegetation im Gebiet der UdSSR. In: *Markov, K. K.,* and *A. I. Popov,* Perigliacial'nye yavleniia na territorii SSSR, 66–100, Moscow 1960

[298] *Grichuk, V. P.,* Zur Vegetationsgeschichte des Europäischen Teiles der UdSSR im Quartär. Trudi Geogr. **37**, 249–66, 1946

[299] ——— Die Vegetation des Russischen Flachlandes im Früh- und Mittelquartär. Trudi Geogr., **46**, 5–202, 1950

[300] ——— Historische Etappen in der Evolution der Vegetation des Südostens des Europäischen Teiles der UdSSR im Quartär. Trudi Geogr., **50**, 5–74, 1951

[301] ——— Beiträge zur paläobotanischen Charakterisierung pleistozäner und pliozäner Sedimente im Nordwestteil der Prikaspijskaja Nizmennost'. Trudi Geogr., **61**, 5–79, 1954

[302] ——— Vorläufige Angaben über eine paläobotanische Erforschung der Sedimente des jungen oberpleistozänen Interglazials an der Balazna. DAN, **137**, 380–83, 1961

[303] ——— Die fossile Flora als paläontologische Grundlage für die stratigraphische Gliederung quartärer Sedimente. In: *Markov, K. K.,* Rel'ef i stratigrafiia chetvertichnych otlozhenii severo-zapada russkoi ravniny. 25–71, Moscow 1961

[304] ——— Über das geologische Alter archäologischer Reste, datiert mit Hilfe paläobotanischen Materials. Trudi Kom. po izuch. chetvert. per., **18**, 146–56, 1961

[305] ——— Comparative Study of the Interglacial and Interstadial Flora of the Russian Plain. Report VI Internat. Congr. Quaternary Warsaw 1961, **2**, 395–406, Łódź 1964

[306] *Grichuk, V. P.*, and *M. P. Grichuk,* Altlakustrine Sedimente bei Pljos. In: *Markov, K. K.*, and *A. I. Popov,* Lednikovyi period na territorii evropeiskoi chasti SSSR i Sibiri. 39–63, Moscow 1959

[307] *Grigor'ev, A. A.*, Die ewige Gefrornis und die ehemalige Vergletscherung. Akad. Nauk SSSR, Kom. po izuch. estestven. proizvod. sil SSSR, Materialy **80**, Sbornik "Vechnaia Merzlota," 43–104, 1930

[308] ——— Die atmosphärische Zirkulation während der maximalen Vergletscherung, als Grundlage zur Rekonstruktion des Klimas der Eiszeit. Trudi Geogr., **37**, 1946

[309] *Gripp, K.*, and *M. Beyle,* Das Interglazial von Billstedt (Öjendorf). Mitt. Geol. Staatsinst. Hamburg, **16**, 19–36, 1937

[310] *Grishchenko, M. N.*, Kurze Mitteilung über die geologischen Lagerungsverhältnisse einer neuentdeckten paläolithischen Station bei Stalingrad. Biull. chetvert., **18**, 87–89, 1953

[311] ——— Die Geologie der Station Suchaja Mechetka bei Volgograd an der Volga, und der Station Rozhok I im Asow-Gebiet. Stratigrafiia i periodizaciia paleolita vostochnoi i central'noi Evropi. 141–56, Moscow 1965

[312] *Gross, H.*, Das Alleröd-Interstadial als Leithorizont der Letzten Vereisung in Europa und Amerika. Eiszeitalter und Gegenwart, **4/5**, 189–209, 1954

[313] ——— Die Fortschritte der Radiocarbon-Methode 1952–1956. Eiszeitalter und Gegenwart, **8**, 141–80, 1957

[314] ——— Die bisherigen Ergebnisse von ^{14}C-Messungen und paläontologischen Untersuchungen für die Gliederung und Chronologie des Jungpleistozäns in Mitteleuropa und den Nachbargebieten. Eiszeitalter und Gegenwart, **9**, 155–87, 1958

[315] ——— Der gegenwärtige Stand der Geochronologie des Spätpleistozäns in Mittel- und Westeuropa. Quartär, **14**, 49–68, 1962/1963

[316] ——— Das Mittelwürm in Mitteleuropa und angrenzenden Gebieten. Eiszeitalter und Gegenwart, **15**, 187–98, 1964

[317] *Gubonina, Z. P.*, Zum Altersproblem der Sedimente der 3. Terrasse oberhalb der Volga-Aue bei Stavropol'. DAN, **135**, 921–24, 1960

[318] ——— Ergebnisse pollenanalytischer Untersuchungen fluviatiler Sedimente der Volga bei Stavropol'. Voprosy paleogeografii i geomorfologii bassejnov Volgi i Urala, 99–121, Moscow 1962

[319] ——— Artbestimmung der Sporomorphen und ihre Bedeutung für die Lösung stratigraphischer Probleme des Quartärs. Sistematika i metodi izucheniia iskopaemich pyl'ci i spor. 190–97, Moscow 1964

[320] *Gudelis, V. K.*, Überblick über die Geologie und Paläogeographie des Quartärs (Anthropogens) in Litauen. Czwartorzęd Europy środkowej i wschodniej, I, 423–97, Warsaw 1961; Engl., Pol. summaries

[321] *Gullentops, F.*, Stratigraphie du pleistocène supérieur en Belgique. Geol. en Mijnb., n. s. **19**, 305, 1957

[322] *Haase, G.*, Die Höhenstufen der Böden im Changai (MVR). Zs. f. Pflanzenernährg., Düngung und Bodenkde., **102**, 113–27, 1963

[323] ——— Stand und Probleme der Lößforschung in Europa. Geogr. Ber., Mitt. d. Geogr. Ges. der DDR, **27**, 97–129, 1963

[324] *Haase, G.*, and *H. Richter,* Fossile Böden im Löß an der Schwarzmeerküste

bei Constanta. Petermanns Geogr. Mitt., **101**, 161–73, 1957
[325] *Hafsten, U.*, Pleistocene Development of Vegetation and Climate in the Southern High Plains as Evidenced by Pollen Analysis. Mus. of New Mexico, Nr. 1, Paleoecology of the Llano Estacado, 59–91, 1961
[326] ——— A Late-glacial Pollen-profile from Lista, South Norway. Grana Palynologica, **4**, 2, 326–37, 1963
[327] *Hahn, O.*, Neuere radioaktive Methoden zu geologischen und biologischen Altersbestimmungen. Naturwiss. Rundschau, **8**, 331–37, 1955
[328] *Hallik, R.*, and *K. Kubitzki*, Über die Vegetationsentwicklung des Weichsel-Interstadials aus Hamburg-Bahrenfeld. Eiszeitalter und Gegenwart, **12**, 92–98, 1962
[329] *Hammen, Th. van der*, The Quaternary Climatic Changes of Northern South America. Ann. New York Acad. Sci., **95**, 676–83, 1961
[330] ——— A Palynological Study on the Quaternary of British Guiana. Leidse Geol. Mededel., **29**, 125–80, 1963
[331] ——— Problems of Quaternary Botany in the Tropics (with Special Reference to South America). Ber. Geobot. Inst. d. ETH, Stiftung Rübel, **34**, 62, 1963
[332] *Hammen, Th. van der*, and *E. Gonzalez*, Upper Pleistocene and Holocene Climate and Vegetation of the "Sabana de Bogotá" (Colombia, South-America). Leidse Geol. Mededel., **25**, 261–315, 1960
[333] ——— A Pollen Diagram from the Quaternary of the Sabana de Bogotá (Colombia) and its Significance for the Geology of the Northern Andes. Geol. en Mijnb., **43**, 113–17, 1964
[334] *Hammen, Th. van der, T. A. Wijmstra*, and *W. H. van der Molen*, Palynological Study of a very Thick Peat Section in Greece, and the Wurm-glacial Vegetation in the Mediterranean Region. Geol. en Mijnb., **44**, 37–39, 1965
[335] *Heck, H. L.*, Zur Fossilführung der Berliner Paludinenschichten, ihrer Beschaffenheit und Verbreitung. Zs. dtsch. geol. Ges., **82**, 385–404, 1930
[336] ——— Die Eem- und ihre begleitenden Junginterglazial-Ablagerungen bei Oldenbüttel in Holstein. Abh. d. Preuß. Geol. Landesanst., N. F., **140**, 1932
[337] *Heide, S. van der*, Correlations of Marine Horizons in the Middle and Upper Pleistocene of the Netherlands. Geol. en Mijnb., n. s. **19**, 272–76, 1957
[338] *Heim, J.*, Recherches sur les relations entre la végétation actuelle et le spectre pollinique récent dans les Ardennes Belges. Bull. Soc. Roy. Bot. Belgique, **96**, 5–92, 1962
[339] *Hellmann, G.*, Über die ägyptischen Witterungsangaben im Kalender von Claudius Ptolemaeus. Sitz.-Ber. d. kgl. preuß. Akad. d. Wiss., 1. Halbbd., 332–41, 1916
[340] *Henriksen, K. L.*, Undersøgelser over Danmark-Skånes kvartaere insektfauna. Vidensk. Meddelels. fra Dansk naturhist. foren. i København, **96**, 77–355, 1933
[341] *Heusser, C. J.*, Late-pleistocene Environments of North Pacific North America. Amer. Geogr. Soc., Special Publ., Nr. **35**, New York 1960

[342] ——— Palynology of Four Bog Sections from the Western Olympic Peninsula, Washington. Ecology, **45**, 23–40, 1964

[343] *Hey, R. W.*, Pleistocene Screes in Cyrenaica (Libya). Eiszeitalter und Gegenwart, **14**, 77–84, 1963

[344] *Heydenreich, S.*, Pollenanalytische Untersuchungen von Flußsedimenten des Mains bei Marktheidenfeld. Staatsexamens-Arbeit Würzburg 1959

[345] *Hibbard, C. W.*, Pleistocene Vertebrate Paleontology in North America. Bull. Geol. Soc., **60**, 1417–28, 1949

[346] *Hiltermann, H.*, Neue Funde von *Azolla* im Pleistozän Deutschlands. Geol. Jahrb., **68**, 653–58, 1954

[347] *Hoinkes, H.*, Die Antarktis und die geophysikalische Erforschung der Erde. Naturwiss., **48**, 354–74, 1961

[348] *Hollingworth, S. E., J. Allison,* and *H. Godwin,* Interglacial Deposits from the Histon Road, Cambridge. Quat. J. Geol. Soc. London, **105**, f. 1949, 495–509, 1950

[348a] *Hollstein, E.*, Über den gegenwärtigen Stand der westdeutschen Eichenchronologie. Kunstchronik, **21**, 159–64, 1968

[349] *Holmsgaard, E.*, Årringanalyser af danske skovtraeer. Det forstl. forsøgsvaesen i Danmark, **22**, 1, 1955

[350] *Hopkins, D. M., F. S. Macneil,* and *E. B. Leopold,* The Coastal Plain at Nome, Alaska: A Late Cenozoic Type Section for the Bering Strait Region. Report Internat. Geol. Congr., 21 sess., 1960, **4**, Chronology and Climatology of the Quaternary, 46–57, Copenhagen 1960

[351] *Hopkins, D. M., F. S. Macneil, R. L. Merklin,* and *O. M. Petrov,* Quaternary Correlations across Bering Strait. Science, **147**, 1107–14, 1965

[352] *Huber, B.*, Die Jahresringe der Bäume als Hilfsmittel der Klimatologie und Chronologie. Naturwiss., **35**, 151–54, 1948

[353] ——— Dendrochronologie. Geol. Rundsch., **49**, 120–31, 1960

[354] *Huber, B.*, and *W. Holdheide*, Jahrringchronologische Untersuchungen an Hölzern der bronzezeitlichen Wasserburg Buchau am Federsee. Ber. D. Bot. Ges., **60**, 261–83, 1942

[355] *Hultén, E.*, Atlas över växternas utbredning i norden. Stockholm 1950

[356] *Huntington, E.*, The Climatic Factor as Illustrated in Arid Arizona. Publ. Nr. 192. Carnegie Inst. of Washingt., 1914

[357] ——— Tree Growth and Climatic Interpretations. Publ. Nr. 352, Carnegie Inst. of Washingt., 155–204, 1925

[358] *Hutchinson, G. E., U. M. Cowgill, W. van Zeist,* and *H. E. Wright,* Preliminary Pollen Studies at Lake Zeribar, Zagros Mountains, Southwestern Iran. Science, **140**, 65–69, 1963

[359] *Ignatenko, I. V.*, Die Böden der arktischen Tundra der Jugor-Halbinsel. Pochvovedenie, **5**, 26–40, 1963

[360] *Ivanova, I. K.*, Die stratigraphische Lage der paläolithischen Station Molodovo am mittleren Dnestr im Lichte allgemeiner Probleme der Stratigraphie und der absoluten Geochronologie des oberen Pleistozäns Europas. Stratigrafiia i periodizaciia paleolita vostochnoi i central'noi Evropi. 123–40, Moscow 1965

[361] *Ivanova, N. G.*, Zum Altersproblem der Terrassen am Mittellauf der Viatka.

Biull. Mosk. Obshch. Ispit. Prir., otd. geol., **37**, 1, 111–19, 1962

[362] *Iversen, J.*, *Viscum, Hedera* and *Ilex* as Climatic Indicators. Geol. Fören. i Stockh. Förhandl., **66**, 463–83, 1944

[363] ——— The Bearing of Glacial and Interglacial Epochs on the Formation and Extinction of Plant Taxa. Uppsala Univ. Årsskrift, 1958, 210–15, 1958

[364] *Jachimovich, V. L.*, Anthropogene (quartäre) Sedimente des südlichen Ural-Vorlandes. Antropogen juzhnogo Urala, 8–54, Moscow 1965

[365] *Jacquiot, C.*, Détermination de bois fossiles provenant de la grotte de Lascaux, Montignac-sur-Vizère (Dordogne). Bull. Soc. Bot. France, **107**, 15–17, 1960

[366] *Jahn, A.*, Zjawiska krioturbacyjne współczesnej i plejstoceńskiej strefy peryglacjalnej. Acta Geol. Polon., **2**, 159–290, 1951

[367] *Jamnov, A. A.*, Über die Ursachen der Überschwemmung des Ssarykamysch-Beckens im Mittelalter und über das Alter der Ssarykamysch-Sedimente mit *Cardium edule*. Izv. AN, ser. geogr., 1953, 4, 61–63, 1953

[368] *Jamnov, A. A.*, and *V. N. Kunin*, Einige theoretische Ergebnisse der neuesten Forschungen im Gebiet des Usboi zur Paläogeographie und Geomorphologie. Izv. AN, ser. geogr., 1953, 3, 21–29, 1953

[369] *Jancik-Kopikowa, S.*, Die Flora des Eem-Interglazials im Gebiet von Kalisch. Report VI Internat. Congr. Quaternary Warsaw 1961, **2**, 429–31, Łódź 1964 (Russ.)

[370] *Jánossy, D.*, *S. Kretzoi-Varrók*, *M. Herrmann* and *L. Vértes*, Forschungen in der Bivakhöhle, Ungarn. Eiszeitalter und Gegenwart, **8**, 18–36, 1957

[371] *Jaroń, B.*, Analiza pyłkowa interglacjału z Żydowszczyzny koło Grodno. Rocznik Polsk. Towarzystwa Geol., **9**, 147–83, Cracow 1933; Ger. summary

[372] *Jatajkin, L. M.*, Flora und Vegetation des Kinel' am Unterlauf der Kama. DAN, **136**, 911–14, 1961

[373] ——— Umgelagerter Pollen und die Anwendung der Korrelations-Analyse für seine Erkennung. Sistematika i metody izucheniia iskopaemich pil'ci i spor. 169–76, Moscow 1964

[374] *Jazewitsch, W. von*, Zur klimatologischen Auswertung von Jahrringkurven. Forstwiss. Centralbl., **80**, 175–90, 1961

[375] *Jazewitsch, W. von*, *H. Siebenlist* and *G. Bettag*, Eine Synchronisiermaschine zum Vergleich von Jahrringkurven und eine langjährige Eichenchronologie. Ber. D. Bot. Ges., **69**, 128–42, 1956

[376] *Jelgersma, S.*, A. Late-glacial Pollendiagram from Madelia, South-Central Minnesota. Amer. J. Sci., **260**, 522–29, 1962

[377] *Jentys-Szaferowa, J.*, Importance of Quaternary Materials for Research on the Historical Evolution of Plants. Veröff. Geobot. Inst. Rübel in Zürich, **34**, 67–73, 1958

[378] ——— Morphological Investigations of the Fossil *Carpinus* Nutlets from Poland. Acta Palaeobotanica, **1**, Nr. 1, Cracow 1960

[379] ——— Anatomical Investigations on Fossil Fruits of the Genus *Carpinus* in Poland. Acta Palaeobotanica, **2**, Nr. 1, Cracow 1961

[380] ——— Metody biometryczne w badaniu ewolucji historyczney roślin. Acta Soc. Bot. Polon., **33**, 77–94, 1964

[381] *Jessen, K.*, and *V. Milthers*, Stratigraphical and Palaeobotanical Studies of

Interglacial Fresh-water Deposits in Jutland and Northwest Germany. Danmarks Geol. Unders., II. raekke, Nr. 48, 1928
[382] *Jessen, K., S. Th. Andersen,* and *A. Farrington,* The Interglacial Deposit near Gort, Co. Galway, Ireland. Proc. Roy. Irish Acad., **60**, Sect. B, Nr. 1, 1959
[383] *Jimbô, T.,* Die heutige palynologische Kenntnis von Ablagerungen der Vergangenheit in Japan. 1.) Zusammenfassende Darstellung über die Angaben, die tertiäre Ablagerungen betreffen. Ecological Review, **14**, 329–41, 1958
[384] *Kac, N. Y.* Die Entwicklung der Wälder und das Klima im mittleren und späten Pleistozän und im Holozän. Biull. Mosk. Obshch. Ispit. Prirodi, otd. biol., **60**, Nr. 3, 49–69, 1955
[385] ——— Ein geologisches Denkmal der Riß-Würm-Zeit mit heute in Europa ausgestorbenen Pflanzen; der Aufschluß bei Korenevo in der Moskau-Oblast'. Nauchn. sess. posviashchen. 100-letiju rozhdenija G. I. *Tanfil'eva,* Tezisi dokladov, 27–28, Odessa 1957
[386] ——— Über den Parallelismus klimatischer und biologischer Erscheinungen. Nauchn. sess. posvjashchen. 100-letiju rozhdenija *G. I. Tanfil'eva,* Tezisi dokladov, 29–30, Odessa 1957
[387] ——— Die Wärmewelle des späten Pleistozäns und die Entwicklung der Vegetation. Izv AN, ser. geogr., 1959, Nr. 6, 77–81, 1959
[388] *Kac, N. Y.* and *S. V. Kac,* Fossile Flora und Vegetation der mindel-riß-interglazialen Sedimente bei Żydowszczyzna nahe Grodno. Biull. chetvert., **25**, 35–49, 1960
[389] ——— Über interglaziale Sedimente bei Rozdol der Drogobychsker Oblast'. Biull chetvert., **26**, 61–73, 1961
[390] ——— Neue Beiträge zur Kenntnis der Flora des Quartärs und ihrer Veränderungen im Ablauf der Zeit. Materialy chetvert., **1**, 331–35, 1961
[391] *Kaiser, K. H.,* Klimazeugen des periglazialen Dauerfrostbodens in Mittel- und Westeuropa. Ein Beitrag zur Rekonstruktion des Klimas der Glaziale des quartären Eiszeitalters. Eiszeitalter und Gegenwart, **11**, 121–41, 1960
[392] *Kapp, R.,* and *A. M. Gooding,* Pleistocene Vegetational Studies in the Whitewater Basin, Southeastern Indiana. Journ. of Geol., **72**, 307–26, 1964
[393] *Karavaev, M. N.,* Konspekt Flori Yakutii. Moscow-Leningrad 1958
[394] *Karavaev, M. N.,* and *A. I. Popova,* Neue Angaben über Pollenspektren neogener Sedimente Nordost-Asiens. Biull. Mosk. Obshch. Ispit. Prirodi, otd. biol., **60**, Nr. 6, 107–13, 1955
[395] *Kashmenskaia, O. V.,* Quartäre Sedimente im Flußgebiet des Berelech (Oberlauf der Kolyma). Materialy chetvert., **3**, 144–46, 1961
[396] *Kempton, I. P.,* and *R. P. Goldthwait,* Glacial Outwash Terraces of the Hocking and Scioto River Valleys, Ohio. Ohio J. of Sci., **59**, 135–51, 1959
[397] *Kes',* A. S., Die Entstehung des Usboi. Izv AN, ser. geogr., 1952, H. 1, 14–26, 1952
[398] ——— Paläogeographische Probleme der Unterläufe des Amu darja und des Syr darja. Tezisi; Sekc. Kazachstana i Srednej Azii, 7–9, 1957
[399] *Kessler, P.,* Das eiszeitliche Klima und seine geologischen Wirkungen im nicht vereisten Gebiet. Stuttgart 1925

[400] *Kind, N. V.*, Einige Bemerkungen zur Position der Karginsker Zeit im chronologischen Schema des Jung-Pleistozäns. Biull. chetvert., **28**, 169–70, 1963

[401] ——— Absolute Chronologie der wichtigsten Etappen in der Geschichte der Letzten Eiszeit und des Postglazials in Sibirien (auf Grund der ^{14}C-Methode). Chetvertichni period i ego istoriia, 157–74, Moscow 1965; Engl. summary

[402] *Kinzl, H.*, Beiträge zur Geschichte der Gletscherschwankungen in den Ostalpen. Zs. f. Gletscherkunde, **17**, 66–121, 1929

[403] ——— Die größten nacheiszeitlichen Gletschervortöße in den Schweizer Alpen und in der Mont Blanc-Gruppe. Zs. f. Gletscherkunde, **20**, 269–397, 1932

[404] *Kirk, W.*, and *H. Godwin*, A Late-glacial Site at Loch Droma, Ross and Cromarty. Transact. Roy. Soc. Edinburgh, **65**, Nr. 11, 225–49, 1963

[405] *Klaus, W.*, Zur pollenanalytischen Datierung von Quartärsedimenten im Stadtgebiet von Wien, südlichen Wiener Becken und Burgenland. Verhandl. d. Geol. Bundesanstalt, 1962, 20–38, Vienna 1962

[406] *Klein, A.*, Die Niederschläge in Europa im Maximum der Letzten Eiszeit. Versuch einer Rekonstrucktion aus dem Höhenunterschied zwischen damaliger und heutiger Schneegrenze. Petermanns Geogr. Mitt., **97**, 98–104, 1953

[407] *Klíma, B.*, *J. Kuklá*, *V. Lozhek* and *H. de Vries*, Stratigraphie des Pleistozäns und Alter des paläolithischen Rastplatzes in der Ziegelei von Dolní Věstonice (Unter-Wisternitz). Anthropozoikum, **11**, 93–145, Prague 1961

[408] *Klute, F.*, Das Klima Europas während des Maximum der Weichsel-Würmeiszeit und die Änderungen bis zur Jetztzeit. Erdkunde, **5**, 273–83, 1951

[409] *Knebelsberg, R. v.*, Handbuch der Gletscherkunde und Glazialgeologie. **1**, 1948; **2**, 1949

[410] *Kneblová, V.*, Fytopaleontologický rozbor uhlíků z paleolítického sídliště v Dolních Věstonicích. Anthropozoikum, **3**, 297–99, Prague 1953

[411] ——— Die paläobotanische Erforschung der Travertine des „Hrádok" in Gánovce. Veröff. Geobot. Inst. ETH Zürich, Stiftung Rübel, **36**, 164–70, 1960

[412] ——— Paleobotanický výzkum interglaciálních travertinů v Gánovcích. Biologické práce, **6**, Nr. 4, 1960

[413] *Kneblová-Vodičková, V.*, Zpráva o paleobotanickém výzkumu sedimentů v okolí Č. Těšína. Zprávy o geologických výzkumech v r. 1960, 170, 1960

[414] ——— Entwicklung der Vegetation des Elster-Saale-Interglazials im Suchá-Stonava-Gebiet (Ostrava-Gebiet). Anthropozoikum, **9**, 129–74, Prague 1961

[415] ——— Flora. Czwartorzed Europy środkowej i wschodniej, **1**, 125–32, Warsaw 1961; Russ., Pol. summaries

[416] ——— Die jungpleistozäne Flora aus Sedimenten bei Český Těšin (letztes Glazial). Preslia, **35**, 52–64, 1963

[417] *Knetsch, G.*, *A. Shata*, *E. Degens*, *K. O. Münnich*, *J. C. Vogel* and *M. M. Shazly*, Untersuchungen an Grundwässern der Ost-Sahara. Geol. Rundsch., **52**, 587–610, 1963

[418] *Knoth, W.*, Zur Kenntnis der pleistozänen Mittelterrassen der Saale und Mulde nördlich von Halle. Geologie, **13**, 598–616, 1964

[419] *Knox, A. S.*, Pollen from the Pleistocene Terrace Deposits of Washington, D.C., Pollen et Spores, **4**, 357–58, 1962

[420] *Kokawa, Sh.*, Morphometric Reconstruction of the Compressed Seed Remains of *Menyanthes* in Japan. Journ. of the Inst. Polytechn., Osaka City Univ., Ser. D, **11**, 79–89, 1960

[421] ——— Distribution and Phytostratigraphy of *Menyanthes* Remains in Japan. Journ. of Biol., Osaka City Univ., **12**, 123–51, 1961

[422] ——— Age Effect on the Morphometric Values of the Fossil *Menyanthes* Seeds in Japan Represented by the Szaferowa's Graphic Method. Journ. of Biol., Osaka City Univ., **13**, 87–97, 1962

[423] *Kolakowskii, A. A.*, Vorläufige Mitteilung über den Fund einer pliozänen Flora in West-Grusinien. Bot. Zhurn., **36**, 408–10, 1951

[424] *Kolumbe, E.*, Neue Untersuchungen an interglazialen Torfen von Hamburg und Burg in Dithmarschen. Mitt. Geol. Staatsinst. Hamburg, **21**, 46–58, 1952

[425] *Kolumbe, E.*, and *M. Beyle*, Neue Interglaziale aus Schleswig-Holstein und Hamburg. Mitt. Geol. Staatsinst. Hamburg, **17**, 59–74, 1940

[426] *Kondratene, O. P., N. S. Chebotareva, A. L. Devirc* and *E. I. Dobkina*, Interstadiale Sedimente im südlichen Litauen. Paleogeografiia i chronologiia verchnego pleistocena i golocena po dannym radiouglerodnogo metoda. 42–51, Moscow 1965

[427] *Konior, K.*, Z badań nad czwartorzędem przedgórza karpackiego między Tarnowem a Dębica. Rocznik Polsk. Towarzystwa Geol., **12**, 353–81, Cracow 1936

[428] *Koperowa, W.*, Późnoglacjalna i holoceńska historia roślinności kotliny Nowotarskiej. Acta Paleobotanica, **2**, Nr. 3, Cracow 1962

[429] *Kopp, K. O.*, Schneegrenze und Klima der Würmeiszeit an der baskischen Küste. Eiszeitalter und Gegenwart, **14**, 188–207, 1963

[430] *Korchagina, I. A.*, Frühquartäre Samenfloren am Unterlauf des Irtish. Bot. Zhurn., **43**, 1121–34, 1958

[431] *Kovanda, J.*, Stratigrafická studie mladokvartérních limnických sedimentů v Opavě-Kateřinkách. Sbornik geol. věd, Antropozoikum, řada A, Sv. 2, 85–111, Prague 1964; Ger. summary

[432] *Kozii, G. V.*, Fossile Flora aus Stationen des prähistorischen Menschen in West-Podolien, Biull. chetvert., **20**, 71–76, 1955

[433] *Kozłowska, A.*, Flora międzylodocowa z pod Rakowa. Acta Soc. Bot. Polon., **1**, 213–32, 1923

[434] ——— Zur Frage des Vorkommens der Gattung *Tsuga* im polnischen Interglazial. Öster. Bot. Zs., **75**, 42–46, 1926

[435] *Kräusel, R.*, Pflanzenreste aus den diluvialen Ablagerungen im Ruhr-Emscher-Lippe-Gebiet. Decheniana, **95** A, 207–40, 1937

[436] *Kremp, G.*, Pollenanalytische Braunkohlenuntersuchungen im südlichen Teil Niedersachsens, insbesondere im Solling. Geol. Jahrb., **64**, 489–517 (1943–48), 1950

[437] *Kriger, N. I.*, and *V. P. Kopylova*, Über pleistozäne „Frost"- und „Trocken"-Risse im Balchaschsee-Gebiet. Biull. chetvert., **29**, 183–88, 1964

[438] *Kubiëna, W. L.*, Über die Braunlehmrelikte des Atakor. Erdkunde, **9**, 115–32, 1955

[439] ——— Paleosoils as Indicators of Paleoclimates. Changes of Climate; Proc. Rome Sympos., Organized by UNESCO and the WMO, 207–9, Paris 1963

[440] *Kubitzki, K.*, and *K. O. Münnich*, Neue ^{14}C-Datierungen zur nacheiszeitlichen Waldgeschichte Nordwestdeutschlands. Ber. D. Bot. Ges., **73**, 137–46, 1960

[441] *Kukla, J.*, and *V. Ložek*, Loesses and Related Deposits. Czwartorzęd Europy środkowej i wschodniej, **1**, 11–28, Warsaw 1961; Russ., Pol. summaries

[442] *Kulczyński, St.*, Flora międzylodowcowa z Timoszkowicz w Nowogródzkiem. Sprawozdanie Kom. Fizjograficz., Polska Akad. Umiej, **63**, 241–52, 1929

[443] ——— Die altdiluvialen *Dryas*floren der Gegend von Przemyśl. Acta Soc. Bot. Polon., **9**, 237–99, 1932

[444] *Kulp, J. L.*, Climatic Changes and Radioisotope Dating. In: *Shapley, H.*, Climatic Change. Evidence, Causes, and Effects. 201–8, Cambridge 1953

[445] *Kunica, N. A.*, Über die Ausnutzung der Molluskenfauna für eine Entschleierung der Bildungsverhältnisse und -weisen der Lösse des mittleren Dnepr-Gebietes. Materialy chetvert., **1**, 192–97, 1961

[446] *Kurdiukov, K. V.*, Alte Seebecken des südöstlichen Kazachstan und die klimatischen Verhältnisse zur Zeit, als sie existierten. Izv AN, ser. geogr., 1952, Nr. 2, 11–24, 1952

[447] *Kuznecova, T. A.*, Neue Angaben über das Pliozän im Gebiet von Russkaja Bektjaška am rechten Ufer der Volga. DAN, **148**, 668–71, 1963

[448] *Laatsch, W.*, Dynamik der mitteleuropäischen Mineralböden. 4. Aufl., Dresden and Leipzig 1957

[449] *Lang, A.*, Untersuchungen über einige Verwandtschafts- und Abstammungsfragen in der Gattung *Stachys* L. auf cytogenetischer Grundlage. Bibliotheca Botanica, **118**, Stuttgart 1940

[450] *Lang, G.*, Die spät- und frühpostglaziale Vegetationsentwicklung im Umkreis der Alpen. Bemerkungen zur Arbeit von *H. Zoller:* Pollenanalytische Untersuchungen zur Vegetationsgeschichte der Insubrischen Schweiz, 1960. Eiszeitalter und Gegenwart, **12**, 9–17, 1962

[451] *Lauer, W.*, Humide und aride Jahreszeiten in Afrika und Südamerika und ihre Beziehung zu den Vegetationsgürteln. Bonner Geogr. Abhandl., **9**, 15–98, 1952

[452] *Laurent, L.*, and *P. Marty*, Flore pliocène des cinérites du Pas-de-la-Mougudo et de Saint-Vincent-La Sabie (Cantal). Ann. Mus. d'hist. nat. Marseille, géol., **9**, Marseilles 1904–5

[453] *Lavrova, M. A.*, Über die geographische Verbreitung des „borealen Meeres" und über seine physisch-geographischen Verhältnisse. Trudi Geogr., **37**, 64–79, 1946

[454] *Lavrova, M. A.*, and *M. P. Grichuk*, Neue Beiträge zur Kenntnis der marinen Sedimente der interglazialen Mga-Transgression. DAN, **135**, 1472–75, 1960

[455] *Lavrushin, Y. A.*, Stratigraphie und einige Besonderheiten in der Bildung

der quartären Sedimente am Unterlauf der Indigirka. Izv AN, ser. geol., 1962, Nr. 2, 73–87, 1962

[456] *Lazukov, G. I.*, Über Synchronie und Metachronie quartärer Vergletscherungen und Transgressionen. Paleogeografiia chetvertichnogo perioda SSSR, 139–49, Moscow 1961

[457] *Lemée, G.*, Successions forestières contemporaines du dépôt des lignites quaternaires dans la cluse de Chambéry. C.R. Acad. Sci., **215**, 23–25, 1942

[458] *Lemée, G.*, and *F. Bourdier*, Une flore pollinique tempérée inclue dans les moraines dites wurmiennes d'Armoy près de Thonon (Haute-Savoie). C. R. Acad. Sci., **230**, 2313–14, 1950

[459] *Leonard, A. B.*, and *J. C. Frye*, Ecological Conditions Accompanying Loess Deposition in the Great Plains Region of the United States. Journ. of Geol., **62**, 399–404, 1954

[460] *Leont'ev, O. K.*, and *P. V. Fedorov*, Zur Geschichte des Kaspischen Meeres im Spät- und Postglazial. Izv AN, ser. geogr., 1953, Nr. 4, 64–74, 1953

[461] *Leopold, E. B.*, and *D. R. Crandell*, Pre-Wisconsin Interglacial Pollen Spectra from Washington State, USA. Veröff. Geobot. Inst. Rübel in Zürich, **34**, 76–79, 1958

[462] *Leopold, L. B.*, Pleistocene Climate in New Mexico. Amer. J. Sci., **249**, 152–68, 1951

[463] *Leroi-Gourhan, A.*, Flores et climates du paléolithique récent. Compte Rendu du Congr. préhist. France, 1–6, Le Mans 1960

[464] ——— Chronologie des grottes d'Arcy-sur-Cure (Yonne). Gallia Préhistoire, **7**, 1–64, 1964

[465] *Leschik, G.*, Die Entstehung der Braunkohle der Wetterau und ihre Mikro- und Makroflora. Palaeontographica, **100** B, 26–64, 1956

[466] *Levina, T. P.*, Sporomorphenspektren quartärer Sedimente aus der periglazialen Zone der Samarov-Vereisung (Flußgebiet des Jenisseii). Sistematika i metodi izucheniia iskopaemich pyl'ci i spor., 208–17, Moscow 1964

[467] *Libby, W. F.*, Altersbestimmung mit radioaktivem Kohlenstoff. Endeavour, **13**, Nr. 49, 5–16, 1954

[468] ——— Radiocarbon Dating. 2. Edition, Chicago 1955

[469] *Lieberoth, I.*, Lößsedimentation und Bodenbildung während des Pleistozäns in Sachsen. Geologie, **12**, 149–87, 1963

[470] *Lilpop, J.*, O utworach międzylodowcowych w Olszewicach pod Tomaszowem Mazowieckim. II. Flora utworów międzylodowcowych w Olszewicach. Sprawozdanie Kom. Fizjograf., Polska Akad. Umiej, **64**, 57–75, 1930

[471] *Livingstone, D. A.*, and *B. G. R. Livingstone*, Late-glacial and Postglacial Vegetation from Gillis Lake in Richmond County, Cape Breton Island, Nova Scotia. Amer. J. Sci., **256**, 341–59, 1958

[472] *Löhnert, E.*, Über *Azolla filiculoïdes* LAM. aus dem Holstein-Interglazial von Hamburg-Billstedt. Abh. u. Verhandl. d. Naturwiss. Ver. Hamburg, N. F. **8**, 155–62, 1964

[473] *Lomaeva, E. T.*, Sporomorphenkomplexe der südukrainischen Lösse. Trudi In-ta geol. nauk, Akad. Nauk USSR, ser. geomorf. i chetvert. geol., **1**, 89–94, Kiev 1957

[474] ——— Geschichte der pollenanalytischen Erforschung quartärer (anthropogener) Sedimente der UdSSR. Chetvertichnyj Period, **13**, **14**, **15**, 323–38, Kiev 1961; Engl. summary
[475] ——— Einige Ergebnisse von Pollenanalysen quartärer Sedimente der Ukrainischen SSR. Materialy chetvert., **1**, 304–8, 1961
[476] *Lomonovich, M. I.*, Quartäre Sedimente und die Genese des Lösses im Südostteil Kazachstans, am Beispiel des Trans-Ili-Alatau. Materialy chetvert., **3**, 367–73, 1961
[477] *Lona, F.*, Prime analisi pollinologiche sui depositi terziari-quaternari di Castell'Arquato: reperti di vegetazione da clima freddo sotto le formazioni calcaree ad *Amphistegina*. Boll. della Soc. Geol. Ital., **81**, 3–5, 1963
[478] ———A Cold Oscillation in the Middle of the Pianico-Sellere (Riß-Würm) Series. Ber. d. Geobot. Inst. d. ETH Zürich, Stiftung Rübel, **34**, f. 1962, 69, 1963
[479] *Lopatnikov, M. I.*, Zur Geschichte der Vegetation der Steppenzone des Russischen Flachlandes (nach Untersuchungsergebnissen am Unterlauf des Choper). In: *Markov, K. K.*, and *A. I. Popov*, Lednikovyi period na territorii evropejskoi chasti SSSR i Sibiri, 227–55, 1959
[480] ——— Einige Probleme der Paläogeographie des Flußgebietes des mittleren Don im Neogen und Quartär. Materialy chetvert., **2**, 292–99, 1961
[481] *Lopatnikov, M. I.*, and *S. M. Shik*, Die Lage des jungquartären Eisrandes in der Smolensk- und Kalinin-Oblast'. Materialy po geol. i poleznym iskopaemym centr. raionov evrop. chasti SSSR, **5**, 123–31, 1962
[482] *Lowenstam, H. A.*, and *S. Epstein*, Paleotemperatures of the Post-Aptian Cretaceous as Determined by Oxygen Isotopic Method. Journ. of Geol., **62**, 207–48, 1954
[483] *Ložek, V.*, Měkkýši československého kvartéru. Rozpravy Ustředního ústavu geologického, **17**, Prague 1955
[484] ——— Das Landschaftsbild der Pavlovské vrchy (Pollauer Berge) im Interglazial. Ochrana. Přírody, **12**, 285–88, 1957
[485] ——— Stratigraphische Erforschung des Travertinlagers bei Skřečoň. Anthropozoikum, **9**, 35–45, Prague 1961
[486] ——— Stratigraphische Erforschung der lockeren Sinter- und Moorablagerungen bei Dluhonice im Landstrich Přerov (Prerau). Anthropozoikum, **9**, 65–76, Prague 1961
[487] ——— Interglaciály na Slovensku a jejich význam pro stratigrafii kvartéru. Geologické práce, **64**, 77–92, 1963
[488] ——— Quartärmollusken der Tschechoslowakei. Rozpravy Ustředního ústavu geologického, **31**, Prague 1964
[489] *Ložek, V.*, and *V. Kneblová*, Die paläontologische Erforschung der interglazialen Travertine in Hradiště pod Vrátnom. Anthropozoikum, **6**, 103–17, Prague 1957
[490] *Ložek, V.*, and *J. Kukla*, Das Lößprofil von Leitmeritz an der Elbe, Nordböhmen. Eiszeitalter und Gegenwart, **10**, 81–104, 1959
[491] *Ložek, V.*, and *F. Prošek*, Über Veränderungen des Landschaftsbildes des Südslowakischen Karstes in der jüngsten geologischen Vergangenheit. Ochrana Přírody, **11**, 33–42, 1956

[492] *Lozhkin, A. V.*, Neue pollenanalytische Beobachtungen zur Vegetationsgeschichte des Nordostteiles der UdSSR im Anthropogen. DAN, **152**, 949–52, 1963

[493] *Lüdi, W.*, Interglaziale Vegetation im Schweizerischen Alpenvorland. Veröff. Geobot. Inst. Rübel in Zürich, **34**, 99–107, 1958

[494] *Lukašev, K. I.*, Kurzer Abriß des gegenwärtigen Standes unserer Kenntnis von den quartären (anthropogenen) Sedimenten Weißrußlands. Czwartorzęd Europy środkowej i wschodniej, **1**, 377–422, Warsaw 1961, Engl., Pol. summaries

[495] *Lundqvist, G.*, The Interglacial Ooze at Porsi in Lapland. Sveriges Geol. Unders., ser. C, Avhandl. och uppsatser, Nr. 575, **54**, Nr. 6, 1960

[496] *Lungersgauzen, G. F.*, and *O. A. Rakovec*, Über die Tertiär/Quartär-Grenze im Gebirgs-Altai. Materialy chetvert., **3**, 229–37, 1961

[497] *Maarleveld, G. C.*, Les phénomènes périglaciaires au pléistocène ancien et moyen aux Pays-Bas. Les Congr. et Coloques de l'Univ. d. Liège, **17**; le périglaciaire préwurmien. 135–41, Liège 1960

[498] *Maarleveld, G. C.*, and *R. P. H. P. van der Schans*, De Dekzandmorfologie van de Gelderse Vallei. Tijdsk. Koninkl. Nederl. Aardrijkskund. Genootsch., **78**, 22–34, 1961

[499] *Macoun, J.*, Stratigrafie sprašovích pokryvů na Opavsku. Přírodovědný Časopis Slezský, **22**, 15–24, 1962

[500] *Macoun, J.*, *V. Šibrava*, *J. Tyráček* and *V. Kneblová-Vodičková*, Kvartér Ostravska a Moravské brány. Prague 1965

[501] *Mądalski, J.*, Pleistoceńska flora ze Ściejowic koło Krakówa. Starunia, **10**, Cracow 1935

[502] *Mädler, K.*, Die pliozäne Flora von Frankfurt am Main. Abh. d. Senckenb. Naturf. Ges., Nr. **446**, 1939

[503] *Magnusson, E.*, An Interglacial or Interstadial Deposit at Gallejaure, Northern Sweden. Geol. Fören. i Stockh. Förhandl., **84**, 363–71, 1962

[504] *Maher, L. J.*, *Ephedra* Pollen in Sediments of the Great Lakes Region. Ecology, **45**, 391–95, 1964

[505] *Mal'gina, E. A.*, Zum Problem der Entstehung von Pollenspektren in den Wüsten Mittel-Asiens. Trudi Geogr., **77**, 113–38, 1959

[506] *Maloletko, A. M.*, Paläogeographie der Flachländer im Vorland des Altai im Quartär. Trudi Kom. po izuch. chetvert. per., **22**, 165–82, 1963

[507] *Mancini, F.*, Osservazioni sui loess e sui paleosuoli dell'anfiteatro orientale del Garda e di quelle di Rivoli (Verona). Atti della Soc. Ital. di Sci. Nat. e del Mus. Civ. di Storia Nat. in Milano, **99**, 221–48, 1960

[507a] *Mania, D.*, Zur stratigraphischen Neugliederung des Mittelpleistozäns im Saalegebiet. Petermanns Geogr. Mitteilungen, **114**, 186–94, 1970

[508] *Manton, I.*, The Problem of *Biscutella laevigata* L. Zs. f. Induktive Abstammungs- und Vererbungslehre, **67**, 41–57, 1934

[509] ——— The Problem of *Biscutella laevigata* L. II. The Evidence from Meiosis. Ann. of Bot., n. s. **1**, 439–62, 1937

[510] *Marchesoni, V.*, Lineamenti paleobotanici dell'interglaciale Riss-Würm nella pianura Padana. Nuov. Giorn. Bot. Ital., **67**, 306–11, 1960

[511] *Marchesoni, V.*, and *A. Paganelli*, Ricerche sul quaternario della pianura

Padana. I. Analisi polliniche di sedimente torbo-lacustri di Padova e Sacile. Rendiconti degli Ist. Sci. Univ. Camerino, **1**, Fasc. 1, 47–54, 1960

[512] *Markov, K. K.*, Probleme der Entwicklung der Natur auf dem Territorium der UdSSR im Quartär (Eiszeitalter-Anthropogen). Trudi Kom. po izuch. chetvert. per., **19**, 3–41, 1962

[513] *Markov, K. K., G. I. Lazukov* and *M. P. Grichuk*, Grundlegende Gesetzmäßigkeiten in der Entwicklung der natürlichen Bedingungen der UdSSR im Quatär (Eiszeitalter-Anthropogen). Izv AN, ser. geogr., 1961, Nr. 4, 10–13, 1961

[514] *Markovič-Marjanovič, J.*, Kvartarni sedimenti Zetske pavnice u svetlosti pleistocene klime. 3. Congr. de géol. de Yougosl., **1**, 259–87, Titograd 1961

[515] ——— Sedimenti Metohije i odredivanje donje granice kvartara. Rep. of the 5. Meeting of the geol. of the F. P. R. of Yugosl., 181–92, Belgrade 1962

[516] *Martin, P. S.*, Taiga-Tundra and the Full-glacial Period in Chester County, Pennsylvania. Amer. J. Sci., **256**, 470–502, 1958

[517] ——— How Many Logs Make a Forest? Ohio J. Sci., **59**, 221–22, 1959

[518] ——— The Last 10,000 Years. A Fossil Pollen Record of the American Southwest. Tucson 1963

[519] *Martin, P. S., B. E. Sabels* and *D. Shutler*, Rampart Cave Coprolite and Ecology of the Shasta Ground Sloth. Amer. J. Sci., **259**, 102–27, 1961

[520] *Martynov, V. A.*, and *V. P. Nikitin*, Interglaziale Schichten im Aufschluß von Belogor'e am Ob'. Chetvert. geol. i geomorf. Zapadno-Sibirsk. Nizmennosti, Trudi in-ta geol. i geofiz., **25**, Akad. Nauk SSSR, sibirsk. otdel, 73–81, Novosibirsk 1964

[521] *Maslova, I. V.*, Ergebnisse einer pollenanalytischen Untersuchung pleistozäner Sedimente der Halbinsel Kertsch. DAN, **137**, 387–90, 1961

[522] *Mattfeld, J.*, Über hybridogene Sippen der Tannen, nachgewiesen an den Formen der Balkan-Halbinsel; zugleich ein Beitrag zur Waldgeschichte der Balkan-Halbinsel. Bibliotheca Botanica, **100**, Stuttgart 1930

[523] *Matveeva, O. V.*, Pollenspektren quartärer Sedimente der Vorberge des Altai, des östlichen Altai-Gebirges und West-Tuvas. Trudi Geol., **31**, 85–112, 1960

[524] *Matveeva, O. V.*, and *A. I. Moskvitin*, Über das Alter und die Entstehungsbedingungen der ersten Terrasse der Cna bei Jaltunovo in der Rjazan'-Oblast'. Biull chetvert., **24**, 56–65, 1960

[525] *Mayr, E.*, Systematics and the Origin of Species, from the Viewpoint of a Zoologist. New York 1947

[526] *Mazenot, G.*, La faune malacologique des deux loess d'Ars (Ain). Compte Rendu Somm. Séances Soc. Géol. France, 1951, 199–201, 1951

[527] *Meckelein, W.*, Beobachtungen und Gedanken zu geomorphologischen Konvergenzen in Polar- und Wärmewüsten. Erdkunde **19**, 31–39, 1965

[528] *Menéndez Amor, J.*, and *F. Florschütz*, Un aspect de la végétation en Espagne méridionale durant la dernière glaciation et l'holocène. Geol. en Mijnb., **41**, 131–34, 1962

[529] ——— Results of the Preliminary Palynological Investigation of Samples from a 50 m Boring in Southern Spain. Bol. R. Soc. Española Hist. Nat. (Geol.), **62**, 251–55, 1964

[529a] *Menke, B.*, Beiträge zur Biostratigraphie des Mittelpleistozäns in Norddeutschland (pollenanalytische Untersuchungen aus Westholstein). Meyniana, **18**, 35–42, 1968
[530] *Mente, A.*, Het resultaat van een palynologisch onderzoek van een Eemian-afzetting bij Liessel (N. Br.). Geol. en Mijnb., **40**, n. s. 23, 75–78, 1961
[530a] *Mercer, J. H.*, The Allerød Oscillation: A European Climatic Anomaly? Arctic and Alpine Research, **1**, 227–34, 1969
[531] *Metel'ceva, E. P.*, and *V. N. Sukachev*, Neue Angaben zur pleistozänen Flora des Zentrums der Russischen Tiefebene (interglaziales Moor in der Vladimir-Oblast'). Biull. chetvert., **26**, 50–60, 1961
[532] *Mizerov, B. V.*, and *M. R. Votach*, Zum Problem der Gliederung der quartären Sedimente am Unterlauf des Čulym. Sistematika i metodi izucheniia iskopaemych pyl'ci i spor., 218–22, Moscow 1964
[533] *Mjachina, A. I.*, Zum Problem der Grenze zwischen Tertiär und Quartär in der Amur-Zeja-Senke. Materialy chetvert., **3**, 305–10, 1961
[534] *Mkrtchjan*, S. S., Geologiia Armjanskoi SSR. **1**, Geomorfologiia, Erevan 1962; **2**, Stratigrafiia, Erevan 1964
[535] *Monod, T.*, The Late Tertiary and Pleistocene in the Sahara and Adjacent Southerly Regions. In: *Howell, F. C.*, and *F. Bourlière*, African Ecology and Human Evolution. 117–229, Chicago 1963
[536] *Monoszon, M. Ch.*, Vertreter der Familie der *Chenopodiaceae* in eiszeitlichen Sedimenten des Europäischen Teiles des UdSSR. Tezisi; sekc. istorii iskop. flori i fauni, 14–15, 1957
[537] ——— Über Pollenfunde von Vertretern der *Chenopodiaceae* in quartären Sedimenten des europäischen Teiles der UdSSR. Materialy chetvert., **1**, 317–30, 1961
[538] ——— Pollen of Halophytes and Xerophytes of the *Chenopodiaceae* Family in the Periglacial Zone of the Russian Plain. Pollen et Spores, **6**, 147–55, 1964
[539] *Mortensen, H.*, Temperaturgradient und Eiszeitklima am Beispiel der pleistozänen Schneegrenzdepression in den Rand- und Subtropen. Zs. f. Geomorph., N. F. **1**, 44–56, 1957
[540] ——— Eiszeiten und Gletscher in der Mongolei. Zs. f. Geomorph., N. F. **1**, 315, 1957
[541] *Moschkov, B. S.*, Photoperiodismus und Frosthärte ausdauernder Gewächse. Planta **23**, 774–803, 1935
[542] *Moskvitin, A. I.*, Neues über den Lichvinsker Aufschluß; die Bedeutung des Lichvinsker Aufschlusses für die Stratigraphie der quartären Sedimente im Europäischen Teil der UdSSR. Biull. Mosk. Obshch. Ispyt. Prirodi, otd. geol., **9**, Nr. 1–2, 174–86, 1931
[543] ——— Der Mologo-Sheksninsker interglaziale See. Trudi in-ta geol. nauk AN SSSR, **88**, geol. ser. Nr. 26, 5–18, 1947
[544] ——— Über Lößhorizonte und Ursachen des Überdeckens interglazialer Böden. Trudi in-ta geol. nauk AN SSSR, ser. geomorf. i chetvert. geol., **1**, 125–30, 1957
[545] ——— Pleistozäne Sedimente und die Entstehungsgeschichte des Tales der mittleren Volga. Trudi Geol., **12**, 1958

[546] ——— Neue Beweise für eine älteste Vereisung des Russischen Flachlandes. DAN, **127**, 852–55, 1959
[547] ——— Über warme und kühlere Interglaziale in der UdSSR. Ber. d. Geol. Ges. DDR, **5**, 5–20, 1960
[548] ——— Paläogeographie des Südostens Europas im Pleistozän. Mezhdunar. geol. kongr., 21. sess, 1960. dokladi sovetsk. geol., 41–47, 1960; Engl. summary
[549] ——— „Warme" und „kalte" Interglaziale als Grundlage einer stratigraphischen Gliederung des Pleistozäns. Materialy chetvert., **1**, 41–52, 1961
[550] ——— Über die Untergliederung des Würms und über die Lage des mittleren und oberen Paläoliths innerhalb der Würm-Eiszeit. Izv AN, ser. geol., 1962, Nr. 7, 35–44, 1962
[551] ——— Brief an die Redaktion. Biull. chetvert., **27**, 162, 1962
[552] ——— Über den Bau der Deckschichten der ältesten Dnestr-Terrassen. Biull. chetvert., **28**, 33–55, 1963
[553] *Müller-Beck, H. J.*, Paläolithische Kulturen und pleistozäne Stratigraphie in Süddeutschland. Eiszeitalter und Gegenwart, **8**, 116–40, 1957
[554] *Müller-Stoll, H.*, Vergleichende Untersuchungen über die Abhängigkeit der Jahrringfolge von Holzart, Standort und Klima. Bibliotheca Botanica, **122**, Stuttgart 1951
[555] *Müller-Stoll, W. R.* Die jüngsttertiäre Flora des Eisensteins von Dernbach (Westerwald). Beih. Bot. Centralbl., **58** B, 376–434, 1938
[556] *Münnich, K. O.*, Ist die Altersbestimmung nach der ^{14}C-Methode zuverlässig? Fehlermöglichkeiten bei der Datierung vorgeschichtlicher Funde. Umschau, **58**, 109–11, 1958
[557] ——— Die ^{14}C-Methode. Geol. Rundschau, **49**, 237–44, 1960
[558] *Mullenders, W.*, Les relations entre la végétation et les spectres polliniques en forêt du Mont-Dieu (Département des Ardennes, France). Bull. Soc. Roy. Bot. Belgique, **94**, 131–38, 1962
[559] *Muller, F. H.*, Quaternary Section at Otto, New York. Amer. J. Sci., **262**, 461–78, 1964
[560] *Murr, J.*, Neue Übersicht über die fossile Flora der Höttinger Breccie. Jahrb. Geol. Bundesanst., **76**, 153–70, Vienna 1926
[561] *Nairn, A. E. M.*, Descriptive Paleoclimatology. London 1961
[562] ——— Problems in Paleoclimatology. London, New York, Sydney, 1964
[563] *Naletov, P. I.*, Katalog mestonachozhdenii iskopaemich fauni, flori, pyl'ci i spor central'noi chasti Buriatskoi ASSR. Moscow 1961
[564] *Nehring, A.*, Über Tundren und Steppen der Jetzt- und Vorzeit mit besonderer Berücksichtigung ihrer Fauna. Berlin 1890
[565] *Neishtadt, M. I.*, Istoriia lesov i paleogeografiia SSSR v golocene. Moscow 1957
[566] *Nejshtadt, M. I., N. A. Chotinskii, N. G. Markova* and *A. L. Devirc*, Das Moor Melechovo (Jaroslavl-Oblast'). In: Paleogeografiia i chronologiia verchnego pleistocena i golocena po dannym radiouglerodnogo metoda. 97–99, Moscow 1965
[567] *Nikiforova, K. V.*, Das Känozoikum der Hungersteppe, Zentral-Kazachstan. Trudi Geol., **45**, 1960

[568] *Nikolaev, N. I.*, and S. S. *Shul'c*, Karta noveishei tektoniki SSSR, m. 1:5 000-000. Akad. Nauk SSSR 1959

[569] *Oaks, R. Q.*, and *N. K. Coch*, Pleistocene Sea Levels, Southeastern Virginia. Science, **140**, 979–83, 1963

[570] *Ogden, I. G.*, A Late-glacial Pollen Sequence from Martha's Vineyard, Massachusetts. Amer. J. Sci., **257**, 366–81, 1959

[571] ——— The Squibnocket Cliff Peat: Radiocarbon Dates and Pollen Stratigraphy. Amer. J. Sci., **261**, 344–53, 1963

[571a] *Ogden, III, J. G.*, Radiocarbon and Pollen Evidence for a Sudden Change in Climate in the Great Lakes Region Approximately 10,000 Years ago. In: *Cushing, E. J.*, and *H. E. Wright*, Quaternary Paleoecology, **7**, Proc. VII Congr. INQUA, Yale Univ. Press, 117–27, 1967

[571b] ——— Radiocarbon Determinations of Sedimentation Rates from Hard and Soft-water Lakes in Northeastern North-America. In: *Cushing, E. J.*, and *H. E. Wright*, Quaternary Paleoecology, **7**, Proc. VII Congr. INQUA, Yale Univ. Press, 175–83, 1967

[572] *Oldfield, F.*, Three Pollen-analyses from an Interglacial Mud-bed on the Foreshore near Biarritz, Southwest France. Bull. Centre d'Études et de Recherches Sci., **3**, 53–62, Biarritz 1960

[573] ——— Late Quaternary Vegetational History in South West France. Pollen et Spores, **6**, 157–68, 1964

[573a] *Olsson, I. U.*, edit., Radiocarbon Variations and Absolute Chronology. Almqvist and Wiksell, Stockholm, 656 pages, 1970

[574] *Ording, A.*, Årringanalyser på gran og furu. Meddelel. fra det norske skogsforsøksvesen, Nr. 25, 7, H. 2, 105–354, 1941

[575] *Pady, S. M.*, and *C. D. Kelley*, Aerobiological Studies of Fungi and Bacteria over the Atlantic Ocean. Canad. J. Bot., **32**, 202–12, 1954

[576] *Paganelli, A.*, Ricerche sul quaternario della pianura Padana. II. Analisi polliniche di sedimenti torbolacustri di ca Marcozzi (Delta Padana). Rendiconti Ist. Sci. Univ. Camerino, **2**, Fasc. 1, 83–96, 1961

[577] ——— Il graduale impoverimento della flora forestale nel quaternario della pianura Padana. Nuovo Giorn. Bot. Ital., **68**, 109–17, 1961

[578] *Palamarev, E.*, Die geochronologische Entwicklung der neogenen Floren der Balkanhalbinsel. Ber. Geol. Ges. DDR, **9**, 369–80, 1964

[579] *Pécsi, M.*, Die periglazialen Erscheinungen in Ungarn. Petermanns Geogr. Mitt., 1963, 161–82, 1963

[580] ——— Die chronologischen Probleme der Ungarischen Strukturböden. Földrajzi Értesítö, **13**, Nr. 2, 141–56, Budapest 1964; Hungarian with German summary

[581] *Penck, A.*, Säugetierfauna und Paläolithikum des jüngeren Pleistozäns in Mitteleuropa. Abh. preuß. Akad. Wiss., Jahrg. 1938, phys.-mat. Kl., Nr. 5, 1938

[582] *Penck, A.*, and *E. Brückner*, Die Alpen im Eiszeitalter. 3 Bde., Leipzig 1901–9

[583] *Petrov, O. M.*, Stratigraphie der quartären Sedimente des Süd- und Ostteiles der Tschuktschen-Halbinsel. Biull. chetvert., **28**, 135–52, 1963

[584] *Péwé, T. L.*, Age of Moraines in Victoria Land, Antarctica. Journ. of Glaciol., **4**, 93–100, 1962

[585] *Péwé, T. L.*, and *R. E. Church,* Glacier Regimen in Antarctica as Related by Glacier-Margin Fluctuation in Historic Time with Special Reference to the McMurdo Sound. Publ. Nr. 58 of the I. A. S. H., Commission of Snow and Ice, 295–305, 1962?

[586] *Péwé, T. L., E. H. Muller, N. V. Karlstrom, D. B. Krinsley, A. T. Fernald, C. Wahrhaftig, D. M. Hopkins,* and *R. L. Dettermann,* Multiple Glaciation in Alaska. A progress Report. Geol. Survey Circular 289, Washington, D. C. 1953

[587] *Pfaffenberg, K.*, Das Interglazial von Haren (Emsland). Eine palaeobotanisch-pollenanalytische Untersuchung. Abh. naturwiss. Ver. Bremen, **31**, 360–76, 1939

[588] *Pia, J.*, Übersicht über die fossilen Kalkalgen und die geologischen Ergebnisse ihrer Untersuchung. Mitt. d. Alpenländ. geol. Ver., **33**, 11–34, 1940

[589] ——— Einige geologische Ergebnisse der Untersuchung fossiler Kalkalgen. Natur und Volk, **71**, 84–90, 1941

[590] *Piech, K.*, Utwory międzylodowcowo w Szczercowie (woj. Łódzkie). Rocznik Polsk. Towarzystwa Geol., **8**, Nr. 2, f. 1932, 51–132, 1932

[591] *Pike, K.*, and *H. Godwin,* The Interglacial at Clacton-on-Sea, Essex. Quat. J. Geol. Soc. London, **108**, f. 1952, 261–72, 1953

[592] *Pogozhev, A. G., V. I. Goljakov* and *A. S. Arsanov,* Stratigraphie der palaeogenen und neogenen Sedimente am Ostufer der Penžina-Bucht. In *Egiazarov, B. Ch.*, Geologiia Korjakskogo Nagor'ja. 122–32, Moscow 1963

[593] *Polak, B.*, and *C. Hamming,* A Peat Layer of Early Würm Glacial Age. Geol. en Mijnb., **42**, 202–5, 1963

[594] *Polunin, N.*, Seeking Airborne Botanical Particles about the North Poles. Svensk Bot. Tidskr., **45**, 1951, 320–54

[595] ——— Circumpolar Arctic Flora. Oxford 1959

[596] *Pons, A.*, and *P. Quézel,* Première étude palynologique de quelques paléosols sahariens. Travaux Inst. de Recherches Sahariennes, **16**, Nr. 2, 15–40, 1957

[597] ——— Premières remarques sur l'étude palynologique d'un guano fossile du Hoggar. C. R. Acad. Sci., **246**, 2290–92, 1958

[598] *Pop, E.*, Palynologische Forschungen in Rumänien und ihre wichtigsten Ergebnisse. Bot. Zhurn., **42**, 363–76, 1957; Fr. summary

[599] *Popov, A. I.*, Das Pleistozän in West-Sibirien. In: *Markov, K. K.*, and *A. I. Popov,* Lednikovyi period na territorii evropeiskoi chasti SSSR i Sibiri. 360–84, Moscow 1959

[600] *Poser, H.*, Dauerfrostboden und Temperaturverhältnisse während der Würm-Eiszeit im nicht vereisten Mittel- und Westeuropa. Naturwiss., **34**, 10–18, 1947

[601] ——— Auftautiefe und Frostzerrung im Boden Mitteleuropas während der Würm-Eiszeit. Ein Beitrag zur Bestimmung des Eiszeitklimas. Naturwiss., **34**, 232–38; 262–67, 1947

[602] ——— Äolische Ablagerungen und Klima des Spätglazials in Mittel- und Westeuropa. Naturwiss., **35**, 269–76, 307–12, 1948

[603] ——— Zur Rekonstruktion der spätglazialen Luftdruckverhältnisse in Mit-

tel- und Westeuropa auf Grund der vorzeitlichen Binnendünen. Erdkunde, **4**, 81–88, 1950

[604] ——— Die nördliche Lößgrenze in Mitteleuropa und das spätglaziale Klima. Eiszeitalter und Gegenwart, **1**, 27–55, 1951

[605] *Post, L. v.*, Postarktiska klimattyper i södra Sverige. Geol. Fören. i Stockh. Förhandl., **42**, 231–42, 1920

[606] ——— Gotlands-agen (*Cladium Mariscus* R. Br.) i Sveriges postarktikum. Ymer, **45**, 295–312, 1925

[607] *Potonié, H.*, Die Entstehung der Steinkohle und der Kaustobiolithe überhaupt (wie des Torfs, der Braunkohle, des Petroleums usw.). 6. Aufl., Berlin 1920

[608] *Prjachin, A. I.*, Gefrornisdeformationen in tertiären und pleistozänen Sedimenten der Flußtäler von Kama, Viatka und Belaia. Tezisi; sekc. Centra i iugovostoka Russk. ravnini, 94–96, 1957

[609] ——— Gefrornisdeformationen in tertiären und quartären Sedimenten der Flüsse Kama, Viatka und Belaia. Materialy chetvert., **2**, 221–26, 1961

[610] *Prošek, F.*, and *V. Ložek*, Sprašový profil v Bance u Pieštan (Západni Slovensko). Anthropozoikum, **3**, 301–23, 1953; Russ., Engl. summaries

[611] ——— Výzkum sprašového profilu v Zamarovcích u Trenčína. Anthropozoikum, **4**, 181–211, 1954; Russ., Ger. summaries

[612] *Quézel, P.*, De l'application de technique palynologique à un territoire désertique. Paléoclimatologie du Quaternaire récent au Sahara. Changes of Climate; Proc. Rome Symposium Organized by UNESCO and WMO, 243–48, Paris 1963

[613] *Quézel, P.*, and *Cl. Martinez*, Premiers résultats de l'analyse palynologique de sédiments recueillis au Sahara méridional à l'occasion de la mission Berliet-Tchad. In: Missions Berliet Ténéré Tchad. 313–27, Paris 1962

[614] *Rabien, I.*, Die Vegetationsentwicklung des Interglazials von Wallensen in der Hilsmulde. Eiszeitalter und Gegenwart, **3**, 96–128, 1953

[615] *Ralska-Jasiewiczowa, M.*, Interstadial zlodowacenia środkowopolskiego w Łabędach na górnym Śląsku. Monographiae Botanicae, **7**, 95–105, Cracow 1958

[616] ——— Plejstoceńska flora z Zabłocie nad Bugiem. Folia Quaternaria, **2**, Cracow 1960

[617] *Ramishvili, I. S.*, Pollenanalytische Befunde über die oberpontischen Sedimente West-Grusiniens. DAN, **139**, 685-87, 1961

[618] *Raniecka, J.*, Pollenanalytische Untersuchungen des Interglazials von Żoliborz in Warschau. Acta Soc. Bot. Polon., **7**, 169–82, 1930

[619] *Raniecka-Bobrowska, J.*, Analiza pyłkowa profilów czwartorzędowych Woli i Żoliborza w Warszawie. Z badań czwartorzędu w Polsce, **5**, 107–40, Warsaw 1954; Russ., Engl. summaries

[620] *Ravskij, E. I.*, and *L. V. Golubeva*, Das Eopleistozän der Tunka-Senke. DAN, **135**, 1207–10, 1960

[621] *Ravskij, E. I., L. P. Aleksandrova, E. A. Vangengeim, V. G. Gerbova* and *L. V. Golubeva*, Anthropogene Sedimente des Südteiles Ostsibiriens. Trudi Geol., **105**, 1964

[622] *Reich, H.*, Die Vegetationsentwicklung der Interglaziale von Großweil-

Ohlstadt und Pfefferbichl im Bayerischen Alpenvorland. Flora, **140**, 386–443, 1953

[623] *Reid, E. M.*, A Comparative Review of Pliocene Floras, Based on the Study of Fossil Seeds. Quat. J. Geol. Soc. London, **76**, f. 1920, 145–61, 1921

[624] *Rein, U.*, Die Vegetationsentwicklung des Interglazials von Lehringen. Zs. dtsch. geol. Ges., **90**, 145–47, 1938

[625] *Richter, H., G. Haase* and *H. Barthel,* Die Landschaften im Osten der Mongolischen Volksrepublik. Geogr. Berichte, **23**, 125–41, 1962

[626] *Richter, K.*, Fluorteste quartärer Knochen in ihrer Bedeutung für die absolute Chronologie des Pleistozäns. Eiszeitalter und Gegenwart, **9**, 18–27, 1958

[627] *Riabchenkov, A. S.*, Zum Problem der Entstehung der Ukrainelösse im Lichte mineralogischer Befunde. Biull. chetvert., **20**, 45–59, 1955

[628] ——— Quartäre Sedimente der Meshchersker Niederung, auf Grund neuester Untersuchungen. Materialy chetvert., **2**, 237–45, 1961

[629] *Riasina, V. E.*, Neue Funde einer quartären Säugetierfauna im Oberlaufgebiet des Ob. DAN, **142**, 1153–54, 1962

[630] ——— Über Genese und Stratigraphie der quartären Sedimente auf dem Steppen-Plateau des oberen Obgebietes. Biull. chetvert., **27**, 86–97, 1962

[631] ——— Über begrabene Böden am Oberlauf des Ob. Trudi Kom. po izuch. chetvert. per., **22**,183–89, 1963

[632] *Rochow, M. v.*, *Azolla filiculoïdes* im Interglazial von Wunstorf bei Hannover und das wahrscheinliche Alter dieses Interglazials. Ber. D. Bot. Ges., **65**, 315–18, 1952

[633] *Roosma, A.*, A Climatic Record from Searles Lake, California. Science, **128**, 716, 1958

[634] *Rosendahl, C. O.*, A Contribution to the Knowledge of the Pleistocene Flora of Minnesota. Ecology, **29**, 284–315, 1948

[635] *Rosholt, J. N., C. Emiliani, J. Geiss, F. F. Koczy,* and *P. J. Wangersky,* Absolute Dating of Deep-sea Cores by the Pa^{231}/Th^{230} Method. Journ. of Geol., **69**, 162–85, 1961

[636] *Rossignol, M.*, Analyse pollinique de sédiments marines quaternaires en Israël. II. Sédiments pleistocènes. Pollen et Spores, **4**, 121–48, 1962

[637] *Różycki, St. Z.*, Das Alter präglazialer Sedimente des mittelpolnischen Flachlandes im Lichte pollenanalytischer Untersuchungen in Ochota/Warschau. Prace o plejstocenie polski środkowej. Polska Akad. Nauk, Kom. geol., 35–42; Warsaw 1961; Russ., Engl. summaries

[638] *Rudloff, H. v.*, Die Klimaschwankungen in den Hochalpen seit Beginn der Instrumenten-Beobachtungen. Archiv f. Meteorol., Geophys. und Bioklimat., Ser. B, **13**, 303–51, 1964

[639] ——— Die Schwankungen und Pendelungen des Klimas in Europa seit dem Beginn der regelmäßigen Instrumenten-Beobachtungen (1670). Die Wissenschaft, Brunswick, in press

[640] *Rudolph, K.*, Grundzüge der nacheiszeitlichen Waldgeschichte Mitteleuropas. Beih. Bot. Centralbl., **47** B, 111–76, 1930

[641] *Ruske, R.*, Das Pleistozän zwischen Halle (Saale), Bernburg und Dessau. Geologie, **13**, 570–97, 1964

[642] ——— Mittelpleistozäne Löße und Böden in Mitteleuropa und deren stratigraphische Einstufung. Geologie, **14**, 554–63, 1965

[643] *Ruske, R.*, and *M. Wünsche*, Löße und fossile Böden im mittleren Saale- und unteren Unstruttal. Geologie, **10**, 9–29, 1961

[644] ——— Zur Gliederung des Pleistozäns im Raum der unteren Unstrut. Geologie, **13**, 211–22, 1964

[645] *Rutte, E.*, Die Fundstelle altpleistozäner Säugetiere von Randersacker bei Würzburg. Geol. Jahrb., **73**, 737–54, 1958

[646] *Shancer, E. V.*, Der gegenwärtige Stand des Problems der Grenze Neogen/Quartär (Anthropogen). Materialy chetvert., **1**, 10–19, 1961

[647] ——— Das Problem der Grenze zwischen dem Neogen und dem Quartär (Anthropogen). Trudi Kom. po izuch. chetvert. per., **20**, 5–24, 1962

[648] *Saporta, G. de*, Sur les caractères propres à la végétation pliocène, à propos des découvertes de *M. J. Rames*, dans le Cantal. Bull. Soc. Géol. France, **3**. sér., **1**, f. 1872–73, 212–32, 1873

[649] *Shatilova, I. I.*, Veränderungen der Flora Guriens während der Kujal'nic-Zeit, auf Grund der Pollenanalyse. DAN, **145**, 895–98, 1962

[650] *Schaedel, K.*, and *J. Werner*, Neue Gesichtspunkte zur Stratigraphie des mittleren und älteren Pleistozäns im Rheingletschergebiet. Eiszeitalter und Gegenwart, **14**, 5–26, 1963

[651] *Schaefer, H.*, and *B. Frenzel*, Beiträge zur Kenntnis der Flora des Ostteiles der Großen Samojedentundra. Botan. Jahrb., **78**, 367–434, 1959

[652] *Scharfetter, R.*, Biographien von Pflanzensippen. Vienna 1953

[653] *Scherhag, R.*, Einführung in die Klimatologie. 2. Aufl., Brunswick, 1962

[654] *Schönhals, F.*, Über einige wichtige Lößprofile und begrabene Böden im Rheingau. Notizbl. d. hess. Landesamtes f. Bodenforschg., **6**, 1, 244–59, Wiesbaden 1950

[655] ——— Über fossile Böden im nichtvereisten Gebiet. Eiszeitalter und Gegenwart, **1**, 109–30, 1951

[656] ——— Gesetzmäßigkeiten im Feinaufbau von Talrandlößen mit Bemerkungen über die Entstehung des Lößes. Eiszeitalter und Gegenwart, **3**, 19–36, 1953

[657] *Schönhals, E.*, *H. Rohdenburg* and *A. Semmel*, Ergebnisse neuerer Untersuchungen zur Würmlöß-Gliederung in Hessen. Eiszeitalter und Gegenwart, **15**, 199–206, 1964

[658] *Schofield, W. B.*, and *H. Robinson*, Late-glacial and Postglacial Plant Macrofossils from Gillis Lake, Richmond County, Nova Scotia. Amer. J. Sci., **258**, 518–23, 1960

[659] *Schott, W.*, Zur Klimaschichtung der Tiefseesedimente im äquatorialen Atlantischen Ozean. Geol. Rundsch., **40**, 20–31, 1952

[660] *Schroeder, H.*, and *J. Stoller*, Marine und Süßwasser-Ablagerungen im Diluvium von Uetersen-Schulau. Jahrb. kgl. preuß. geol. Landesanst. u. Bergakademie, **26**, f. 1905, 94–102, 1908

[661] ——— Diluviale marine und Süßwasser-Schichten bei Ütersen-Schulau. Jahrb. kgl. preuß. geol. Landesanst., f. 1906, 455–528, 1909

[662] *Schütrumpf, R.*, Das Interglazialprofil von Lauenburg a. d. Elbe (Kuhgrund II) im Lichte der Pollenanalyse. Mitt. Geol. Staatsinst. Hamburg, **16**, 37–45, 1937

[663] *Schulmann, E.*, Definitive Dendrochronologies: A Progress Report. Tree-ring Bulletin, **18**, 10–18, 1951/52
[664] ——— Tree-ring Evidence for Climatic Changes. In: *Shapley, H.*, Climatic Change. Evidence, Causes, and Effects. 209–19, Cambridge 1953
[665] *Schwabedissen, H.*, Fällt das Aurignacien ins Interstadial oder ins Interglazial? Germania, **34**, 12–41, 1956
[666] *Schwarzbach, M.*, Das Klima der Vorzeit; eine Einführung in die Paläoklimatologie. Stuttgart 1950
[667] ——— Orogenesen und Eiszeiten. Zur Ursache des Klimawechsels in der Erdgeschichte. Naturwiss., **40**, 452–55, 1953
[668] ——— Das Alter der Wüste Sahara. Neues Jahrb. Geol. u. Paläontol., Mh., 1953, 157–74, 1953
[669] *Sears, P. B.*, and *K. H. Clisby*, Pollen Spectra Associated with the Orleton Farm Mastodon Site. Ohio J. Sci., **52**, 9–10, 1952
[670] *Selle, W.*, Beiträge zur Mikrostratigraphie und Paläontologie der nordwestdeutschen Interglaziale. Jahrb. d. Reichst. f. Bodenforschung, **60**, f. 1939, 197–235, 1941
[671] ——— Die Interstadiale der Weichselvereisung. Eiszeitalter und Gegenwart, **2**, 112–19, 1952
[672] ——— Gesetzmäßigkeiten im pleistozänen und holozänen Klima-Ablauf. Abh. naturw. Ver. Bremen, **33**, 259–90, 1953
[673] ——— Die Vegetationsentwicklung des Interglazials von Ober-Ohe in der Lüneburger Heide. Abh. naturw. Ver. Bremen, **33**, 457–63, 1954
[674] ——— Das letzte Interglazial in Niedersachsen. 103. Bericht d. naturhist. Ges. Hannover, 77–89, 1957
[675] ——— Zur Gliederung des Riß/Würm-Interglazials in Nordwestdeutschland und den angrenzenden Gebieten. Zs. dtsch. geol. Ges., **112**, 3. Teil, 519–20, 1961
[676] ——— Geologische und vegetationskundliche Untersuchungen an einigen wichtigen Vorkommen des letzten Interglazials in Nordwestdeutschland. Geol. Jahrb., **79**, 295–352, 1962
[677] *Semken, H. A.*, *B. B. Miller*, and *J. B. Stevens*, Late Wisconsin Woodland Musk Oxen in Association with Pollen and Invertebrates from Michigan. Journ. of Paleont., **38**, 823–35, 1964
[678] *Šercelj, A.*, On Quaternary Vegetation in Slovenia. Razprave Geologija Poročila, **7**, 25–34, Ljubljana 1962
[679] ——— Pelodne analize pleistocenskich sedimentov v Horjulski dolini. Arheološki Vestnik, **13–14**, Biodarjev Zbornik, 273–86, 1962–63; Engl. summary
[680] ——— Razvoj würmske in holocenske gozdne vegetacije v Sloveniji. Razprave, Slov. Akad. znanosti in umetnosti, razred za prirodosl. in medic. vede, oddelek za prirodosl. vede, 7, 363–418, 1963; Engl. summary
[681] *Sheveleva, N. S.*, Alte Gefrorniserscheinungen in mittelpleistozänen fluviatilen Sedimenten bei Krasnojarsk. Problemi paleogeografii i morfogeneza v poljarnych stranach i vysokogor'e. 98–108, Moscow 1964
[682] *Shapley, H.*, Climatic Change. Evidence, Causes, and Effects. Cambridge 1953

[683] *Shimada, M.*, Pollen Analysis of Lignites. IV. Pliocene Lignites from Yamuke Formation of Mogami Group. Ecolog. Review, **14**, 117–19, 1955

[684] ——— Pollen Analysis of Lignites. V. Pliocene Lignites from Sakegawa Formation of Mogami Group. Ecolog. Review, **14**, 265–66, 1957

[685] *Shotton, F. W.*, and *I. Strachan*, The Investigation of a Peat Moor at Rodbaston, Penkridge, Staffordshire. Quart. J. Geol. Soc. London, **115**, 1–15, 1959

[686] *Šibrava, V.*, Sediments at the Southern Margin of the Continental Glaciation in Moravia and Czechoslovakian Silesia. Report VI Internat. Congr. Quaternary Warsaw 1961, **3**, 327–35, 1963

[687] *Shik, S. M.*, Neue Beiträge zur Kenntnis der mittelpleistozänen interglazialen Sedimente des Smolensker Gebietes. Materialy chetvert., **2**, 252–58, 1961

[688] ——— Stratigraphie quartärer Sedimente. In: *Markov, K. K.*, Rel'ef i stratigrafiia chetvertichnych otlozhenii severo-zapadnoi chasti Russkoi ravnini, 151–72, Moscow 1961

[689] *Simpson, I. M.*, and *R. G. West*, On the Stratigraphy and Palaeobotany of a Late-Pleistocene Organic Deposit at Chelford, Cheshire. New Phytologist, **57**, 239–50, 1958

[690] *Sitler, R. F.*, and *J. Baker*, Thickness of Loess in Clark County, Illinois. Ohio J. Sci., **60**, 73–77, 1960

[691] *Slujs, P. van der*, and *G. C. Maarleveld*, Dekzandruggen uit des Jonge *Dryas*tid in Zeeuws-Vlaanderen. Boor en Spade, **13**, 21–27, 1963

[692] *Smith, A. G.*, Threshold and Inertia in British Late Quaternary Paleoecology. Pollen et Spores, **4**, 378–79, 1962

[693] *Smolíková, L.*, Zur Erforschung der terrae calcis-Böden im Murān-Karst (Zentral-Slowakei). Věstnik Ustředního ústavu geolog., **36**, 373–75, 1961

[694] *Smolíková, L.*, and *V. Ložek*, Zur Altersfrage der mitteleuropäischen Terrae calcis. Eiszeitalter und Gegenwart, **13**, 157–77, 1962

[695] *Sobolewska, M.*, Dzika winorośl (*Vitis silvestris* GMEL.) w plejstocenie polskim. Z badań czwartorzędu w Polsce, **5**, 159–66, Warsaw 1954

[696] ——— Roślinność plejstoceńska z Syrnik nad Wieprzem. Z badań czwartorzędu w Polsce, **7**, 143–92, 1956; Russ., Engl. summaries

[697] ——— *Azolla filiculoïdes* LAM. w starszym interglacjale w Polsce. Z badań czwartorzędu w Polsce, **7**, 241–46, 1956; Russ., Engl. summaries

[698] ——— Wyniki analizy pyłkowej osadów interglacjalnych z Olszewic. Z badań czwartorzędu w Polsce, **7**, 271–89, 1956, Russ., Engl. summaries

[699] ——— Eem-Interglazialflora von Góra Kalvarija. Z badań czwartorzędu w Polsce, **10**, 73–90, 1961; Russ., Engl. summaries

[700] *Sobolewska, M.*, and *A. Środoń*, Late-pleistocene Deposits at Białka Tatrzańska (West Carpathians). Folia Quaternaria, **7**, Cracow 1961

[701] *Sobolewska, M.*, *L. Starkel* and *A. Środoń*, Młodoplejstoceńskie osady z florą kopalna w Wadowicach. Folia Quaternaria, **16**, Cracow 1964; Engl., Russ. summaries

[702] *Sohma, K.*, Pollenanalytische Untersuchungen der pliozänen Braunkohlen der Sendai-Gruppe. I. Übersichtliches. Ecolog. Review, **14**, 121–32, 1956

[703] ——— Palynological Studies on a Peaty Lignite and a Peat from the Environs of Nagoya. Ecolog. Review, **14**, 289–90, 1958

[704] ——— Eine pollenanalytische Untersuchung von Braunkohlen in der Provinz Mie, Mittel-Japan. Ecolog. Review, **15**, 9–12, 1959

[705] *Sparks, B. W.*, and *R. G. West,* The Paleoecology of the Interglacial Deposits at Histon Road, Cambridge. Eiszeitalter und Gegenwart, **10**, 123–43, 1959

[706] *Środoń, A.*, Rozvój roślinności pod Grodnem w czasie ostatniego interglacjału. Acta Geol. Polon., **1**, 365–400, 1950

[707] ——— Flory plejstoceńskie z Tarzymiechów nad Wieprzem. Z badań czwartorzędu w Polsce, **5**, 5–78, 1954; Russ., Engl. summaries

[708] ——— W sprawie interglacjału w Szelągu pod Poznaniem. Z badań czwartorzędu w Polsce, **7**, 45–60, 1956; Russ., Engl. summaries

[709] ——— Flora interglacjalna z Gościęcina koło Koźla Inst. Geol., Biul., **118**, 7–60, Warsaw 1957; Russ., Engl. summaries

[710] ——— Pollen Spectra from Spitsbergen. Folia Quaternaria, **3**, Cracow 1960

[711] ——— Palaeobotany and Stratigraphy of the Late-pleistocene Deposits in the Northern Carpathians. Report VI Internat. Congr. Quaternary Warsaw 1961, **2**, 483–86, Łódź 1964

[712] *Środoń, A.*, and *M. Gołąbowa,* Plejstoceńska flora z Bedlna. Z badań czwartorzędu w Polsce, **7**, 7–44, 1956; Engl., Russ. summaries

[713] *Stachurska, A.*, Roślinność interglacjalna z Włodawy nad Bugiem. Inst. Geol., Biul., **118**, 61–89, Warsaw 1957; Russ., Engl. summaries

[714] ——— *Juglandaceae* in the Interglacial Deposits of Suszno. Acta Soc. Bot. Polon., **29**, 495–97, 1960

[715] ——— Das Präglazial-Profil von Ochota in Warschau auf Grund pollenanalytischer Untersuchungen. Prace o plejstocenie polski środkowej. Polska Akad. Nauk, Kom. geol., 43–45, Warsaw 1961; Russ., Engl. summaries

[716] ——— Das Ende des Masovian-Interglazials bei Suszno, nahe Włodawa am Bug, auf Grund palaeobotanischer Untersuchungen. Z badań czwartorzędu w Polsce, **10**, 155–73, 1961; Russ., Engl. summaries

[717] *Stark, P., F. Firbas* and *F. Overbeck,* Die Vegetationsentwicklung des Interglazials von Rinnersdorf in der östlichen Mark Brandenburg. Abh. naturw. Ver. Bremen, Sonderh. zum **28**. Bd., 105–30, 1931/32

[718] *Staszkiewicz, J.*, Histoire d'espèce *Pinus silvestris* en Europe. Report VI Internat. Congr. Quaternary Warsaw 1961, **2**, 473–77, Łódž 1964

[719] *Stefanov, B.*, and *D. Jordanov,* Weitere Materialien zur Kenntnis der fossilen Flora des Pliozäns bei dem Dorfe Kurilo (Bez. Sofia). Godišnik na sofijskija universitet, V Agronomo-lesovden fakultet, **2**, Lesovodstvo, **13**, 1–56, Sofia 1935

[720] *Stevens, L. A.*, The Interglacial of the Nar Valley, Norfolk. Quat. J. Geol. Soc. London, **115**, 291–316, 1960

[721] *Straus, A.*, Vorläufige Mitteilung über den Wald des Oberpliozäns von Willershausen (Westharz). Mitteil. dtsch. dendrol. Ges., **47**, 182–86, 1935

[722] ——— Beiträge zur Pliozänflora von Willershausen III. Die niederen Pflanzengruppen bis zu den Gymnospermen. Palaeontographica, B, **93**, 1–44, 1952

[723] *Stuiver, M.*, Carbon Isotopic Distribution and Correlated Chronology of Searles Lake Sediments. Amer. J. Sci., **262**, 377–92, 1964

[723a] *Stuiver, M.*, Long-term C-14 Variations. In: *Olsson, I. U.*, see [573a], 197–213, 1970

[724] *Suess, H. E.*, Grundlagen und Ergebnisse der Radiokohlenstoff-Datierung. Angew. Chemie, **68**, 540–46, 1956

[724a] *Suess, H. E.*, The Three Causes of the Secular C-14 Fluctuations, Their Amplitudes and Time Constants. In: *Olsson, I. U.*, see [573a], 595–604, 1970

[725] *Suggate, R. P.*, Time-stratigraphic Subdivisions of the Quaternary as Viewed from New-Zealand. Quaternaria, **5**, 5–17, 1958–1961

[726] ——— The Upper Boundary of the Hawera Series. Transact. Roy. Soc. New Zealand, geol., **1**, 4, 11–16, 1961

[727] *Sukatscheff, W.*, Über das Vorkommen der Samen von *Euryale ferox* SALISB. in einer interglazialen Ablagerung in Rußland. Ber. D. Bot. Ges., **26** a, 132–37, 1908

[728] *Sukachev, V. N.*, and *A. K. Nedoseeva*, Über die Vegetationsentwicklung im Verlauf des Riß-Würm-Interglazials. DAN, **94**, 1171–74, 1954

[729] *Szafer, W.*, Über den Charakter der Flora und des Klimas der letzten Interglazialzeit bei Grodno in Polen. Bull. Internat., Acad. Polon. Sci. Lettres, Cl. sci. math. nat., sér. B: sci. nat., 1925, 277–314, 1926

[730] ——— Flora plejstoceńska w Jarosławiu. Rocznik Polsk. Towarzystwa Geol., **9**, 237–43, 1933; Ger. summary

[731] ——— Plioceńska flora okolic Czorsztyna i jej stosunek do Plejstocenu. Inst. Geol., Prace **11**, Warsaw 1954; Russ., Engl. summaries

[732] ——— Über die Zweiteilung des Riß-Glazials. Veröff. Geobot. Inst. Rübel in Zürich, **34**, 126–31, 1957

[733] *Szafer, W.*, and *J. Trela*, Interglacjał w Szelągu pod Poznaniem. III. Flora międzylodowcowa z Szelagu pod Poznaniem ze szczególnem uwzględnieniem wyników analizy pyłkowej. Sprawozdanie Kom. Fizjograf., Polska Akad. Umiej, **63**, 71–82, 1929

[734] *Szafer, W.*, *J. Trela* and *M. Ziembianka*, Flora interglacjalna z Bedlna koło Końskich. Rocznik Polsk. Towarzystwa Geol., **7**, 402–14, 1931

[735] *Terasmaë, J.*, Palynological Study of Pleistocene Deposits on Banks Island, Northwest Territories, Canada. Science, **123**, 801–2, 1956

[736] ——— Palaeobotanical Studies of Canadian Pleistocene Nonglacial Deposits. Science, **126**, 351–52, 1957

[737] ——— A Palynological Study of Pleistocene Interglacial Beds at Toronto, Ontario. Geol. Survey Canada, Bull., **62**, 23–41, 1960

[738] *Ter Wee, M. W.*, The Saalian Glaciation in the Netherlands. Mededel. van de Geol. Stichting, n. s. **15**, 57–76, 1962

[739] *Theobald, N.*, Les climats de l'Europe occidentale au cours des temps tertiaires d'après l'étude des insectes fossiles. Geol. Rundschau, **40**, 89–92, 1952

[740] *Thienemann, A.*, Verbreitungsgeschichte der Süßwassertierwelt Europas. Versuch einer historischen Tiergeographie der europäischen Binnengewässer. Stuttgart 1950

[741] *Thiergart, F.*, Pollen und Sporen aus dem Pliozän von Willershausen.

Geologie, **3**, 536–47, 1954

[742] *Thomson, P. W.*, Das Interglazial von Wallensen im Hils. Eiszeitalter und Gegenwart, **1**, 96–102, 1951

[743] ——— Zur Frage des Alters des Braunkohlenlagers vom Roten Kliff auf der Insel Sylt. Neues Jahrb. Geol. u. Paläontol., Mh., 1955, 69–71, 1955

[744] *Thoral, M.*, and *H. Gauthier*, Loess et limons anciens du Lyonnais. C. R. Acad. Sci., **236**, 1182–84, 1953

[745] *Tichomirov, B. A.*, Beiträge zur Kenntnis des Ferntransportes von Baumpollen über die polare Waldgrenze hinaus. DAN, **71**, 753–55, 1950

[746] *Timoféeff-Ressovsky, N. W.*, Genetik und Evolution. Zs. f. indukt. Abstammungs- und Vererbungslehre, **76**, 158–219, 1939

[747] *Tjurina, L. S.*, Palynologische Charakteristik quartärer und oberpliozäner Sedimente am Unterlauf der Volga. Materialy chetvert., **1**, 288–95, 1961

[748] *Tołpa, St.*, Interglazialflora von Sławno nahe Radom. Z badań czwartorzędu w Polsce, **10**, 15–56, Warsaw 1961; Russ., Engl. summaries

[749] *Tolstov, A. N.*, Über das Auftreten von Eisadern in der Tundra, deren Chemismus dem des Meerwassers nahe steht. Problemi paleogeografii i morfogeneza v poliarnych stranach i vysokogor'e. 182–84, Moscow 1964

[750] *Tolstov, S. P.*, Nizov'ja Amu-dar'i, Sarykamysh, Uzboi. Istoriia formirovaniia i zaseleniia. Materialy Chorezmskoi Ekspedicii, **3**, Moscow 1960

[751] *Tolstov, S. P., A. S. Kes'* and *T. A. Zhdanko*, Geschichte des Sarykamysch-Sees im Mittelalter. Izv AN, ser. geogr., 1954, Nr. 1, 41–50, 1954

[752] *Tongiorgi, E.*, La flora fossile di Saccopastore. Suo significato nella storia della vegetazione nella regione laziale. Rivista di Antropologia, **32**, 237–42, 1938–39

[753] *Trela, J.*, O utworach międzylodowcowych w Olszewicach pod Tomaszowem Mazowieckim. III. Analiza pyłkowa utworów międzylodowcowych w Olszewicach. Sprawozdanie Kom. Fizjogr., Polska Akad. Umiej, **64**, 77–86, 1930

[754] *Trela, J.*, Interglacjał w Samostrzelnikach pod Grodnem (wyniki analizy pyłkowej). Starunia, **9**, Cracow 1935

[755] *Tricart, J.*, Paléoclimats quaternaires et morphologie climatique dans le midi méditerranéen. Eiszeitalter und Gegenwart, **2**, 172–88, 1952

[756] *Tumel', V. F.*, Zur Geschichte der ewigen Gefrornis in der UdSSR. Trudi Geogr., **37**, 124–32, 1946

[757] *Urey, H. C., H. A. Lowenstam, S. Epstein,* and *C. R. McKinney*, Measurement of Paleotemperatures and Temperatures of the Upper Cretaceous of England, Denmark, and the Southeastern United States. Bull. Geol. Soc., **62**, 399–416, 1951

[758] *Urry, W. D.*, Radioactivity of Ocean Sediments. VI. Concentrations of the Radioelements in Marine Sediments of the Southern Hemisphere. Amer. J. Sci., **247**, 257–75, 1949

[759] *Ushko, K. A.*, Der Aufschluß interglazialer lakustriner Sedimente bei Lichvin (Chekalin). In: *Markov, K. K.* and *A. I. Popov*. Lednikovyi period na territorii evropeiskoi chasti SSSR i Sibiri. 148–226, Moscow 1959

[760] *Vangengeim, E. A.*, Paläontologische Begründung der Stratigraphie der anthropogenen Sedimente im Norden Ost-Sibiriens. Trudi Geol., **48**, 1961

[761] *Vangengeim, E. A.*, and *V. G. Gerbova,* Einige Angaben über Zeit und Umstände der Akkumulation der Sande in Transbaikalien. Trudi Kom. po izuch. chetvert. per., **19**, 268–74, 1962

[762] *Vasil'ev Y. M.* and *P. V. Fedorov,* Zum Problem der Beziehungen zwischen marinen und kontinentalen Sedimenten des unteren und mittleren Volgagebietes. Izv AN, ser. geol., 1961, Nr. 9, 91–99, 1961

[763] *Vasil'ev, V. N.*, Klimaverhältnisse Ost-Sibiriens im Pleistozän. Trudi Kom. po izuch. chetvert. per., **12**, 22–53, 1955

[764] ——— Entstehung der Flora und der Vegetation des Fernen Ostens und Ost-Sibiriens. Materialy po istorii flori i rastitel'nosti SSSR, **3**, 361–457, 1958

[765] *Vas'kovskii, A. P.*, Kurzer Abriß der Vegetation, des Klimas und der Chronologie des Quartärs am Oberlauf der Flüsse Kolyma, Indigirka und am Nordufer des Ochotskischen Meeres. In: *Markov, K. K.*, and *A. I. Popov*, Lednikovyi period na territorii evropeiskoi chasti SSSR i Sibiri, 510–45, Moscow 1959

[766] *Veklich, M. F.*, Fossile Böden in quartären (anthropogenen) Sedimenten des Südwestteiles des Russischen Flachlandes. Chetvertichnyi Period, **13**, **14**, **15**, 87–106, Kiev 1961; Engl. summary

[767] ——— Mollusken quartärer kontinentaler Sedimente der USSR. Materialy chetvert., **1**, 342–46, 1961

[768] *Veklich, M. F.*, and *N. A. Kunica,* Die Malakofauna quartärer (anthropogener) kontinentaler Sedimente der Ukrainischen SSR. Chetvertichnyi Period, **13**, **14**, **15**, 280–93, Kiev 1961; Engl. summary

[769] *Velichko, A. A.*, Geologicheskii vozrast verchnego paleolita central'nych raionov Russkoi ravnini. Moscow 1961

[770] ——— Stratigraphie der Lößsedimente in der periglazialen Zone der Valdai- und Moskau-Vereisung. Paleografiia chetvertichnogo perioda SSSR, 93–108, Moscow 1961; Fr. summary

[771] ——— Das vom Frost verursachte Relief der spätpleistozänen Periglazialzone (Kryolithozone) Ost-Europas. In: Chetvertichnyi period i ego istoriia, 104–20, Moscow 1965; Fr. summary

[772] *Vent, W.*, Die Pflanzenwelt der Ilmtaltravertine von Weimar-Ehringsdorf zur Unstrut-Warmzeit. Alt-Thüringen, **3**, 16–28, 1958

[773] *Villaret-von Rochow, M.*, Vergleichende Beobachtungen an rezenten und fossilen *Euryale*-Samen. Veröff. Geobot. Inst. ETH Zürich, Stiftung Rübel, **37**, 303–14, 1962

[774] *Vinogradov, A. P., V. M. Kutiurin, M. V. Ulubekova* and *I. K. Zadorozhnyi,* The Isotopic Composition of Photosynthetic Oxygen. DAN, **125**, 1151–53, 1959

[775] *Van der Vlerk, I. M.*, and *F. Florschütz,* Nederland in het ijstijdvak. Utrecht 1950

[776] ——— The Palaeontological Base of the Subdivision of the Pleistocene in the Netherlands. Verhandl. Koninkl. Nederl., Acad. van Wetensch., afd. natuurk. 1. reeks, **20**, Nr. 2, 1953

[777] *Voskresenskii, S. S.*, and *M. P. Grichuk,* Wesentliche Etappen in der Geschichte des Südteiles Ostsibiriens im Quartär. Tezisi; sekc. vost. Sibiri i Dal'n.

Vostoka, 15, 1957
[778] *Vostrjakov, A. V., I. V. Mizinov, A. I. Moskvitin* and *A. A. Chigurjaeva,* Die klimatischen Verhältnisse des Akchagyls nach neuen lithologischen und paläobotanischen Untersuchungen im südlichen Transvolga-Gebiet. DAN, **105**, 144–46, 1955
[779] *Vries, H. de,* and *A. Dreimanis,* Finite Radiocarbon Dates of the Port Talbot Interstadial Deposits in Southern Ontario. Science, **131**, 1738–39, 1960
[780] *Vries, H. de, F. Florschütz* and *J. Menéndez Amor,* Un diagramme pollinique simplifié d'une couche de «Gyttja», située à Poueyferré près de Lourdes (Pyrénées françaises centrales), daté par la méthode du radio-carbone. Koninkl. Nederl. Akad. van Wetensch., Proc., ser. B, **63**, Nr. 4, 498–500, 1960
[781] *Wasylikowa, K.,* Pollen Analysis of the Late-glacial Sediments in Witów near Łęczyca, Middle Poland. Report VI Internat. Congr. Quaternary Warsaw 1961, **2**, 497–502, Łódź 1964
[782] *Watts, W. A.,* Interglacial Deposits at Baggotstown, near Bruff, Co. Limerick. Proc. Roy. Irish Acad., **63**, Sect. B, Nr. 9, 167–89, 1964
[783] *Weber, C. A.,* Über die diluviale Vegetation von Klinge in Brandenburg und über ihre Herkunft. Beibl. zu den Botan. Jahrb., **17**, Nr. 40, 1–20, 1893
[784] ——— Über die diluviale Flora von Fahrenkrug in Holstein. Beibl. zu den Bot. Jahrb., **18**, Nr. 43, 1–13, 1894
[785] ——— Über die fossile Flora von Honerdingen und das nordwestdeutsche Diluvium. Abh. naturw. Ver. Bremen, **13**, 413–68, 1896
[786] *Weischet, W.,* Zum Problem der Stabilität der Klimabedingungen in Westsibirien während der Glaziale und Interglaziale. Eiszeitalter und Gegenwart, **11**, 77–87, 1960
[787] *Welten, M.,* Die spätglaziale und postglaziale Vegetationsentwicklung der Berner Alpen und -Voralpen und des Walliser Haupttales. Veröff. d. Geobot. Inst. Rübel in Zürich, **34**, 150–58, 1958
[788] *West, R. G.,* The Ice Age. The Advancement of Science, Nr. 64, 428–40, 1960
[789] ——— The Glacial and Interglacial Deposits of Norfolk. Transact. Norfolk and Norwich Naturalists' Soc., **19**, 365–75, 1961
[790] ——— Vegetational History of the Early Pleistocene of the Royal Society Borehole at Ludham, Norfolk. Proc. Roy. Soc. London, Ser. B, Biol. Sci., **155**, 437–53, 1962
[791] ——— A Note on *Taxus* Pollen in the Hoxnian Interglacial. New Phytologist, **61**, 189–90, 1962
[792] ——— Problems of the British Quaternary. Proc. Geol. Assoc., **74**, pt. 2, 147–86, 1963
[793] *West, R. G.,* and *B. W. Sparks,* Coastal Interglacial Deposits of the English Channel. Philos. Transact. Roy. Soc. London, Ser. B, Biol. Sci., **243**, 95–133, 1960
[794] *Weyl, R., U. Rein* and *M. Teichmüller,* Das Alter des Sylter Kaolinsandes. Eiszeitalter und Gegenwart, **6**, 5–15, 1955
[795] *Whitehead, D. R.,* and *D. R. Bentley,* A Post-glacial Pollen Diagram from Southeastern Vermont. Pollen et Spores, **5**, 115–27, 1963

[796] *Willett, H. C.*, The General Circulation at the Last (Würm) Glacial Maximum. Geogr. Annaler, **32**, 179–87, 1950

[797] ——— Atmospheric and Oceanic Circulation as Factors in Glacial-Interglacial Changes of Climate. In: *Shapley, H.*, Climatic Change. Evidence, Causes, and Effects. 51–71, Cambridge 1953

[798] *Willett, H. C.*, and *F. Sanders*, Descriptive Meteorology. 2. Aufl., New York 1959

[799] *Willis, E. H.*, *H. Tauber*, and *K. O. Münnich*, Variations in the Atmospheric Radiocarbon Concentration over the Past 1300 Years. Amer. J. Sci., Radiocarbon Suppl., **2**, 1960, 1–4

[800] *Winter, Th. C.*, Pollen Sequence at Kirchner Marsh, Minnesota. Science, **138**, 526–28, 1962

[801] *Wissmann, H. v.*, Die heutige Vergletscherung und Schneegrenze in Hochasien, mit Hinweisen auf die Vergletscherung der Letzten Eiszeit. Abh. d. Akad. d. Wiss. u. d. Lit., math.-nat. Kl., Jahrg. 1959, Nr. 14, Mainz 1960

[802] *Woldstedt, P.*, Norddeutschland und angrenzende Gebiete im Eiszeitalter. Stuttgart 1950

[803] ——— Die Klimakurve der Tertiärs und Quartärs in Mitteleuropa. Eiszeitalter und Gegenwart, **4/5**, 5–9, 1954

[804] ——— Das Eiszeitalter, Grundlinien einer Geologie des Quartärs. 1. Bd., Stuttgart 1954

[805] ——— Das Eiszeitalter, Grundlinien einer Geologie des Quartärs. 2. Bd., Stuttgart 1958

[806] ——— Das Eiszeitalter, Grundlinien einer Geologie des Quartärs. 3. Bd., Stuttgart 1965

[807] ——— Die interglazialen marinen Strände und der Aufbau des antarktischen Inlandeises. Eiszeitalter und Gegenwart, **16**, 31–36, 1965

[808] *Wolters, R.*, Nachweis der Günz-Eiszeit und der Günz-Mindel-Wärmezwischenzeit am Niederrhein. Geol. Jahrb., **65**, 769–72, 1949

[809] ——— Ausbildung und Lagerung der pliozän/pleistozänen Grenzschichten im niederrheinischen Grenzgebiet von Niederkrüchten/Brüggen. Geol. Jahrb., **69**, 339–48, 1954

[810] *Wright, H. E.*, Late Pleistocene Soil Development, Glaciation, and Cultural Change in the Eastern Mediterranean Region. Ann. New York Acad. Sci., **95**, 718–28, 1961

[811] ——— Late Pleistocene Geology of Coastal Lebanon. Quaternaria, **6**, 525–39, 1962

[812] ——— Aspects of the Early Postglacial Forest Succession in the Great Lakes Region. Ecology, **45**, 439–48, 1964

[813] *Wright, H. E.*, and *H. L. Platten*, The Pollen Sum. Pollen et Spores, **5**, 445–50, 1963

[814] *Yamagata, O.*, A Palynological Study of a *Menyanthes* Bed from Nagano Prefecture, Japan. Ecolog. Review, **14**, 267–68, 1957

[815] *Zagwijn, W. H.*, Vegetation, Climate, and Time-Correlations in the Early Pleistocene of Europe. Geol. en Mijnb., n. s. **19**, 233–44, 1957

[816] ——— Aspects of the Pliocene and Early Pleistocene Vegetation in the

Netherlands. Mededel. Geol. Stichting, ser. C-III-1, Nr. 5, 1960
[817] ——— Vegetation, Climate, and Radiocarbon Datings in the Late Pleistocene of the Netherlands. I: Eemian and Early Weichselian. Mededel. Geol. Stichting, n. s. Nr. **14**, 15–45, 1961
[818] ——— Pleistocene Stratigraphy in the Netherlands, Based on Changes in Vegetation and Climate. Verhandel. Koninkl. Nederl. Geol. Mijnbouwk. Genootsch., geol. ser., deel 21–2, 173–96, 1963
[819] ——— Pollen-analytic Investigations in the Tiglian of the Netherlands. Mededel. Geol. Stichting, n. s. Nr. **16**, 49–71, 1963
[820] *Zarina, E. P., F. A. Kaplanskaya, I. I. Krasnov, I. M. Michankov,* and *W. D. Tarnogradski,* Periglaziale Formation von Westsibirien. Report VI Internat. Congr. Quaternary Warsaw 1961, **4**, 199–215, Łódź 1964
[821] *Zemcov, A. A.,* and *S. B. Shackii,* Zur Geologie und Stratigraphie der quartären Sedimente im Nordostteil der Westsibirischen Tiefebene. Materialy chetvert., **3**, 32–38, 1961
[822] *Zinderen-Bakker, E. M. v.,* Pollen Analysis and its Contribution to the Palaeoecology of the Quaternary in Southern Africa. In: *Davis, D. H. S.,* Ecology in South Africa. Amsterdam 1961
[823] ——— Botanical Evidence for Quaternary Climates in Africa. Annals of the Cape Provincial Museums, **2**, 16–31, 1962
[824] ——— A Late-glacial and Post-glacial Climatic Correlation between East-Africa and Europa. Nature (L.), **194**, 201–3, 1962
[825] ——— Symposium on Early Man and His Environments in Southern Africa. Paleobotanical Studies. South Afr. J. Sci., **59**, 332–40, 1963
[826] ——— A Pollen Diagram from Equatorial Africa, Cherangani, Kenya. Geol. en Mijnb., **43**, 123–28, 1964
[827] *Zinderen-Bakker, E. M. v.,* and *J. D. Clark,* Pleistocene Climates and Cultures in North-Eastern Angola. Nature (L.), **196**, 639–42, 1962
[828] *Zoller, H.,* Pollenanalytische Untersuchungen zur Vegetationsgeschichte der insubrischen Schweiz. Denksch. Schweiz. Naturf. Ges., **83**, 45–156, 1960
[829] ——— Die wärmezeitliche Verbreitung von Haselstrauch, Eichenmischwald, Fichte und Weißtanne in den Alpenländern. Bauhinia, **1**, 189–207, 1960
[830] *Zol'nikov, V. G.,* Über die Entdeckung eines Skeletts von *Mammonteus primigenius* (BL.) im Megino-Kangalassker Gebiet der Jakutischen ASSR. Biull. chetvert., **24**, 104–9, 1960
[831] ——— Über Funde von Resten der Tiere des Mammutkomplexes an der Suola im Megino-Kangalassker Gebiet der JaASSR. Izv AN, ser. geogr., 1961, Nr. 3, 81–85, 1961
[832] *Zorin, L. V., E. M. Malaeva* and *N. G. Sudakova,* Zur Paläogeographie Ost-Transbaikaliens im Quartär. Paleogeografiia chetvertichnogo perioda SSSR, 174–88, Moscow 1961
[833] *Zubakov, V. A.,* Über das Vorhandensein interglazialer Verhältnisse in der Umgebung des Jenisseij in der Samburg-Zeit. DAN, **131**, 628–31, 1960
[834] ——— Stratigraphie und Paläogeographie der quartären Sedimente im Flußgebiet des Jenisseij. Materialy chetvert., **3**, 157–66, 1961

[835] *Zhuze, A. P.*, Wichtige Etappen in der Entwicklung der marinen Diatomeenflora des Fernen Ostens während des Tertiärs und des Quartärs. Bot. Zhurn., **44**, 44–55, 1959; Engl. summary
[836] ——— Diatomeen und ihre Bedeutung für die Entschleierung der Geschichte der Ozeane. Izv AN, ser. geogr., 1961, Nr. 2, 13–20, 1961
[837] ——— Stratigrafiia i paleogeograficheskie issledovaniia v severo-zapadnoi chasti Tichogo Okeana. Moscow 1962
[838] *Zhuze, A. P.*, and *E. V. Koreneva*, Zur Paläogeographie des Ochotskischen Meeres. Izv AN, ser. geogr., 1959, Nr. 2, 12–24, 1959
[839] *Walter, H.*, Die Jahresringe der Bäume als Mittel zur Feststellung der Niederschlagsverhältnisse in der Vergangenheit, insbesondere in Deutsch-Südwestafrika. Naturwissenschaften, **28**, 607–12, 1940
[840] *Hedin, S.*, Der wandernde See. 5. Aufl., Leipzig 1940
[841] *Solger, F., P. Graebner, J. Thienemann, P. Speiser*, and *F. W. O. Schulze*, Dünenbuch. Stuttgart 1910
[842] *Devjatkin, E. V.*, Über tertiäre Sedimente in der Džulukul-Senke (Ost-Altai). DAN, **135**, 1457–60, 1960
[843] *Dokturowsky, W.*, Neue Angaben über die interglaziale Flora in der UdSSR. Abh. d. naturw. Ver. Bremen, Sonderheft zum **28**. Bd., 246–61, 1931/32
[844] *Bertsch, K.*, Die diluviale Flora des Cannstatter Sauerwasserkalkes. Zs. f. Botanik, **19**, 641–59, 1926/1927
[845] *Anufriev, G. I.*, Einige Angaben über die Pflanzenreste aus dem Kos'kovsker Fundort fossilen Sapropelits. Izv. Sapropelevogo Komiteta, **2**, 84–90, Leningrad 1925
[846] *Adam, K. D.*, Die Bedeutung der altpleistozänen Säugetier-Faunen Südwestdeutschlands für die Gliederung des Eiszeitalters. Geologica Bavarica, **19**, 357–63, 1953
[847] *Toepfer, V.*, Tierwelt des Eiszeitalters. Leipzig 1963
[848] *Emiliani, C.*, Temperature and Age Analysis of Deep-Sea Cores. Science, **125**, 383–87, 1957
[849] *Schenk, E.*, Pleistozäne Eiskeil-Polygonnetze im Vogelsberg im Vergleich mit heutzeitlichen Vorkommen in Alaska. Natur und Museum, **95**, 8–16, 1965
[850] *Hofer, E.*, Arktische Riviera. Ein Bildband über die Schönheit Nordost Grönlands. Berne 1957
[851] *Hultén, E.*, The Circumpolar Plants. I Vascular Cryptogams, Conifers, Monocotyledons. Kungl. Svenska Vetenskaps akad. Handl., fjärde ser., **8**, Nr. 5, 275 S., 1962
[852] ——— The Amphi-Atlantic Plants and their Phytogeographical Connections. Kungl. Svenska Vetenskaps-akad. Handl. fjärde ser., **7**, Nr. 1, 340 S., 1958
[853] *Meusel, H.*, Vergleichende Arealkunde. Einführung in die Lehre von der Verbreitung der Gewächse mit besonderer Berücksichtigung der mitteleuropäischen Flora. **1**, Berlin 1943
[854] *Walter, H.*, Grundlagen der Pflanzenverbreitung. I. Teil: Standortslehre. Stuttgart 1960
[855] *Clark, J. D.*, and *E. M. van Zinderen Bakker*, Prehistoric Culture and

Pleistocene Vegetation at the Kalambo Falls, Northern Rhodesia. Nature (L.), **201**, 971–75, 1964

[856] *Emiliani, C.*, Oxygen Isotopes and Paleotemperature Determinations. Actes du IV. Congr. Internat. du Quaternaire, **2**, 831–44, 1956

Index